ALLE ZEIT WACH 1842

M.N.O. Davies P.R. Green (Eds.)

Perception and Motor Control in Birds

An Ecological Approach

With 129 Figures and 3 Tables

Springer-Verlag
Berlin Heidelberg New York
London Paris Tokyo
Hong Kong Barcelona
Budapest

Dr. MARK N.O. DAVIES
University College London
Department of Psychology
Gower St
London, WC1E 6BT
UK

Dr. PATRICK R. GREEN
University of Nottingham
Department of Psychology
Nottingham NG7 2RD
UK

ISBN-13:978-3-642-75871-3 e-ISBN-13:978-3-642-75869-0
DOI: 10.1007/978-3-642-75869-0

Library of Congress Cataloging-in-Publication Data. Davies, Mark N.O., 1960– Perception and motor control in birds: an ecological approach / Mark N.O. Davies, Patrick R. Green. p. cm. Includes bibliographical references and index. ISBN-13:978-3-642-75871-3 1. Birds – Sense organs. 2. Birds – Behavior. 3. Perceptual-motor processes. I. Green, Patrick R. II. Title. QL698.D22 1994 598.2'18 – dc20 93-33022

Softcover reprint of the hardcover 1st edition 1994

Typesetting: Macmillan India Ltd., Bangalore - 25
31/3145/SPS – 5 4 3 2 1 0 – Printed on acid-free paper

Preface

The scope of the chapters in this book will seem at once broad and narrow. They encompass topics in both perception and motor organization, but are restricted to just one group of animals. Our reason for bringing the areas of perception and motor control together is to make the case that each should be considered in the context of the other, rather than separately. Psychologists will recognize the provenance of this view in Gibsonian or "ecological" theory, and our subtitle reflects this influence on our selection of research topics. Many features of "ecological psychology" are familiar to biologists as principles which should underlie any good biology – the analysis of structure and function in the context of the whole animal and its ecological niche – and these principles guide all the research described in this book.

By restricting the scope of the book to research on birds, we hope to provide enough points of contact between different areas of research to convince readers that fruitful interactions between them are a practical possibility. Such topics as the control of flight manoeuvres or pecking provide concrete examples of how problems in perception and motor organization can be treated together.

The chapters fall into three sections, and each begins with a short introduction drawing out some of the links between the chapters it contains. Section I contains reviews of recent research on a variety of sensory and perceptual processes in birds, which all involve subtle analyses of the relationships between species' perceptual mechanisms and their ecology and behaviour. Chapters in Section II describe research using a variety of methods – behavioural, neurophysiological, anatomical and comparative – but all dealing with the common problem of understanding how the activities of large numbers of muscles are co-ordinated to generate adaptive behaviour. In Section III, chapters are concerned with a range of approaches to analyzing the links between perceptual and motor processes, through cybernetic modelling, neurophysiological analysis, or behavioural methods.

We are grateful to Dr Dieter Czeschlik and his colleagues at Springer-Verlag for their support and advice throughout the planning and preparation of this book. We would also like to thank Carlos Martinoya, who was prevented by illness from contributing a chapter, for his advice on the selection of topics. The compilation and editing of the chapters was made far easier by using manuscripts on disc, and we thank Karen Sherlock and Howard Martin for their help in handling different disc and document formats. We are also grateful to Charlotte Dewey for handling all our editorial correspondence so efficiently.

Nottingham, UK
January 1994

Mark N.O. Davies
Patrick R. Green

Contents

List of Contributors

Addresses are given at the beginning of the respective contribution

Introduction to Section I

The analysis of sensory processes in an adaptive and comparative context is a distinguished tradition in ethology and neurobiology. A classic example is work on the adaptive radiation of the vertebrate eye, and, in the first chapter of this section, Martin takes up the themes pioneered by Walls (1942), describing the great strides made in recent years in understanding how the designs of eyes of different bird species can be related to their ecological niches. These advances have come about in part through new methods, particularly the construction of schematic eyes from optical measurements, which have provided insights into the optical design of birds' eyes impossible to obtain from anatomical descriptions alone.

Advances in ecological analysis have been equally important. Martin shows that old distinctions, such as those between diurnal and nocturnal ways of life, or terrestrial and aquatic habitats, must be replaced by more detailed descriptions of the *ranges* of ecological conditions in which the eyes of particular species operate. Further, different adaptive demands on birds' eyes interact and cannot be treated independently; the relative contributions of nocturnal and amphibious habits to the evolution of a cornea with low refractive power in the Manx shearwater provide an example.

Another theme developed in Martin's chapter is the importance of behavioural differences, and particularly differences in foraging strategies, for understanding variations between species in eye design. The two chapters which follow deal with different aspects of bird vision and also stress the importance of relating visual processes to the particular behaviour patterns which they control.

In Chapter 2, Schaeffel analyzes the problems which must be solved in using the state of accommodation of the eye as a source of distance information, and considers whether it is used for this purpose in birds. In species where the speed of accommodation has been measured (chickens and two owl species), it is strikingly fast, suggesting a role in providing distance information for the timing of rapid acts such as pecking. There is evidence that accommodation does play a part in controlling pecking in chickens and barn owls, but the mechanisms involved differ in subtle ways between these species.

Several other themes in the adaptive radiation of birds' eyes are also taken up in Chapter 2. These include the mechanisms of corneal accommodation in pigeons and chickens, and the evidence for lower field myopia in pigeons and

M.N.O. Davies and P.R. Green (Eds.)
Perception and Motor Control in Birds

other ground-foraging birds. Schaeffel discusses critically the evidence for this possible means of "keeping the ground in focus", and considers some implications of the hypothesis for eye growth and accommodation.

In Chapter 3, McFadden considers another "physiological cue" to depth, binocular disparity. She stresses that binocular depth perception is by no means a single process, but can take a variety of forms, and also interacts in complex ways with the control of convergence. There is behavioural evidence for local and global stereopsis in birds, but physiological evidence only for the former.

McFadden argues that local and global stereopsis have different roles in the control of behaviour. Global stereopsis is unlikely to be used in the timing of actions such as pecking or grasping, but is an important means of revealing differences in relative depth of surfaces and so in breaking the camouflage of prey or predators. Local stereopsis, in contrast, can play a role in providing distance information for controlling the extension of a body part such as a foot or beak towards a target.

The theme of visual depth perception addressed in Chapters 2 and 3 is taken up again in Section III, where Chapter 13 deals with sources of distance information in optic flow, and Chapter 16 with the integration of distance cues to achieve maximum accuracy in the control of behaviour. The remaining chapters in Section I move away from an emphasis on vision to consider other senses of birds. In Chapter 4, McGregor discusses recent research on one aspect of hearing in birds; their ability to use acoustic cues in the vocalizations of other birds to determine their distance, or "range". This information is important to territorial birds in discriminating between neighbours and intruders, and so in making adaptive decisions about whether to challenge another bird.

McGregor describes how sophisticated behavioural methods relying on field playback of recorded songs have identified the acoustic cues involved and have demonstrated the roles of learning and song familiarity in the ranging of song. The results which McGregor describes also have important implications for more general theories of the evolution of animal communication. In particular, the mechanisms of range detection imply that songbirds may not be able to withhold range information from, or convey false information to, their territorial neighbours, as some theoretical arguments have predicted. Auditory processes in birds are discussed again in Chapter 14, which deals with directional hearing in owls.

In Chapter 5, Wiltschko and Wiltschko describe the integration of a variety of different kinds of sensory information – visual, magnetic, auditory and olfactory – in the navigation of migratory and homing birds. They describe the major advance made in understanding bird navigation when it was discovered that multiple, redundant cues are used both to determine the home direction and to locate a compass heading. Since then, a wealth of experimental results obtained by their own group and by other researchers has demonstrated that multiple cues for navigation are not arranged in a fixed hierarchy. Instead, the ways in which they are used vary in subtle ways with ecological context and with the life histories of individual birds. (This theme is also developed in Chapter 16 in the context of depth perception.)

In particular, Wiltschko and Wiltschko describe the progress which has been made in understanding how different navigational cues interact during development. A magnetic compass ability is used to calibrate sun and star compasses, and also to enable young homing birds to use route-specific information to determine a homeward course. These calibration mechanisms continue to operate in adult birds, and provide means of adjusting compass mechanisms to environmental fluctuations, such as magnetic storms and seasonal changes in the sun's arc. Like Chapter 1, Wiltschko and Wiltschko's discussion stresses the importance of environmental variation *within* a species' niche when considering the ecological factors influencing sensory processes.

Reference

Walls GL (1942) The vertebrate eye and its adaptive radiation. Cranbrook Institute of Science, Michigan

1 Form and Function in the Optical Structure of Bird Eyes

G.R. Martin

1.1 Introduction

The anatomy of bird eyes has often been described in near eulogistic terms. Polyak (1957, p. 852), for example, reported that the eyes of swallows (*Hirundo rustica*) exhibit "extraordinary development" and asserted (in the absence of detailed descriptions) that they are "a model of structural and functional refinement". Ornithologists have tended to accept such strong assertions as confirmatory evidence of the general primacy of vision in the control of much bird behaviour (e.g. Welty and Baptista 1988; Gill 1990). However, knowledge of eye structures which can be explicitly related to particular visual tasks is limited to relatively few of the presently extant bird species (numbering between 9200–9700). Since bird species occupy a wide diversity of habitats and exhibit a great diversity of life styles (Welty and Baptista 1988; Gill 1990; Sibley and Ahlquist 1990; Sibley and Monroe 1990; Brooke and Birkhead 1991; Howard and Moore 1991) there is a clear need to broaden the comparative base if general principles concerning the form and function of birds' eyes and their evolutionary origins are to be understood.

Something of the diversity of avian eye structures has been captured in a number of reviews (Wood 1917; Walls 1942; Rochon-Duvigneaud 1943; Polyak 1957; Duke-Elder 1958; Meyer 1977). However, the scope of such works for shedding light on factors which may have led to the evolution of present day structures is somewhat limited. This is because they are based solely upon anatomical descriptions. A firmer base for interpreting links between form and function in bird eyes has been provided by more recent quantitative descriptions of optical structure and performance. Although these descriptions are also based upon few species, they have led to the questioning of certain older assumptions concerning relationships between form and function in avian eyes. Examples of these findings are the principal topics discussed in this chapter.

School of Continuing Studies, The University of Birmingham, Edgbaston, Birmingham B15 2TT, UK

M.N.O. Davies and P.R. Green (Eds.)
Perception and Motor Control in Birds

1.2 The Bases of Diversity in Avian Eye Structure

In all vertebrate eyes the image of a segment of space is projected onto the retina by a simple optical system consisting of just two principal refractive components, cornea and lens, and between these lies the iris which controls the pupil aperture of the system (Fig. 1.1). For a general account of avian eye anatomy, see Martin (1985) and Evans and Martin (1993). This simplicity of optical structure has a particular strength since it embodies many degrees of freedom. These can produce eyes which, while sharing the same basic design, have markedly different optical performances.

Variations in optical design can involve: (1) the shape and relative positions of all principal refractive surfaces, (2) the effective refractive index of the lens,

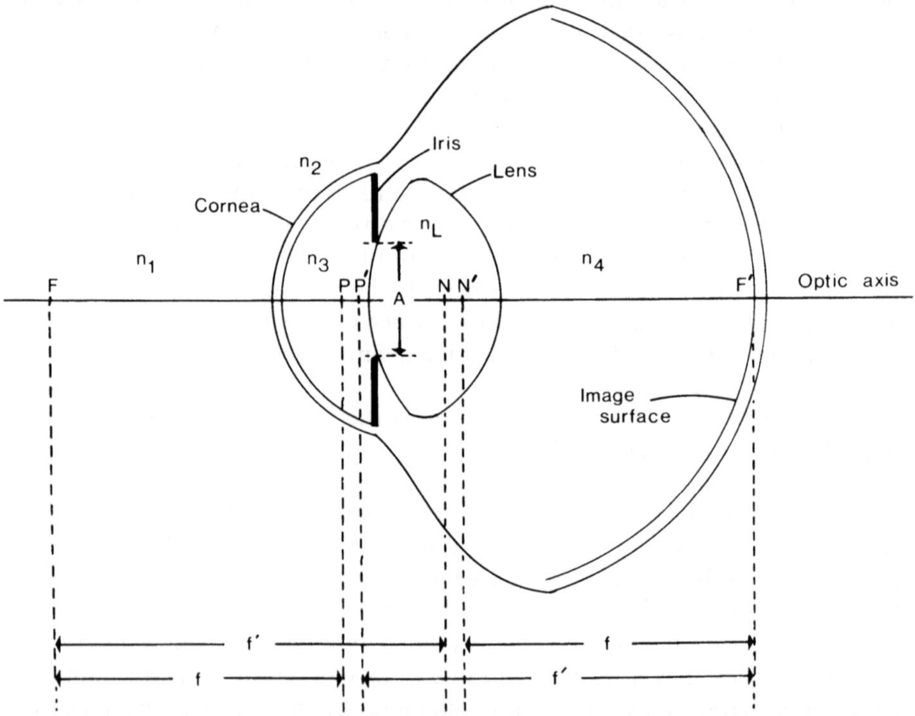

Fig. 1.1. Principal optical components of a bird eye showing the main parameters described by a schematic eye model. These are used in interspecific comparisons of optical structure and performance. *N,N'* nodal points; *P,P'* principal points; *F,F'* focal points; *f* is the anterior focal length (= PND), *f'* the posterior focal length; n_1 refractive index of the medium outside the eye (either air or water); n_2 refractive index of the cornea; n_3 refractive index of the aqueous humour filling the anterior chamber; n_4 refractive index of the vitreous humour filling the posterior chamber; n_L the calculated equivalent bulk refractive index of the lens. *A* is the diameter of the Real Pupil defined by the margins of the Iris. A full specification of the eye's optical structure also requires the radii of curvature of the two lens surfaces and the two corneal surfaces. In this diagram the optical components and the image surface are assumed to be symmetrical about the optic axis, but this is unlikely to be the case in a real eye where marked asymmetries are known to occur

(3) the relative and absolute sizes of optical components and (4) the size of pupil aperture. The outcome of these variations in optical design is that when viewing the same scene, retinal images in different eye types can have quite different characteristics. In particular, retinal images may differ in absolute size (Sect. 1.4), brightness (Sect. 1.5), and the angular extent of the portion of space imaged upon the retina (Sect. 1.6). Furthermore, within an eye, optical structures are plastic and can be altered (within limits) to focus clearly objects at different distances (Sect. 1.5.3 and Chap. 2), or to take account of changes in the ambient luminance of a scene (Sect. 1.5.1).

It has long been assumed (e.g. Walls 1942; Polyak 1957; Duke-Elder 1958) that differences in eye design, and the resultant differences in retinal image properties, are not the "neutral" result of random variations but that they can be interpreted by "adaptionist" arguments. That is, differences in eye design may be seen as correlated with, or adapted to, aspects of the sensory problems faced by individual species in completing their annual life cycles. This assumption underpins the discussions which follow.

1.3 Quantitative Descriptions of Eye Structures and Their Properties

Rochon-Duvigneaud (1943) showed clearly the value of anatomical descriptions in presenting intriguing functional questions about diversity in the optical structures of bird eyes. Since the function of these optical structures is the production of an image of an animal's world, answers to these questions ideally require information on the properties of images formed in different eyes. Such information is best obtained by direct measurement in the intact eye. While this can be achieved, it does require elaborate procedures and yields information principally concerned with image quality rather than, for example, details of image brightness and size (e.g. Shlaer 1972; Westheimer 1972).

Information of more direct relevance to these latter properties, which can be related to interspecific differences in eye structure and the sensory problems posed by particular life-styles, can be achieved by computations based upon *schematic eye models*. These are mathematical descriptions of the average eye's optical structure and properties. The advantage of such models is that they replace comparative anatomical descriptions of optical structures with quantitative differences which can be readily compared between species.

In constructing schematic eyes, optical characteristics of the components and the overall properties of the system are usually described in terms similar to those used to characterize any other linear magnification system, such as a camera or projector lens. Six parameters are most commonly used: focal length, pupil aperture, *f*-number, magnification, field of view and depth of field (see Fig. 1.1). Just two of these parameters, focal length and pupil aperture, are of primary importance, since they define the limits of the other parameters. For descriptions of procedures for constructing schematic eyes, see Fig. 1.1 and Davson (1962), Fincham and Freeman (1980), Martin (1983) and Hughes (1977, 1986).

Most schematic models of vertebrate eyes have important limitations, discussed more fully in Martin (1983). Many apply strictly to the paraxial region of the optical system and therefore do not describe fully image formation in the periphery. There is likely to be marked intraspecific variation in the optical properties of individual components of the eye and also in the total refractive properties of individual eyes. This has been clearly demonstrated in human eyes where significant variations in surface curvatures and lens thickness have been recorded (Howcroft and Parker 1977), together with marked differences between the eyes in total refractive power (Sorsby et al. 1957, 1961). While these considerations do not invalidate the construction of an "average" schematic eye for a species, they do restrict the subtlety of interspecific differences in optical properties which can be sensibly discussed. They also limit the degree of correlation which is possible between parameters of schematic eyes, behaviour and sensory problems posed by life in particular habitats.

One further parameter which has been derived from a schematic eye model is the optical visual field (Vakkur and Bishop 1963; Martin 1982, 1986a). However, since these computations ignore the assumptions of paraxial optics, it is better that data on visual fields are derived by direct measurement (see Sect. 1.6).

1.4 Interpretations of Diversity

Framing hypotheses which relate form to function within an adaptionist framework is always a difficult exercise (e.g. King and King 1980) and this is no less problematic when discussing visual systems (Ali 1978; Lythgoe 1979). This is because such an exercise not only requires quantitative descriptions of eye structure, but also needs a detailed understanding of the ecology and behaviour of the species concerned. Progress made in understanding the sensory aspects of differing life styles has indicated the complexity of factors which must be considered (see also Chaps. 4, 14 and 15). For example, older certainties (Walls 1942; Tansley 1965) about the differing visual demands of nocturnal versus diurnal, or aquatic versus terrestrial life-styles, have now been questioned (Lythgoe 1979; Martin 1990). Furthermore, there is now some understanding of the interdependence in birds of different perceptual systems (including touch, audition and olfaction) in conducting a seemingly straightforward "visual" task such as foraging (Martin 1990).

Reasonable progress in understanding the form and function of eye structure can only be expected if conducted within a comparative context in which data from species with a range of different sensory problems and behavioural repertoires can be contrasted. Such a comparative approach reduces the likelihood that analyses lead only to naive adaptionist arguments which cannot account in general terms for the diversity of structures seen.

1.4.1 Shape and Size of Eyes

Figure 1.2 illustrates two extremes of eye size and shape found in birds. In owls (Strigiformes) eyes are tubular and absolutely large. An owl eye is so large relative to the skull that much of it extends outside its orbit. Such an arrangement prohibits eye movements of any significance (Steinbach and Money 1973) since the usual three pairs of muscles used to move bird eyes (Martin 1985) are employed principally to hold the eye in the orbit. In the case of the tit (Passeriformes) the eyes are flatter and well concealed in the orbit within which they readily move. In both species eye size relative to the skull is so great that the eyes almost meet and are separated by only a thin bone septum. The absolute difference in eye size between these species is made clear by the observation that the whole tit skull would fit comfortably within a single owl eye.

The largest eyes of any vertebrates are reputed to be those of ostriches (*Struthio camelus*; Struthioniformes) which are estimated to have axial lengths of approximately 50 mm (Duke-Elder 1958); for comparison, the average axial length of the human eye is 24 mm. By contrast, eyes in many small passerines are between 6–8 mm in length (Walls 1942; Donner 1951; Martin 1986a).

1.4.1.1 Eye Size and Weight Considerations

A particularly important factor in the evolution of all avian anatomical structures has been weight reduction, especially of peripheral body structures

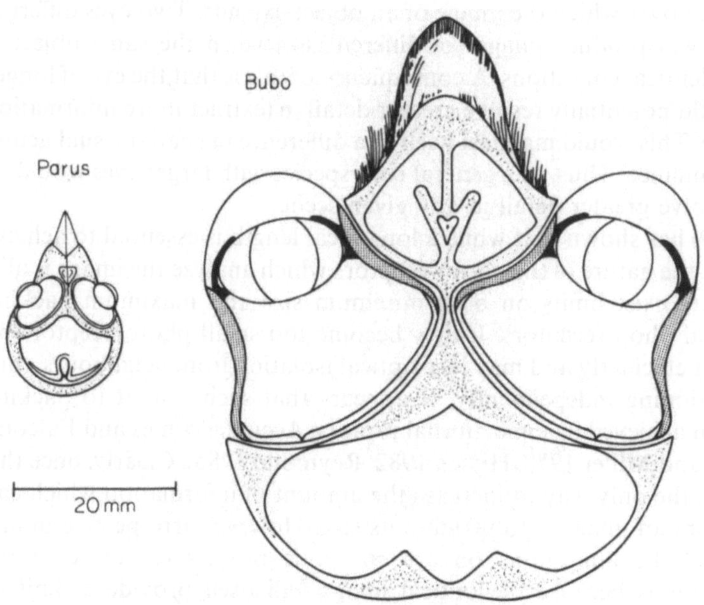

Fig. 1.2. Horizontal sections through the head of the black-capped chickadee (*Parus atricapillus*) and the great horned owl (*Bubo virginianus*). Redrawn to scale from Wood (1917)

including the head (King and King 1980). Regarded as approximately spherical water-filled bodies, it is clear that eyes are, by avian standards, absolutely heavy structures whose weight increases as the cube of their axial length. Thus, doubling eye size will increase its weight approximately 8-fold, and the difference in weight between typical passerine eyes and those of ostriches will be almost 240-fold. It is not surprising, therefore, that the largest eye is in a flightless bird where weight considerations may no longer be of prime importance. However, flightlessness is not a sufficient condition for the evolution of large eyes. The flightless kiwis, Apterygiformes (body weight 1200–3500 g), have eyes similar in size to those of passerine species weighing less than 75 g (Dawson 1978; Martin 1986a).

The different overall shapes of bird eyes (e.g. Fig. 1.2) are not independent of absolute size; only the larger eyes are tubular and smaller eyes tend to be more spherical or flat. The tubular shape may have little to do with visual functioning, although it is not without visual consequences since it restricts the extent of the functional visual field (Sect. 1.6). Tubular eyes are perhaps satisfactorily viewed as a solution to the demands of weight reduction within an aviform skull. Thus owl eyes can be viewed as absolutely large eyes in which peripheral portions have been removed to effect weight reduction.

1.4.1.2 Eye Size and Visual Resolution

In terms of visual functioning the most important parameter associated with eye size is focal length (Fig. 1.1). As in any other lens system focal length determines the area of retina over which the image of an object extends. Two eyes differing in focal length will produce images of different size when the same object is viewed under identical conditions. A consequence of this is that the eye of longer focal length could potentially resolve greater detail in (extract more information from) the image. This would manifest itself in a difference in spatial visual acuity at the same luminance. Thus, as a general rule, species with larger eyes would be expected to resolve greater detail in any given scene.

Miller (1979) has shown that while a long focal length is essential to achieve high resolution, the nature of the photoreceptors which analyze the image is also important. There exist limits on both minimum size and maximum packing density of retinal photoreceptors. If they become too small photoreceptors no longer trap light efficiently and may lose optical isolation from neighbours, thus no longer functioning independently. It appears that such a limit to packing density has been achieved in some diurnal raptors, Accipitriformes and Falconiformes (Snyder and Miller 1978; Hirsch 1982; Reymond 1985). Clearly, once this limit is reached, the only way to increase the amount of information which can be extracted from an image is to increase its size. However, irrespective of any weight or metabolic limitations on eye size, this process cannot continue ad infinitum. This is because the optical image will itself provide a limit to resolution due to the increased likelihood of aberrations within the optical system (Miller 1979; Sect. 1.5.2).

1.4.1.3 Eye Size and Ambient Light Levels

Natural light levels in terrestrial habitats vary over a truly vast range. Within its daily cycle of activity a mobile animal could experience light levels which span 11 orders of magnitude (Martin 1990). Factors which influence this wide range of natural illuminance include: elevation of the sun, elevation and phase of the moon, presence of starlight and airglow, degree of cloud cover, attenuation produced by a vegetation canopy and reflectance of adjacent surfaces.

Vertebrate visual systems must have evolved within the context of this widely varying light regime and so it is pertinent to consider how this range of light levels might have influenced the evolution of eye structure. One approach has been to consider those features which would be necessary in any image analyzing system which is designed to extract the maximum amount of information from an image at any particular light level, and then to consider how these features might be combined in a system that functions through a range of light levels.

An analysis of this kind (Snyder et al. 1977) has shown that in order to maximize the amount of information which can be extracted from a retinal image, both the effective size and density of photoreceptors must alter with luminance level. In essence, larger and more widely spaced receptors are required at lower light levels, and smaller, densely packed receptors are required at high light levels. Thus for an eye to function adequately over a wide luminance range the effective size and spacing of the photoreceptors analyzing the image must vary. Snyder et al. (1977) suggest that this is achieved by the pooling of photoreceptor outputs at the bipolar cells to achieve increasingly large "effective receptors" as light levels decrease.

The degree of flexibility which can be achieved in this way will be greater if the image is spread over many photoreceptors whose outputs can be summed in various ways by the nervous system as light levels decrease. The large eyes of owls (Fig. 1.2) can be interpreted on this basis rather than simply as eyes designed to produce a bright retinal image (Martin 1982; Sect. 1.5.1). Owl eyes can be viewed as designed to function throughout the wide range of natural light levels which can occur between dusk and dawn (9 orders of magnitude), rather than just within the lower ranges of night-time light levels. Indeed there is evidence that although some owls adopt a strictly nocturnal lifestyle (restricting all regular activity to between dusk and dawn) their visual system functions adequately by both night and day. For example, at high daytime light levels the visual acuity of tawny owls (*Strix aluco*) is similar to that of strictly diurnal pigeons (*Columba livia*), yet at low light levels absolute sensitivity in this owl exceeds that of pigeons 100-fold and this difference can be attributed to retinal mechanisms (Martin 1977, 1986b, 1990).

Species whose life-style exposes them only to the lowest naturally occurring light levels might be expected to have smaller eyes. This is because at such light levels only low spatial resolution is possible and this can be adequately achieved with a small image analyzed by large effective receptors (Snyder et al. 1977;

Martin 1983). Birds which are exclusively active at night on a forest floor, such as the flightless kiwis (Reid and Williams 1975; Fuller et al. 1991) are the most likely to experience only these lowest natural light levels (Martin 1990). As noted above (Sect. 1.4.1.1), kiwi eyes are less than one-third the size of owl eyes, with an axial length of approximately 8 mm. Although Walls (1942) suggested that kiwi eyes are degenerate, it has recently been shown that they are indeed optically functional (Sivak and Howland 1987). However, as in nocturnal rodents, which live in similar habitat types and also have small eyes (e.g. the rat, Hughes 1979), kiwis seem to depend in their foraging on a sense other than vision. Among birds, these species are unusually reliant on olfaction (Bang and Wenzel 1985), and this could be partly attributed to the low visual spatial resolution which is possible at the nocturnal light levels of forest floors.

1.4.2 The Optical Design of Eyes

The above discussions suggest that bird eyes are not simply large or small as a result of allometric scaling with body size, but are influenced by ecological and behavioural factors. In this section it is also seen that optical systems within eyes of different sizes are not just allometrically scaled models of each other. Differences in optical design can also be interpreted with reference to ecological and behavioural parameters.

One convenient way of comparing optical structure is to examine the relative contributions of the principle refractive elements, lens and cornea, to the total refractive power of the eye (F_E). Refractive power of a cornea (F_C) is determined principally by its radius of curvature, while the refractive power of a lens (F_L) is a function of surface curvatures, thickness and refractive index gradient. The latter is particularly important since two lenses with identical physical dimensions could differ in refractive power if their internal structures are different.

The marked ways in which optical designs differ between species are made clear in Table 1.1. Although the absolute sizes of Manx shearwater (*Puffinus puffinus*) and pigeon eyes are almost identical, the ratio $F_L{:}F_C$ is quite different. In pigeons the cornea does most of the refraction, contributing about 2.5 times the refractive power of the lens, while in shearwaters the situation is almost reversed, with the lens contributing considerably more than the cornea to total refraction. By contrast, comparison between starlings (*Sturnus vulgaris*) and tawny owls show that although these two eyes differ in absolute size by a factor of 3.6, the ratio $F_L{:}F_C$ is similar, thus indicating that their optical systems are virtually scaled versions of each other.

These differences in optical structure are not functionally trivial. Thus Table 1.1 shows that while pigeon and shearwater eyes are of the same absolute size, differences in their optical design means that their focal lengths (PND) are different. In pigeons the cornea contributes most to total refraction and this brings the nodal point more forward giving the eye a longer focal length compared with shearwater eyes. Functionally this means that if these two

Table 1.1. Selected eye parameters for four bird species[a]

Species	F_E (*D*)	PND (mm)	F_L (*D*)	F_C (*D*)	Ratio F_L:F_C	Axial length (mm)	Ratio PND:axial length
Starling *Sturnus vulgaris*	209	4.78	112.5	124.6	0.90	7.92	0.60
Shearwater *Puffinus puffinus*	154	6.49	108.6	68.1	1.60	11.82	0.55
Pigeon *Columba livia*	126	7.91	38.7	95.9	0.40	11.62	0.68
Tawny owl *Strix aluco*	58	17.24	29.9	35.7	0.84	28.50	0.60

[a] *D* dioptres; F_E, F_L and F_C refractive powers of eye, lens and cornea respectively; PND posterior nodal distance (or anterior focal length). Data taken from schematic eye models of the European starling (Martin 1986a), tawny owl (Martin 1982), pigeon (Marshall et al. 1973; Martin and Brooke 1991) and Manx shearwater (Martin and Brooke 1991)

species were looking at the same scene, pigeons would, by virtue of a larger image, have the potential for higher spatial resolution.

It has long been assumed that at least some interspecific differences in optical design can be attributed to differences in the sensory problems posed by particular ecological conditions and life-styles (e.g. Walls 1942; Tansley 1965; Lythgoe 1979). Two factors which have been considered to be particularly important in this regard are ambient light levels and an amphibious life-style. In analyses of how these factors might be correlated with eye design, data from mammalian as well as avian species have been used, on the assumption that if differences in eye design are correlated with fundamental ecological or life-style parameters then the same general rules might apply to the optical design of both mammalian and avian eyes.

1.4.2.1 Optical Design and Ambient Light Levels

To simplify discussion, it has often been convenient to describe species as exhibiting regular nocturnal or diurnal activity. However, as noted above (Sect. 1.4.1.3), many "nocturnal" species may be active over the full range of light levels (9 orders of magnitude) which can occur naturally after sunset. Diurnally active species are typically active over a smaller range (4 orders of magnitude) of higher light levels. Figure 1.3 shows the ratio of the lens to corneal refractive power (F_L:F_C) as a function of eye size (axial length) in four bird and six mammal species which have been allotted to either nocturnal or diurnally active categories.

The straight line of Fig. 1.3 is the linear regression of data points for nocturnal species. It indicates that among nocturnal forms lens power becomes relatively greater as eye size decreases. No such relationship is apparent among diurnally active species. In the latter group, optical structure (indicated by the

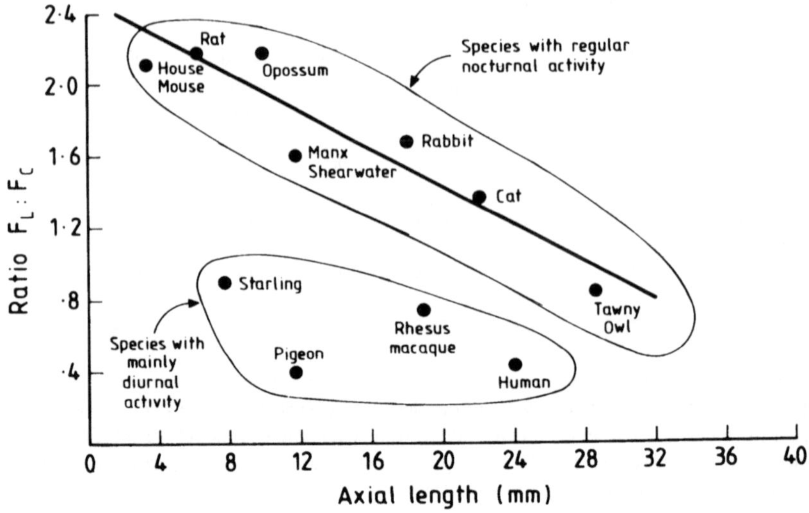

Fig. 1.3. Ratio of lens: cornea refractive powers (F_L:F_C) as a function of axial length in the schematic eye models of seven species of mammals and four species of birds. The *straight line* is the linear regression of the data points for the nocturnally active species. Data from the following sources: tawny owl (Martin 1982); pigeon (Marshall et al. 1973); European starling (Martin 1986a); rat (Hughes 1979); house mouse (Remtulla and Hallet 1985); opossum (Oswaldo-Cruz et al. 1979); cat (Vakkur and Bishop 1963); human (Hughes 1977); rabbit (Hughes 1972); rhesus macaque monkey (Hughes 1977)

F_L:F_C ratio), is relatively constant across a three-fold range of eye size. In these eyes the cornea always contributes the greater part of total refractive power, thus bringing the nodal point forward within the eye. This has the result that, in eyes of the same axial length, "diurnal" forms, by virtue of their larger image size, have the potential for higher visual acuity while "nocturnal" eyes produce a brighter retinal image at a given pupil size (also see the discussion of image brightness in Sect. 1.5.1).

One particular assumption about the optical structure of bird eyes has been that focal length (PND) can be deduced simply from knowledge of an eye's axial length (AL), and it has been suggested that the PND:AL ratio has a constant value of 0.6 across species (e.g. Fite and Rosenfield-Wessels 1975; Hughes 1977). Such a rule implies that all bird eyes have the same optical design, but the above discussion makes clear that this is not the case. Figure 1.4 indicates that the PND:AL ratio is also a function of eye size and of the degree to which a species is nocturnally or diurnally active. In the small nocturnal eyes PND:AL is approximately 0.52, whereas in the largest owl eyes it is about 0.65. As expected from the noted differences in optical structure, PND:AL is higher in diurnal than in nocturnal eyes of the same size but the ratio PND:AL is not constant. However, it does not appear to be a simple function of eye size as in the nocturnal forms.

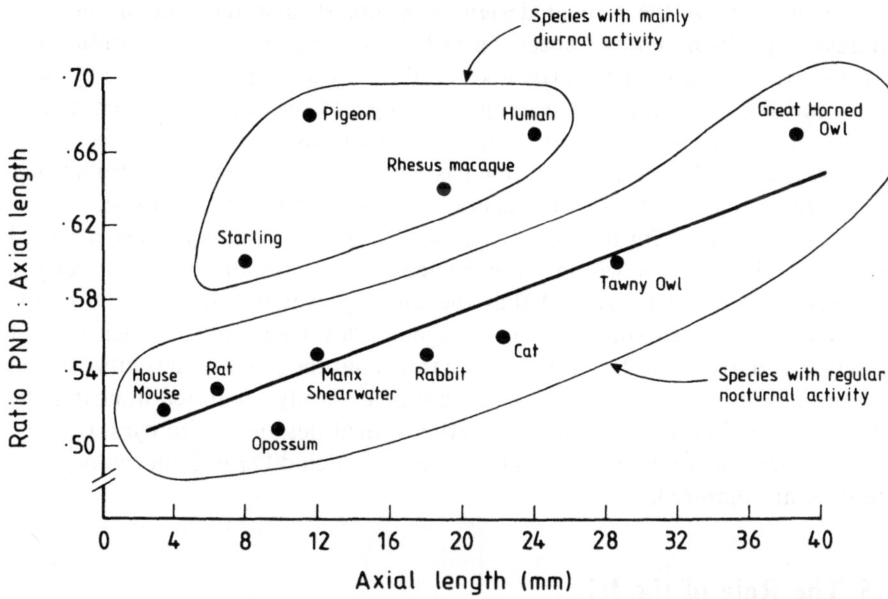

Fig. 1.4. Ratio of focal length (Posterior Nodal Distance): Axial length as a function of axial length in the eyes of five bird species and seven species of mammals. The *straight line* is the linear regression of the data points for species which exhibit regular nocturnal activity. All data are based upon schematic eye models from the same sources as listed in Fig. 1.3 plus data for great horned owl (Murphy et al. 1985)

1.4.2.2 Optical Design and Amphibious Habits

Species which employ vision in both air and under water are faced with a specific optical problem. Upon immersion the refractive power of the cornea is abolished because the corneal surface now separates media of near identical refractive index. If the eye of an amphibious animal is to produce a well-focused image in both air and water, this loss of corneal refractive power must be compensated for, probably by an accommodative increase in the refractive power of the lens (see Sect. 15.1). Table 1.1 indicates, for example, that should an eye like that of a starling be immersed in water, its lens would need to provide 124 D of additional refractive power in order to produce a clearly focused image.

These considerations led Sivak (1976) to propose that an eye designed for emmetropic vision in both air and water would benefit from a cornea which is relatively flat, and hence of low refractive power. The lower the refractive power of the cornea in air, the less accommodation is required upon immersion. Such an optical design is found in the eyes of penguins Sphenisciformes (Sivak 1976), probably the most aquatic of all birds (Davis and Darby 1990). In Humboldt penguins (*Spheniscus humboldti*) the cornea has a refractive power of 29 D, the loss of which upon immersion can perhaps be compensated for by an accommodative mechanism (Martin and Young 1984; Sivak et al. 1987).

Accounting for the optical design of Manx shearwater eyes presents an interesting problem in this respect. These birds catch prey at the sea surface and a metre, perhaps more, below (Brooke 1990). They are regularly active by both night and day, entering and leaving breeding colonies at night when vision, among other senses, is used to locate individual nest burrows (Brooke 1990). Shearwaters can therefore be described as both semi-nocturnal and amphibious.

If Figs. 1.3 and 1.4 indicate general features in the optical design of vertebrate eyes adapted for use at low light levels, then the relatively low-powered cornea of shearwaters would be expected for an eye of that axial length. Conversely, it could be argued that the lower powered cornea (68 D) of the shearwater compared with that of the pigeon (96 D) is a feature which is most closely correlated with the amphibious habit. That the possible amphibious and nocturnal features of shearwater eyes cannot be easily separated indicates the complexity of factors which may underlie optical design in bird eyes. Clearly, further comparative data on a range of species which differ in both ecology and life-style are required.

1.5 The Role of the Iris

There are few studies of the role of the iris specifically in bird eyes and most of its functions can be discussed in general terms only. These functions are, however, of central importance to visual performance since they influence the brightness and quality of the retinal image, and the depth of focus of the visual system. In amphibious birds there is evidence that the iris may also serve an accommodatory function by influencing the refractive power of the lens (Hess 1910; Levy and Sivak 1980).

1.5.1 Pupil Size and Image Brightness

The diameter of the entrance pupil in any optical system plays a critical role in controlling image brightness. When viewing a point source of light, such as a distant star, image brightness is directly related to the diameter of the entrance pupil irrespective of the characteristics of the optical system. When viewing extended light sources (the majority of natural stimuli) image brightness is inversely related to the square of the *f*-number [the anterior focal length (PND)/entrance pupil diameter (*A*); see Fig. 1.1], and so the smaller the *f*-number, the relatively brighter the image. For a detailed discussion of the relationship between *f*-number and retinal image brightness, see Davson (1962) and Kirschfeld (1974).

It might be expected that the eyes of nocturnally active species would have lower minimum *f*-numbers (higher maximum image brightness) than diurnal species. Figure 1.5 shows that this is indeed the case, although the average difference in image brightness between animals with eyes of equal size which are primarily nocturnal or diurnally active equals only four- to five-fold. Figure 1.5 also shows that maximum retinal image brightness increases as eye size decreases.

In an earlier survey of estimated retinal image brightness in a range of vertebrate species, Hughes (1977) concluded that nocturnal and diurnal forms differ in their maximum image brightness by a maximum of five-fold. In a direct comparison between the eyes of nocturnal and diurnal forms of a lizard, Citron and Pinto (1973) found only a two-fold difference in image brightness.

These differences in maximum image brightness between nocturnal and diurnal forms are very small compared with the actual difference in light levels which occur by night and by day (see Sect. 1.4.1.3), and so differences in image brightness cannot be viewed as compensating for them in any way. However, there is evidence that these relatively small differences in maximum image brightness are indeed functional. Thus, the approximately 2.5-fold difference in absolute visual sensitivity between human and tawny owl eyes is accounted for by the difference in image brightness between the two eyes (Martin 1977, 1982); tawny owls have higher visual sensitivity because their retinal images are brighter, not because of differences in the sensitivity of their retinas, which in both species appears to be close to a theoretical maximum limit (Barlow 1981; Martin 1982). However, it is also clear from comparison of maximum image brightness in pigeons and tawny owls that the 100-fold difference in absolute visual sensitivity between these diurnal and nocturnal birds (Martin 1977, 1982) is due to a difference in their retinal sensitivities.

The mobile, circular shaped, pupil of pigeons is capable of altering retinal image brightness approximately 16-fold (Marshall et al. 1973). This is insufficient to equalize image brightness over other than a small part of the natural

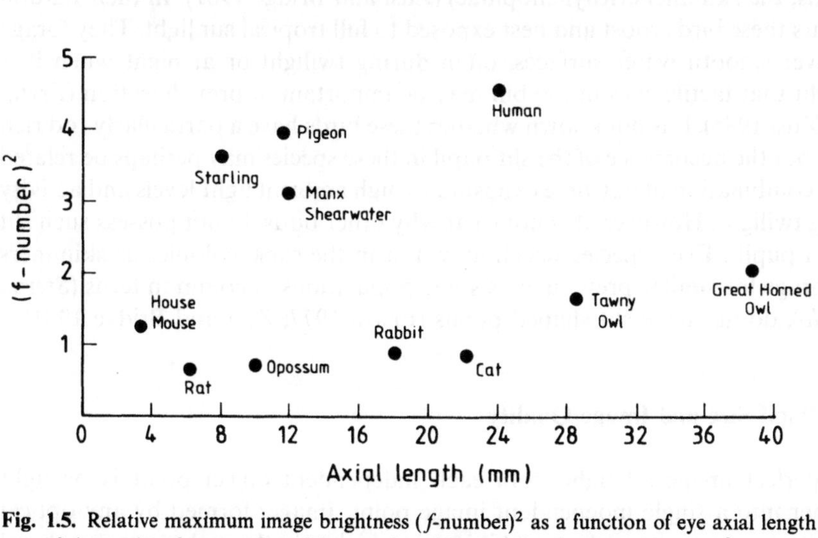

Fig. 1.5. Relative maximum image brightness (f-number)^2 as a function of eye axial length in the eyes of six species of mammals and five species of birds. All values for (f-number)^2 are based upon data from schematic eye models. Same sources as in Figs. 1.3 and 1.4

range of light levels. Within a particular habitat, however, natural light levels usually change by only a small amount. Even when thick clouds quickly cover the sun, illuminance levels change by a maximum of only ten-fold (Natural Illumination Charts 1952). A variable aperture pupil which can alter image brightness over a 16-fold range could therefore serve to equalize retinal illumination as ambient light levels change throughout the day or as the animal moves in and out of cover within a particular habitat type.

It has been proposed that one particular function of a variable pupil is in maintaining the retina in a partially dark-adapted state (Woodhouse and Campbell 1975). The adaptive significance of this is that retinal mechanisms of dark adaptation are relatively slow compared with the speed of response of the pupil (Barlow 1972). With a fixed pupil a rapid change in ambient light level, which could be produced by movement through a structurally complex habitat, would result at times in the eye being inadequately adapted to a new light level. However, a partially closed mobile pupil could be rapidly opened or closed and thus maintain retinal adaptation near the optimum as ambient light level changes.

Slit shaped pupils can be closed to a smaller aperture than circular pupils of the same maximum aperture (Walls 1942). Although such small apertures may function particularly in controlling the depth of focus of the eye in some vertebrate species (see Sect. 1.5.3) they may also serve to protect retinas from excessive bleaching of rod visual pigment. This may be particularly important in species which have a predominantly rod retina, especially if they are likely to be exposed to the highest levels of ambient light during the day, as cats are when they bask in the sun. While slit shaped pupils are relatively common amongst mammals and reptiles (Walls 1942), they occur in only three congeneric species of birds, the skimmers (Rhynchopidae) (Zusi and Bridge 1981). In their natural habitats these birds roost and nest exposed to full tropical sunlight. They forage low over smooth water surfaces, often during twilight or at night when it is thought that tactile cues in the bill may be important in prey detection (Erwin 1977; Zusi 1985). It is not known whether these birds have a particularly rod rich retina but the occurrence of the slit pupil in these species may perhaps be related to the combination of day time exposure to high ambient light levels and activity during twilight. However, it is not clear why other birds do not possess such slit shaped pupils. Even species which may nest in the same colonies as skimmers and forage for similar prey, such as some populations of common terns (*Sterna hirundo*), do not have slit shaped pupils (Erwin 1977; Zusi and Bridge 1981).

1.5.2 Pupil Size and Image Quality

In a perfect image all light from each independent object point is brought together into a single independent image point. Images formed by any optical system are never quite perfect, and it is rays which pass through more peripheral parts of the system which are the principal source of image imperfections. Image quality can therefore be enhanced by removal of these peripheral rays by the iris. Such "optical surgery" (Weale 1974) has been demonstrated convincingly in humans and indirectly in birds.

In humans, retinal image quality is high as long as pupil diameter is below 2.4 mm. Above that size image quality deteriorates as increasingly aberrant rays come to form part of the image (Campbell and Gregory 1960; Campbell and Gubisch 1966). However, it is theoretically understood that in a perfect optical system the larger the pupil aperture, the finer the detail which can be resolved in the image (Miller 1979). Thus optical systems which are closer to perfection than human eyes may become limited in their optical performance at larger pupil apertures. It is predicted therefore that optical systems designed for maximum resolution of detail would have larger pupils when exposed to bright light conditions.

This has been clearly demonstrated in the eyes of some diurnal raptors (Accipitriformes). In Congo snake eagles (*Dryotriorchis spectabilis*), image quality does not deteriorate until a pupil diameter in excess of 6.25 mm is reached. At this diameter image quality is considerably superior to that of human eyes (Shlaer 1972). Miller (1979) studied pupil size in eight species of raptors in stable bright-light conditions. In all of these birds, stable pupil diameter was greater than in human eyes, and Miller concluded that these birds' eyes have superior optical quality to those of humans.

1.5.3 Pupil Size and Depth of Field

Like most optical systems, vertebrate eyes have a fixed surface (the retina) onto which images are projected for analysis by the nervous system. If the optical system has a fixed focal length, only objects at one distance would be clearly focused. Images of objects both closer and further away from that object plane would be defocused to varying degrees. However, whether these defocused images result in a reduction in the amount of information which the visual system can extract from them depends upon properties of both the optical system and of the system which analyses the image (retina and higher brain centres).

The *depth of field* of an optical system is the distance which a focused object can be moved towards or away from the system without appreciably decreasing the amount of information which can be extracted from its image. It has only been measured in detail in humans (Campbell 1957). In experimental terms the criterion used in determining the depth of field is the amount that an object can be moved without causing perceptible defocus of its image. A problem arises since the degree of perceptible defocus will depend upon the resolution of the image analyzing system at any particular image brightness, but in general it can be understood that the poorer the resolution of the visual system, the greater the depth of field. The *depth of focus* of an optical system is the conjugate parameter of the depth of field in the image plane. Formulae have been developed which permit the depth of focus in eyes with different characteristics to be compared (Green et al. 1980; Land 1981; Martin 1982).

These formulae show that depth of focus is related to both focal length (f) and pupil diameter (A) of the eye as well as resolution of the visual system. If z = the maximum diameter that a blur circle (produced by the image of a point source of light) can be before it will be detected as blurred, then depth of focus $= 2z \times f/A = 2z \times f$-number. At high day time light levels it is assumed that z is equivalent to the diameter of a cone photoreceptor outer segment (approximately 2 μm), while at lower light levels z will increase and perhaps be equivalent to the diameter of the "effective receptors" into which rod photoreceptor outputs are grouped by summation at the retinal bipolar cells (Sect. 1.4.1.3).

The influence of pupil diameter upon depth of focus, and consequently upon depth of field, is far from trivial. The importance of this influence is seen most clearly in Fig. 1.6. Here, the calculated depth of focus has been used to determine depth of field in eyes of two focal lengths (5 and 17 mm, approximately equal to those of starling and tawny owl eyes) focused on objects 500, 1000 and 2000 mm from the eye. Depth of field is shown as a function of f-number. It is assumed that the diameter of the permissible blur circle is the same in both eyes at 4 μm, equivalent to twice the typical diameter of an avian photoreceptor outer segment (Bowmaker 1977).

The following general points can be made:

1. Depth of field is not symmetrical about the object upon which the eye is focused. There is a greater distance beyond the plane of focus than before it in which objects will be perceived as focused.

2. The larger the eye (longer focal length), the narrower the depth of field at any given f-number. For example, at f-number = 13, depth of focus of the smaller eye when focused on an object at 500 mm, stretches from about 250 mm to infinity, while in the larger eye only objects lying between 460 and 545 mm from the eye are in focus.

3. The distance from the eye to the object has an important influence upon depth of field. Around close objects the depth of field is smaller at a given f-number than around more distant objects and this effect is larger in smaller eyes.

4. F-number has a marked influence on the depth of field. At a large pupil aperture (low f-number, brighter image) depth of field is very narrow. As pupil aperture decreases, depth of field expands rapidly, and this happens more quickly, the smaller the eye.

It should also be noted that these marked changes in depth of field with f-number are produced by relatively small changes in absolute size of pupil aperture. Furthermore the influence of pupil aperture becomes increasingly marked as eye size decreases. For example, Fig. 1.7 shows that, in the smaller eye, the 3-fold change in f-number between 5 and 15 (which would produce a 9-fold change in image brightness) is produced by altering the diameter of the

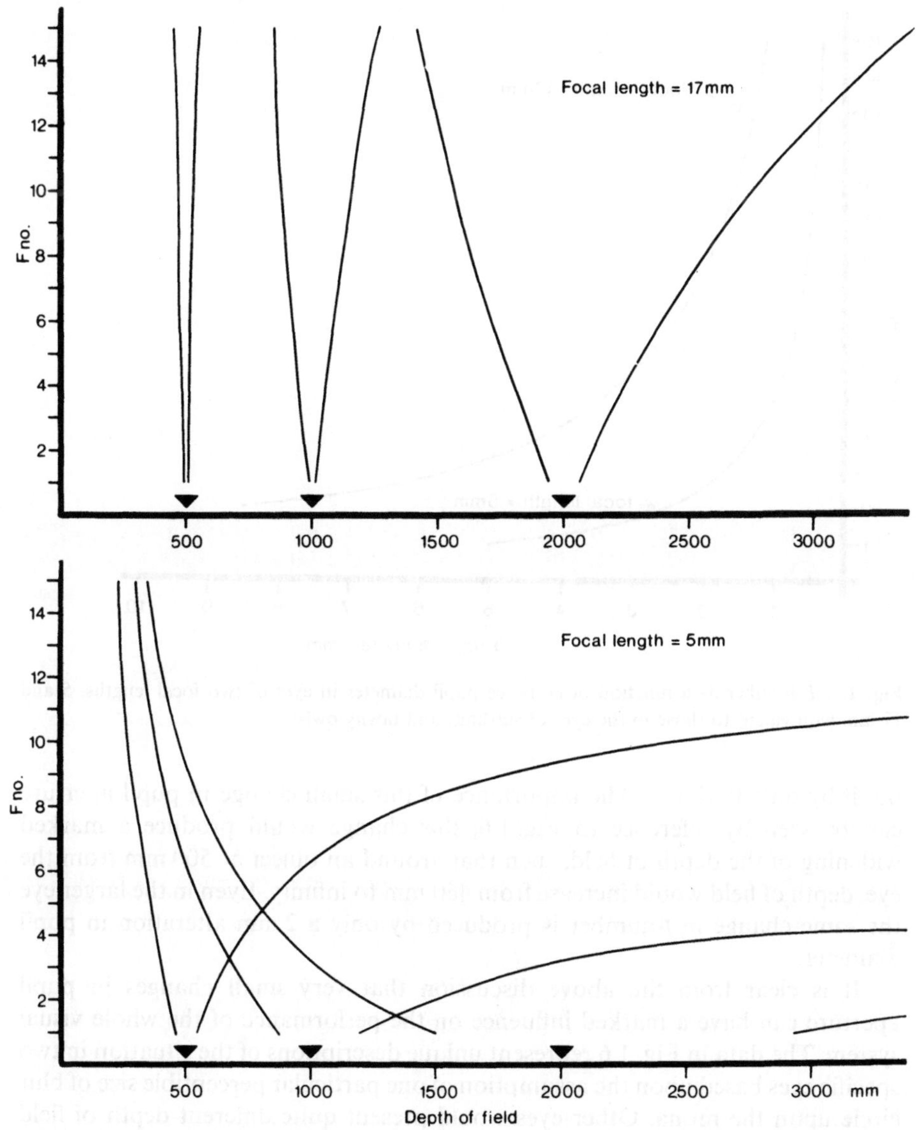

Fig. 1.6. Depth of field as a function of *f*-number in eyes of two focal lengths, 5 and 17 mm (equivalent to the focal lengths in the eyes of starlings and tawny owls) when focused on objects 500, 1000 and 2000 mm from the eye. The *pairs of lines* radiating from each point of focus (indicated by *arrows*) show, at any given *f*-number, the nearest and furthest points from the eye at which another object could be placed and its image still be considered to be in acceptable focus. These pairs of lines indicate clearly that depth of field increases with *f*-number, while the absolute size of depth of field is a complex function both of eye size and of the distance at which the eye is focused. Depth of field was calculated for each eye using the formula described in the text to estimate depth of focus at each *f*-number and depth of field calculated from these values using standard lens equations

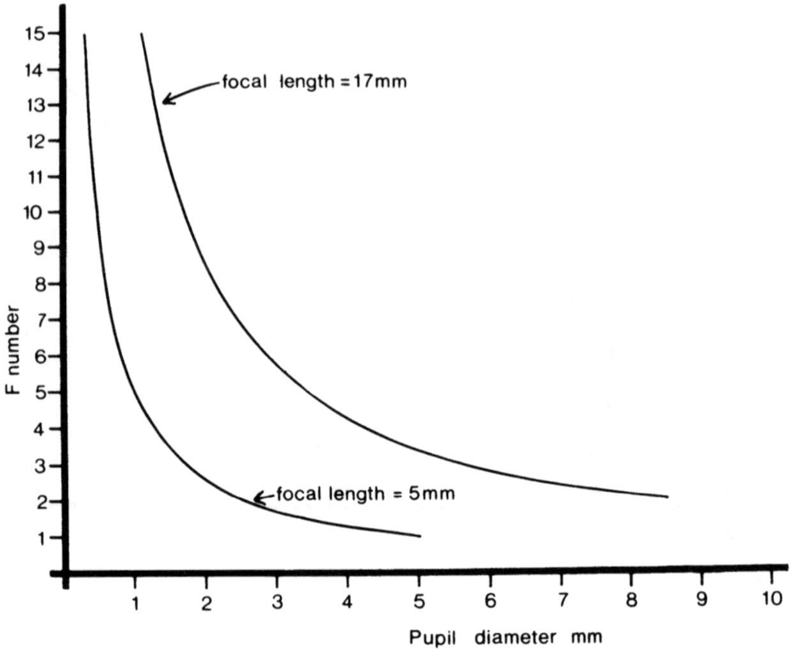

Fig. 1.7. *F*-number as a function of entrance pupil diameter in eyes of two focal lengths, 5 and 17 mm (equivalent to those in the eyes of starlings and tawny owls)

pupil by only 0.67 mm. The importance of this small change in pupil aperture can be seen by reference to Fig. 1.6; this change would produce a marked widening of the depth of field, such that around an object at 500 mm from the eye, depth of field would increase from 460 mm to infinity. Even in the larger eye the same change in *f*-number is produced by only a 2 mm alteration in pupil diameter.

It is clear from the above discussion that very small changes in pupil aperture can have a marked influence on the performance of the whole visual system. The data in Fig. 1.6 represent unique descriptions of the situation in two specific eyes based upon the assumption of one particular perceptible size of blur circle upon the retina. Other eyes would present quite different depth of field values. However, the general principles concerning the influence of *f*-number on depth of focus and depth of field are the same across all eye types. As light levels decrease and resolution of the eye necessarily deteriorates (Snyder et al. 1977), then depth of field will increase as the diameter of the acceptable blur circle increases. Thus, the depth of field of a dark adapted owl is likely to be much larger than that indicated in Fig. 1.6, which uses a criterion for perceptible defocus that would apply to high light levels.

Clearly, a full understanding of the factors which may have led to the evolution of any one particular eye design in birds must take account of the ways that pupil size can influence visual performance. Small changes in the absolute size of the pupil diameter can have marked effects which are perhaps easily overlooked in considering the visual problems of birds and the ways in which they have been resolved during evolution.

1.6 Visual Fields

The visual fields of birds show a high degree of complexity and interspecific variability. This results from differences in the optical structure of eyes, their placement in the skull and the ways in which they can be moved. As in the case of optical structure, functional interpretations of visual fields are dependent upon detailed analyses of very few bird species. However, these species differ in ecology, life-style and evolutionary origins, and this has permitted the postulation of some general functional principles which are discussed below.

It is necessary to distinguish between *optical* and *retinal* visual fields. The optical field is the angular extent of space which is imaged by the optical system. It is approximately indicated by the angular limit from which the pupil can be seen. The retinal field is that portion of the optical field which is served by the retina. Hence it defines the functional visual field; the segment of space in which stimuli must lie if they are to elicit a behavioural response. The difference between the boundaries of the optical and retinal fields may be quite large (up to 20°). Hence, if conclusions are based solely upon observation of the bird's pupil, the casual observer may be led to conclude that a particular bird's functional visual field is considerably larger than it actually is.

1.6.1 Monocular Fields

Ways of calculating the extent of optical fields from knowledge of an eye's optical structure have been proposed (Martin 1983, 1984). In these calculations it is assumed that optical fields are likely to be symmetrical about the optic axis, but this may not be true (see Sect. 1.3). Certainly, in the bird species so far examined, the functional *retinal* fields are highly asymmetric with respect to the optic axis. This has been most fully described in tawny owls (Martin 1984). In all species so far examined (Table 1.2) the nasal hemifield (the portion of the visual field in an approximately horizontal plane between the optic axis and the direction of the bill) is smallest.

The function of such asymmetry is not clear. One consequence is that the frontal binocular field is narrower than it could be if full use were made of the monocular optical field. For example, in tawny owls, frontal binocular field width could be nearly double that actually observed (Martin 1984) and recent observations in four different species of herons, Ciconiiformes, and in the European woodcock (*Scolopax rusticola*; Charadriiformes), show that binocular fields "appear" from casual observation of the pupils to be much larger than

Table 1.2. Comparison of monocular retinal field widths in five species of birds[a]

Species	Total	Nasal	Temporal	Nasal:temporal
Pigeon *Columba livia*	169	77	92	0.84
Mallard duck *Anas platyrhynchos*	191	92	99	0.93
Tawny owl *Strix aluco*	124	51	73	0.70
European starling *Sturnus vulgaris*	161	70	91	0.77
Shearwater *Puffinus puffinus*	148	65	83	0.78

[a] Field widths measured in degrees in an approximately horizontal plane which contains the optic axis. Sources: tawny owl and pigeon (Martin 1982); mallard duck (Martin 1986c); European starling (Martin 1986a); Manx shearwater (Martin and Brooke 1991)

they functionally are because full use is not made of the available optical field (Martin and Katzir 1994).

One possible explanation of this failure to use the full potential width of the nasal optical field may be the need for a high quality optical image in frontal vision, which is invariably used to guide flight and foraging. The assumption is that image quality necessarily deteriorates with increasing eccentricity from the optical axis. While this assumption is generally true for man-made optical systems it may not, however, be a necessary property of all vertebrate eyes. Data on image quality throughout the entire visual field in a bird are lacking, but in the rabbit high quality images are produced in the far periphery (Hughes and Vaney 1978) and there is some evidence that this might also be the case in some owl species (Murphy and Howland 1983).

Comparison of monocular retinal field widths in tawny owls and starlings (Table 1.2) shows how eye shape and the extent of the retina may restrict the functional visual field. Thus, although these two eyes differ in size, their optical designs are nearly identical (Sect. 1.4.2) and so they are likely to have optical fields of the same size. Table 1.2 shows, however, that the owl's monocular retinal field is nearly 40° narrower than that of starlings. This difference can be attributed to the tubular shape of owl eyes (Fig. 1.2), which results in a less extensive retina than in the flatter starling eyes.

1.6.2 Binocular and Panoramic Fields

The eyes of all birds are placed laterally in the skull. No bird species has been described in which the eyes face forward, parallel to each other. Figures 1.2 and 1.8 indicate that even in owls, which are frequently described as having frontally directed eyes, the optic axes diverge by 55°. In some species, such as woodcocks, mallard ducks (*Anas playtrhynchos*) and northern shovelers (*Anas clypeata*), the eyes may diverge by nearly 180° (Martin 1986c, 1994).

The position and movements of eyes in the skull determines the extent of the binocular and panoramic visual fields. These fields are complex three-dimensional structures which are difficult to present in illustrations on a flat surface. Something of the diversity of visual field shapes can be appreciated in Fig. 1.8 which presents the visual fields of four species in an approximately horizontal

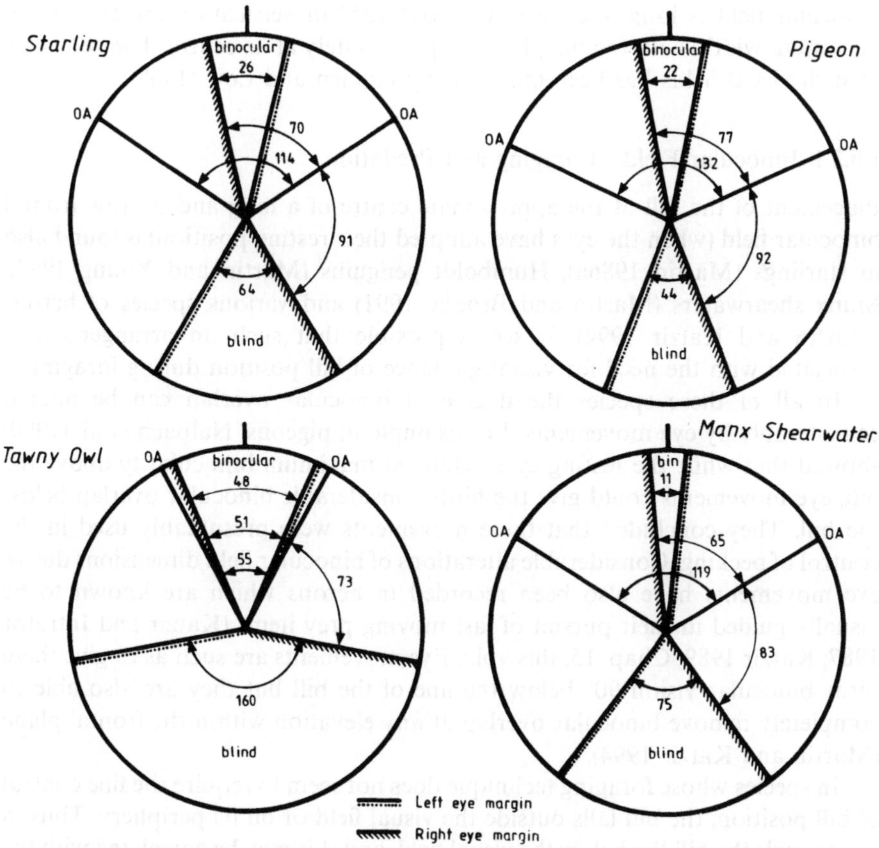

Fig. 1.8. Diagrammatic representations of the retinal visual fields of the European starling, pigeon, tawny owl and Manx shearwater in an approximately horizontal plane containing the optic axes (*OA*). The *bar* at the top of each diagram indicates the direction of the bill. Visual field widths are in degrees expressed with reference to a hypothetical cyclopean projection centre which lies at the centre of the bird's head midway between the nodal points of the two eyes. *Numerical values* indicate for each species the width of the binocular field, the blind area behind the head, the divergence of the optic axes and the widths of the nasal and temporal hemi-fields in a single eye. These data refer to the visual fields when the eyes have adopted their resting position. In all species, except the owl, the cyclopean and binocular visual fields can be altered by eye movements. Data from the following sources: European starling (Martin 1986a); pigeon (Martin and Young 1983); tawny owl (Martin 1984); Manx shearwater (Martin and Brooke 1991)

plane when the head and eyes have adopted a resting position. These comparisons reveal marked interspecific differences in the placement of the eyes in the skull and in the ways in which the two monocular fields combine to give different panoramic and binocular fields (see Sect. 3.4).

Only the visual fields of domestic pigeons have been studied extensively (Martinoya et al. 1981; Martin and Young 1983; Jahnke 1984; McFadden and Reymond 1985; Nalbach et al. 1990), and subtle differences have been found which probably result from differences in investigative procedures and variation between different domestic strains. All investigations show that the frontal binocular field is long and narrow (about 135° in vertical extent and 30° in maximum width) with the bill placed approximately at its centre. The geometry of such visual fields has been modelled by Holden and Low (1989).

1.6.2.1 Binocular Fields, Foraging and Predation

Placement of the bill at the approximate centre of a long and narrow frontal binocular field (when the eyes have adopted their resting position) is found also in starlings (Martin 1986a), Humboldt penguins (Martin and Young 1984), Manx shearwaters (Martin and Brooke 1991) and various species of herons (Martin and Katzir 1994). It seems possible that such an arrangement is associated with the need for visual guidance of bill position during foraging.

In all of these species the degree of binocular overlap can be altered considerably by eye movements. For example in pigeons, Nalbach et al. (1990) showed that while the resting eyes exhibited maximum binocularity above the bill, eye movements could give the birds considerable binocular overlap below the bill. They concluded that these movements were presumably used in the control of pecking. Considerable alterations of binocular field dimensions due to eye movements have also been recorded in herons which are known to be visually guided in their pursuit of fast moving prey items (Katzir and Intrator 1987; Katzir 1989; Chap. 15, this vol.). Eye movements are such as to give these birds binocular vision 90° below the line of the bill but they are also able to completely remove binocular overlap at any elevation within the frontal plane (Martin and Katzir 1994).

In species whose foraging technique does not seem to require the fine control of bill position, the bill falls outside the visual field or on its periphery. Thus in tawny owls the bill lies below the visual field, and this may be correlated with the owl's habit of using its feet to catch prey in a "perch and pounce" feeding technique (Martin 1984). The feet are swung forward to lie within the region of binocular vision just prior to prey capture (Payne 1971).

In mallard ducks (Martin 1986c), northern shovelers and woodcocks (Martin 1994) the bill also falls on the very periphery of the visual field. This can be correlated with findings that the foraging of these birds may be guided exclusively by tactile and taste cues rather than vision (Gottschaldt 1985; Martin 1991a; see Sect. 10.5.2). In these species eye movements are of small amplitude and the eyes are placed so high in the skull that binocular overlap extends from just above the bill through approximately 180° to directly behind the head. Thus these birds have no blind area above or behind them and can see the whole of the celestial hemisphere, which can be scanned for the presence of predators whilst their foraging is guided by tactile cues.

Such comprehensive visual coverage is potentially important in many birds, especially ground feeding species which may be particularly vulnerable to predation by birds of prey. However, it is perhaps only achieved by those species such as woodcocks and ducks which feed by "dabbling" and do not need to monitor their bill position visually during foraging. In visually guided ground feeding species some degree of comprehensive vision must be sacrificed so that the eyes are placed sufficiently forwards for the bill to fall within the frontal binocular field.

1.6.2.2 Binocular Fields and Eye Movements

Even in those species in which the bill is placed near the centre of the binocular field, near complete visual coverage of the space around the bird may be achieved through eye movements. This is exemplified in European starlings, in which the eyes can be swung forward and downwards to achieve a binocular field about the bill as wide as that of the owl. However, unlike owls, whose eyes exhibit no appreciable movement (Steinbach and Money 1973; Sect. 1.4.1), starling eyes can be swung backwards and upwards to give the bird almost complete coverage of the celestial hemisphere, but at the same time abolishing the frontal binocular field (Martin 1986a). Eye movements have their largest amplitude in a plane which passes approximately 20° below the bill, which is similar to the situation in pigeons (Martinoya et al. 1984; Nalbach et al. 1990).

These alterations to the starling's visual field can be functionally interpreted as enabling the birds to use a large binocular field when searching for prey at or near the bill tip, using a specialized visually guided foraging technique termed open-bill probing (Beecher 1978), whilst momentarily being able to scan the celestial hemisphere for conspecifics and/or predators. Starlings are highly sociable species and their habit of feeding in open habitats makes them particularly vulnerable to predatory birds (Feare 1984). It is also noteworthy that visual coverage of the celestial hemisphere could be of particular importance in the detection of navigational and orientational cues (Martin 1991b).

A similar demonstration of how eye movements can considerably alter the visual fields of birds has been provided by tawny frogmouths (*Podargus strigoides*; Caprimugiformes) (Wallman and Pettigrew 1985). In these birds, which take prey with the bill, eye movements can provide extensive visual coverage of the lateral visual field, or the eyes can be swung forward to provide a large frontal binocular field.

1.6.2.3 Binocular Fields and Nocturnal Habits

It is nocteworthy that frontal binocular field width in the nocturnal tawny owls (48°; Martin 1984) is similar to that achieved through eye movements in diurnally active starlings (43°; Martin 1986a) and considerably less than that in humans (120–140°; Emsley 1948; Weale 1960; Vakkur and Bishop 1963). Thus the presumed association of large binocular fields with a predominantly nocturnal life-style (e.g. Walls 1942; Tansley 1965) is brought into question.

Pettigrew (1979) raised the possibility that the more frontal placement of the eyes in both owls and diurnal birds of prey is an example of ecological convergence in two groups of birds of quite different evolutionary origins (Sibley and Ahlquist 1990). He suggested that increased binocular field width was the result of selection to bring more central portions of the eye's optics to point in a forward direction. However, as argued in Sect. 1.6.1, it cannot be assumed that optical quality in peripheral parts of the visual field is necessarily inferior to that near the optical axis.

A further approach to questions concerning the extent of frontal positioning of the eye and binocular field width in nocturnally active birds is suggested by the superiority, in man, of binocular over monocular viewing in the absence of stereoscopic depth cues (Jones and Lee 1981). This superiority in the execution of visually guided tasks becomes apparent when subjects are free to move their heads, and is more marked as luminance levels decrease. Jones and Lee (1981) suggested that stereopsis is not the principal function of binocular vision as supposed, for example by Walls (1942) and Fox (1979), but rather a secondary consequence of increasing binocularity in order to extract the maximum amount of information from the optic flow field (Lee 1980; Chap. 13). Jones and Lee's results suggest that binocular, as opposed to monocular viewing, becomes more important as light levels decrease, not because of increased sensitivity, which in man amounts to approximately 0.15 $\log_{10}$ units (Thorn and Boynton 1974), but rather because of the increase in redundancy of information which can be extracted from the identical optic flow fields presented in the binocular portion of each monocular field. Such redundancy, and hence increased binocularity, is likely to become more important with decreasing light levels, since these inevitably result in a reduction in visual resolution (Snyder et al. 1977), which in turn results in a reduction in the number of optic elements comprising the optic flow field (see Chap. 13). According to such an analysis, it might be expected that the eye axes of owls would be parallel. However, given that an absolutely large size may be an important feature of owl eyes (Sect. 1.4.1.3), the anatomical constraints of placing such eyes within an aviform skull (Sect. 1.4.1.1, Fig. 1.2) may prohibit more frontal placement and hence greater binocular overlap. It should be noted, however, that asymmetry of the monocular visual field does suggest that full use is not in fact made of the nasal hemifield (Sect. 1.6.1) and hence binocular field width is not maximized, even given the anatomical constraints of an aviform skull.

1.6.3 Visual Fields and Amphibious Habits

One further consideration of how ecological factors may influence visual field geometry is provided by penguins. These birds forage almost exclusively below water (Davis and Darby 1990). As in other apparently visually guided foragers, they have a long vertical frontal binocular field with the bill placed approximately at its centre when the eyes have adopted their resting position (Fig. 1.9).

Upon immersion, the refractive power of the cornea is abolished (Sect. 1.4.2.2), narrowing the monocular fields and therefore reducing the width

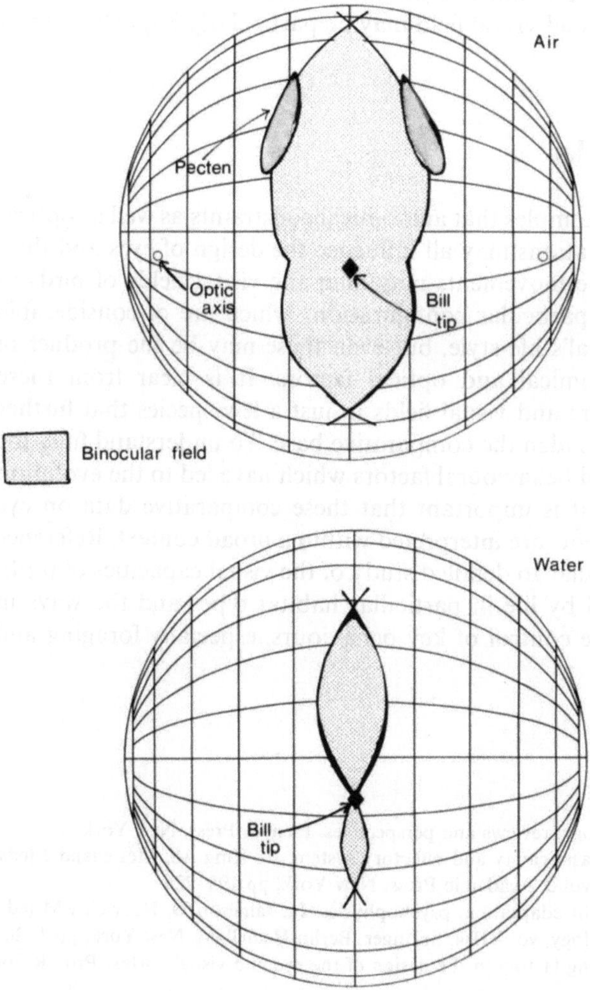

Fig. 1.9. Diagrammatic presentation of the frontal visual fields of the Humboldt penguin (*Spheniscus humboldti*) in air and in water. An equatorial zenithal orthographic projection is used which permits the direct reading of binocular field widths at any elevation about the median sagittal plane of the birds. In these projections the bird is facing the reader with the bill pointing perpendicularly from the plane of the page. It can be imagined that the head is surrounded by a transparent sphere onto which the margins of the left and right retinal visual fields have been drawn. Also shown are the projections of the optic axes and the pectens. The co-ordinates of the projection are equivalent to those of conventional latitude and longitude with the equator in the bird's median sagittal plane

of the binocular and panoramic fields (Fig. 1.9). However, resting eye position is such that the underwater binocular field is not completely abolished, and hence comprehensive visual coverage of the frontal field is maintained. This suggests that resting eye position is such as to maximize the panoramic field underwater. That is, the blind area behind the bird is kept to a minimum given the constraint of maintaining comprehensive frontal vision upon immersion. Penguins are vulnerable to predation by a number of marine mammals and fish (Davis and Darby 1990), and thus a broad visual field may be particularly important when they are underwater.

1.7 Concluding Remarks

It is clear from the above examples that anatomical constraints as well as optical factors and visual considerations may all influence the design of eyes and their placement in the skull. Eye movements may alter the visual fields of birds in complex ways to provide particular configurations which are of considerable importance in an individual's life-style, but even these may be the product of interaction between anatomical and optical factors. It is clear from these analyses of optical structure and visual fields in just a few species that further studies are necessary to broaden the comparative base. To understand fully the possible environmental and behavioural factors which have led to the evolution of present-day structures, it is important that these comparative data on eye structures and eye movements are interpreted within a broad context. Reference needs to be made in particular to detailed study of the visual capacities of birds, the visual problems posed by life in particular habitat types and the ways in which vision is used in the control of key behaviours, especially foraging and orientation.

References

Ali MA (ed) (1978) Sensory ecology: reviews and perspectives. Plenum Press, New York

Bang BG, Wenzel BM (1985) Nasal cavity and olfactory system. In: King AS, McLelland J (eds) Form and function in birds, vol 3. Academic Press, New York, pp 195–225

Barlow HB (1972) Dark and light adaptation: psychophysics. In: Jameson D, Hurvich LM (eds) Handbook of sensory physiology, vol VII/4. Springer, Berlin Heidelberg New York, pp 1–28

Barlow HB (1981) Critical limiting factors in the design of the eye and visual cortex. Proc R Soc Lond B 212:1–34

Beecher WJ (1978) Feeding adaptations and evolution in the starlings. Bull Chic Acad Sci 11:269–298

Bowmaker JK (1977) The visual pigments, oil droplets and spectral sensitivity of the pigeon. Vision Res 17:1129–1138

Brooke M (1990) The Manx shearwater. Poyser, London

Brooke M, Birkhead T (eds) (1991) The Cambridge encyclopedia of ornithology. Cambridge University Press, Cambridge

Campbell FW (1957) The depth of field of the human eye. Opt Acta 4:157–164

Campbell FW, Gregory AH (1960) Effect of size of pupil on visual acuity. Nature (Lond) 187:1121–1123

Campbell FW, Gubisch RW (1966) Optical quality of the human eye. J Physiol Lond 186:558–578

Citron MC, Pinto LH (1973) Retinal image larger and more luminous for a nocturnal than a diurnal lizard. Vision Res 13:873–876

Davis LS, Darby JT (eds) (1990) Penguin biology. Academic Press, London

Davson H (1962) Visual optics and the optical space sense. In: Davson H (ed) The eye, vol 4. Academic Press, New York, pp 3–131

Dawson EW (1978) Apterygiformes. In: Harrison CJO (ed) Bird families of the world. Elsevier, Oxford, pp 25–26

Donner KO (1951) The visual acuity of some passerine birds. Acta Zool Fenn 66:1–40

Duke-Elder S (1958) System of ophthalmology, vol 1. The eye in evolution. Henry Kimpton, London

Emsley HH (1948) Visual optics, 4th edn. Hatton Press, London

Erwin RM (1977) Black Skimmer breeding ecology and behaviour. Auk 94:709–717

Evans HE, Martin GR (1993) Organa sensuum. In: Baumel JJ (ed) Handbook of avian anatomy: Nomina anatomica avium, 2nd edn. Nuttall Ornithological Club, Cambridge, Mass, pp 585–611

Feare CJ (1984) The starling. Oxford University Press, Oxford

Fincham WHA, Freeman MH (1980) Optics, 9th edn. Butterworths, London

Fite KV, Rosenfield-Wessels (1975) A comparative study of deep avian foveas. Brain Behav Evol 12:97–115

Fox R (1979) Binocularity and stereopsis in the evolution of vertebrate vision. In: Cool SJ, Smith EL (eds) Frontiers of visual science. Springer, Berlin Heidelberg New York, pp 336–341

Fuller E, Ching R, Andrews JRH (1991) A monograph of the family Apterygidae Kiwis. Swan Hill, Shrewsbury

Gill FB (1990) Ornithology. Freeman, New York

Gottschaldt K-M (1985) Structure and function of avian somatosensory receptors. In: King AS, McLelland J (eds) Form and function in birds, vol 3. Academic Press, New York, pp 375–461

Green DG, Powers MK, Banks MS (1980) Depth of focus, eye size and visual acuity. Vision Res 10:827–836

Hess C (1910) Die Akkommodation bei Tauchervögeln. Arch Vergl Ophthalmol 1:153–164

Hirsch J (1982) Falcon visual sensitivity to grating contrasts. Nature (Lond) 300:57–58

Holden AL, Low JC (1989) Binocular fields with lateral-eyed vision. Vision Res 29:361–367

Howard R, Moore A (1991) A complete checklist of the birds of the world, 2nd edn. Academic Press, London

Howcroft MJ, Parker JA (1977) Aspheric curvature of the human lens. Vision Res 17:1217–1223

Hughes A (1972) A schematic eye for the rabbit. Vision Res 12:123–138

Hughes A (1977) The topography of vision in mammals of contrasting life style: comparative optics and retinal organization. In: Cresitelli F (ed) Handbook of sensory physiology, vol VII/5. Springer, Berlin Heidelberg New York, pp 613–756

Hughes A (1979) A schematic eye for the rat. Vision Res 19:569–588

Hughes A (1986) The schematic eye comes of age. In: Pettigrew J, Sanderson KJ, Levick WR (eds) Visual neuroscience. Cambridge University Press, Cambridge, pp 60–89

Hughes A, Vaney DI (1978) The refractive state of the rabbit eye: variation with eccentricity and correction of oblique astigmatism. Vision Res 18:1351–1355

Jahnke HJ (1984) Binocular visual field differences among various breeds of pigeons. Bird Behav 5:96–102

Jones RK, Lee DN (1981) Why two eyes are better than one: the two views of binocular vision. J Exp Psychol Hum Percept Perform 7:30–40

Katzir G, Intrator N (1987) Striking of underwater prey by reef herons, *Egretta gularis schistacea.* J Comp Physiol A 160:517–523

Katzir G, Lotem A, Intrator N (1989) Stationary underwater prey missed by reef herons, *Egretta gularis*: head position and light refraction at the moment of strike. J Comp Physiol A 165:573–576

King AS, King DZ (1980) Avian Morphology: general principles. In: King AS, McLelland J (eds) Form and function in birds, vol 1. Academic Press, London, pp 1–38
Kirschfeld K (1974) Absolute sensitivity of lens and compound eyes. Z Naturforsch 29c:592–596
Land MF (1981) Optics and vision in invertebrates. In: Dartnall HJA (ed) Handbook of sensory physiology, vol VII/6b. Springer, Berlin Heidelberg New York, pp 471–594
Lee DN (1980) The optic flow field: the foundation of vision. Philos Trans R Soc Lond B 290:169–179
Levy B, Sivak JG (1980) Mechanisms of accommodation in the bird eye. J Comp Physiol 137:267–272
Lythgoe JN (1979) The ecology of vision. Clarendon, Oxford
Marshall J, Mellerio J, Palmer DA (1973) A schematic eye for the pigeon (*Columba livia*). Vision Res 13:2449–2453
Martin GR (1977) Absolutte visual threshold and scotopic spectral sensitivity in the tawny owl, *Strix aluco*. Nature (Lond) 268:636–638
Martin GR (1982) An owl's eye: schematic optics and visual performance in *Strix aluco* L. J Comp Physiol 145:341–349
Martin GR (1983) Schematic eye models in vertebrates. In: Ottoson D (ed) Progress in sensory physiology, vol 4. Springer, Berlin Heidelberg New York, pp 43–81
Martin GR (1984) The visual fields of the tawny owl (*Strix aluco*). Vision Res 24:1739–1751
Martin GR (1985) Eye. In: King AS, McLelland J (eds) Form and function in birds, vol 3. Academic Press, London, pp 311–373
Martin GR (1986a) The eye of a passeriform bird, the European starling (*Sturnus vulgaris*): eye movement amplitude, visual fields and schematic optics. J Comp Physiol A 159:545–557
Martin GR (1986b) Sensory capacities and the nocturnal habit in owls. Ibis 128:266–277
Martin GR (1986c) Total panoramic vision in the mallard duck, *Anas platyrhynchos*. Vision Res 26:1303–1305
Martin GR (1990) Birds by night. Poyser, London
Martin GR (1994) Visual fields in woodcocks *Scolopax rusticola* (Scolopacidae: Charadriiformes). J Comp Physiol A (in press)
Martin GR (1991a) The sensory bases of nocturnal foraging. Acta XX Congr Int Ornithol, New Zealand Ornithological Congress Trust Board, Wellington, pp 1130–1135
Martin GR (1991b) Aspects of avian vision and orientation. Acta XX Congr Int Ornithol, New Zealand Ornithological Congress Trust Board, Wellington, pp 1830–1836
Martin GR, Brooke M (1991) The eye of a Procellariiform seabird, the Manx Shearwater, *Puffinus puffinus*: visual fields and optical structure. Brain Behav Evol 37:65–78
Martin GR, Katzir G (1994) Visual fields and eye movements in herons (Ardeidae). Brain Behav Evol (in press)
Martin GR, Young SR (1983) The retinal binocular field of the pigeon (*Columba livia*): English racing homer. Vision Res 23:911–915
Martin GR, Young SR (1984) The eye of the Humboldt penguin *Spheniscus humdoldti*: visual fields and schematic optics. Proc R Soc Lond B 223:197–222
Martinoya C, Rey J, Bloch S (1981) Limits of the pigeon's binocular field and direction for best binocular viewing. Vision Res 21:1197–1200
Martinoya C, Le Houezec J, Bloch S (1984) Pigeons' eyes converge during feeding: evidence for frontal binocular fixation in a lateral eyed bird. Neurosci Lett 45:335–339
McFadden SA, Reymond L (1985) A further look at the binocular visual field of the pigeon (*Columba livia*), Vision Res 25:1741–1746
Meyer DB (1977) The avian eye and its adaptations. In: Crescitelli F (ed) Handbook of sensory physiology, vol VII/5. Springer, Berlin Heidelberg New York, pp 560–611
Miller WH (1979) Ocular optical filtering. In: Autrum H (ed) Handbook of sensory physiology, vol VII/6A. Springer, Berlin Heidelberg New York, pp 69–143
Murphy CJ, Howland HC (1983) Owl eyes: accommodation, corneal curvature and refractive state. J Comp Physiol A 151:277–284

Murphy CJ, Evans HE, Howland JC (1985) Towards a schematic eye for the great horned owl. Fortschr Zool 30:703–706
Nalbach H-O, Wolf-Oberhollenzer F, Kirschfeld K (1990) The pigeon's eye viewed through an ophthalmoscopic microscope: orientation of retinal landmarks and significance of eye movements. Vision Res 30:529–540
Natural illumination charts (1952) US Navy research and development project NS 714-100 Rep 374-1 (September), US Navy, Washington, DC
Oswaldo-Cruz E, Hokoc JN, Sousa APB (1979) A schematic eye for the opossum. Vision Res 19:263–278
Payne RS (1971) Acoustic location of prey by barn owls (*Tyto alba*). J Exp Biol 54:535–573
Pettigrew JD (1979) Comparison of the retinotopic organization of the visual wulst in nocturnal and diurnal raptors, with a note on the evolution of frontal vision. In: Cool SJ, Smith EL (eds) Frontiers of visual science. Springer, Berlin Heidelberg New York, pp 328–335
Polyak S (1957) The vertebrate visual system. University of Chicago Press, Chicago
Reid B, Williams GR (1975) The Kiwi. In: G Kuschell (ed) Biogeography and ecology in New Zealand. Junk, The Hague
Remtulla S, Hallet PE (1985) A schematic eye for the mouse, and comparisons with the rat. Vision Res 25:21–31
Reymond L (1985) Spatial visual acuity of the eagle *Aquila audax*: a behavioural, optical and anatomical investigation. Vision Res 25:1477–1491
Rochon-Duvigneaud A (1943) Les yeux et la vision des vertébrés. Masson, Paris
Shlaer R (1972) An eagle's eye: quality of retinal image. Science (Wash DC) 176:920–922
Sibley CG, Ahlquist JE (1990) Phylogeny and classification of birds. A study in molecular evolution. Yale University Press, New Haven
Sibley CG, Monroe BL (1990) Distribution and taxonomy of birds of the world. Yale University Press, New Haven
Sivak JG (1976) The role of a flat cornea in the amphibious behaviour of the blackfoot penguin (*Spheniscus demersus*). Can J Zool 54:1341–1346
Sivak JG, Howland HC (1987) Refractive state of the eye of the brown kiwi (*Apteryx australis*). Can J Zool 65:2833–2835
Sivak JG, Howland HC, McGill-Harelstad P (1987) Vision of the Humboldt penguin (*Spheniscus humboldti*) in air and water. Proc R Soc Lond B 229:467–472
Snyder AW, Miller WH (1978) Telephoto lens system of the falconiform eyes. Nature (Lond) 275:127–129
Snyder AW, Laughlin SB, Stavenga DG (1977) Information capacity of eyes. Vision Res 17:1163–1175
Sorsby A, Benjamin B, Davey JB, Sheridan M, Tanner JM (1957) Emmetropia and its aberrations. Med Res Counc (GB) Spec Rep Ser 293
Sorsby A, Benjamin B, Sheridan M (1961) Refraction and its components during the growth of the eye from the age of three. Med Res Counc (GB) Spec Rep Ser 301
Steinbach MJ, Money KE (1973) Eye movements of the owl. Vision Res 13:889–891
Tansley K (1965) Vision in vertebrates. Chapman Hall, London
Thorn F, Boynton RM (1974) Human binocular summation at absolute threshold. Vision Res 14:445–458
Vakkur GJ, Bishop PO (1963) The schematic eye in the cat. Vision Res 3:357–381
Wallman J, Pettigrew J (1985) Conjugate and disjunctive saccades in two avian species with contrasting oculomotor strategies. J Neurosci 5:1418–1428
Walls GL (1942) The vertebrate eye and its adaptive radiation. Cranbrook Institute of Science, Michigan
Weale RA (1960) The eye and its function. Hatton Press, London
Weale RA (1974) Natural history of optics. In: Davson H (ed) The Eye, vol 6. Academic Press, New York, pp 328–356
Welty JC, Baptista L (1988) The life of birds, 4th edn. Saunders, Philadelphia

Westheimer G (1972) Optical properties of vertebrate eyes. In: Fuortes MGF (ed) Handbook of sensory physiology, vol VII/2. Springer, Berlin Heidelberg New York, pp 449–482
Wood CA (1917) The fundus oculi of birds especially as viewed by the ophthalmoscope. Lakeside Press, Chicago
Woodhouse JM, Campbell FW (1975) The role of the pupil light reflex in aiding adaptation to the dark. Vision Res 15:649–653
Zusi RL (1985) Skimmer. In: Campbell B, Lack E (eds) A dictionary of birds. Poyser, Calton, pp 546–547
Zusi RL, Bridge D (1981) On the slit pupil of the black skimmer. J Field Ornithol 52:338–340

2 Functional Accommodation in Birds

F. Schaeffel

2.1 The Power and Precision of Accommodation as a Distance Cue

Distances can be estimated using a number of monocular and binocular visual cues. Under binocular viewing conditions, distances relative to the plane of fixation can be extracted from the disparity in the retinal images of the two eyes (see Sect. 3.2.1). This mechanism requires elaborate neural operations but has evolved independently in birds and mammals (Pettigrew 1989). Its presence has been shown in a number of birds, such as owls (Pettigrew and Konishi 1976) and kestrels (Fox et al. 1977). There is much variability in its occurrence among species (Pettigrew 1989) and its presence in granivorous birds is still a subject of discussion. Evidence for the use of binocular disparity by pigeons has been presented by McFadden and Wild (1986) and McFadden (1987; see also Chap. 3), while opposing arguments are made by Martinoya et al. (1988) and Pettigrew (1989). A second binocular mechanism would be to extract distance information from an extraretinal cue, the tension of the extraocular muscles, and determine the angle of convergence of the eyes during fixation. The resolution of such a mechanism, however, is limited (Foley 1978).

There are also several ways to measure relative and absolute distances under monocular viewing conditions: if the eye is moving tangentially with respect to the target, the relative retinal angular velocities of images of objects at different distances provide information on relative depth ("motion parallax"). Also retinal image size can provide distance information if information is available concerning the object's size, either from experience (familiarity) or through genetic preprogramming. Finally, the accommodative effort to see a target clearly can provide distance information to the viewer. For all the distance cues listed above, absolute resolution in depth declines with increasing target distance. The resolution of accommodative distance estimation is limited by the depth of focus of the eye (Fig. 2.1; Green et al. 1980; Sect. 1.5.3).

For example, a chicken's eye with a depth of focus of about 1 D (Schaeffel and Howland 1988) has a maximal possible precision, when locating a target at

Department of Experimental Ophthalmology, University Eye Hospital, Röntgenweg 11, D-72076 Tübingen, Germany

M.N.O. Davies and P.R. Green (Eds.)
Perception and Motor Control in Birds

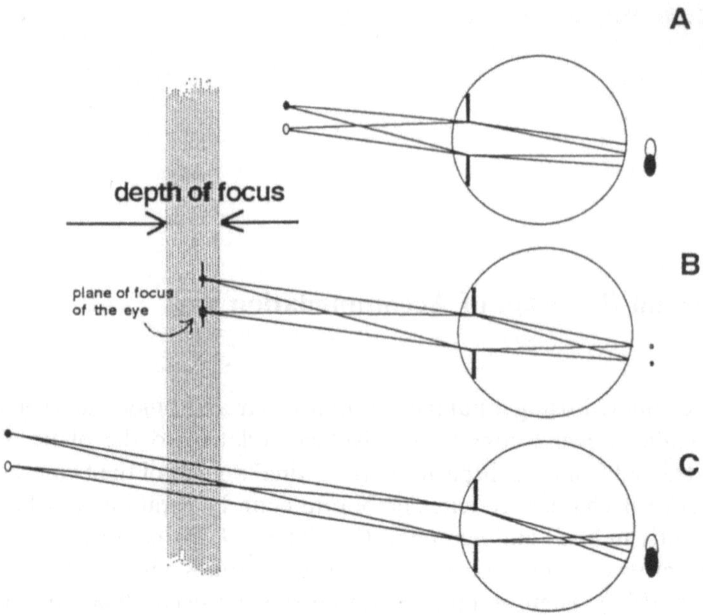

Fig. 2.1. Depth of focus of the eye. Points on the surface of an object closer **A** or more distant **C** than the plane of focus of the eye **B** give rise to blurred images on the retina. The diameter of the blur circle depends on the amount of defocus and pupil size. However, there is a certain range (depth of focus) through which the object can move in depth without causing a detectable change in the diameter of the blur circle. Depth of focus declines with increasing spatial resolution of an eye, because an eye with higher visual acuity can distinguish smaller changes in image sharpness. The quality of the optics of the eye, diffraction of light at the pupil aperture (limited visual acuity) and neuronal factors are responsible for the major part of depth of focus. Depth of focus is most appropriately measured in dioptres because its dioptric equivalent is constant for all viewing distances

8 cm distance (equivalent to 12.5 D) solely by accommodative cues, of about 100/12.5–100/13.5, equivalent to 6 mm. On the other hand, if the target is 10 m away, its distance cannot be estimated more accurately than to 50 m, indicating that accommodative cues are of no value. Therefore, accommodative distance estimation can be useful during pecking behaviour or prey capture from an immobile posture but not for example during flight. As depth of focus is related to visual acuity (Green et al. 1980), the precision of accommodative distance estimation is also correlated with visual acuity. Furthermore, a large pupil reduces depth of focus and, therefore, improves precision as well (see Sect. 1.5.3).

The extraction of distance information from accommodative tonus is not trivial. It requires that the information from the accommodative feedback loop is carried on to motor centres in the brain which guide responses such as pecking. Also, the refractive state of the eye must be known by the brain. Like stereopsis, accommodative distance estimation has the major advantage that it also works from an immobile posture. This may be the reason why it is used by predators which lie in wait for prey, like the chameleon (Harkness 1977), the toad (Collett 1977), and possibly the gecko (Murphy and Howland 1986).

Accommodative distance estimation is useful only under some conditions, and so other visual cues are commonly used simultaneously (see Sect. 16.2.3). Nevertheless, recent behavioural experiments in birds confirm that under monocular and even under binocular viewing conditions, errors in distance estimation can be induced by experimental changes in the accommodative tonus.

There are at least two requirements that must be satisfied if accommodation is to be used as a distance cue. First, the range of accommodation must be large enough that objects can be focused at close range where a reasonable resolution in depth can be obtained even in the presence of some depth of focus of the eye. Second, the speed of accommodation must be high enough to ensure a fast adjustment of focus during a behavioural pattern like pecking where viewing distances change quickly. It is clear that birds with a low amplitude of accommodation, like many large raptors (Walls 1942) or large owls (Murphy and Howland 1983), cannot rely on accommodation when they estimate distances. On the other hand, granivorous birds with extended monocular visual fields and large ranges of accommodation could gain valuable distance information from accommodation. In this chapter, as examples of granivorous birds, data from the chicken (*Gallus domesticus*) and the pigeon (*Columba livia*) are presented. Research on the barn owl (*Tyto alba pratincola*) has recently demonstrated that this small owl also uses accommodative cues (Wagner and Schaeffel 1991), despite its extended binocular visual fields and its ability to use stereopsis (Pettigrew and Konishi 1976).

2.2 A Technique to Measure Accommodation in Unrestrained, Alert Birds

One of the most powerful techniques for recording refractive state and accommodation in freely moving animals is infrared photoretinoscopy (Schaeffel et al. 1987). Because infrared light is used, the animals are not aware that they are being measured and because the photoretinoscope works at 1–2 m distance, they are not disturbed by the measurement. The procedure involves birds being video-taped while the infrared photoretinoscope is operating. The infrared photoretinoscope consists of five horizontal rows of infrared LEDs (light emitting diodes) positioned at different distances from the edge of a shield which occludes about half of the aperture of the video camera lens (Fig. 2.2A).

If light from the LEDs is projected into the eye, a small proportion is also reflected from the fundus (the layers of tissue at the back of the eye, behind the vitreous humour) and returns to the camera. If the eye is focused onto the camera, the returning light is refocused back on the LEDs. As a result, for a well focused eye, the pupil appears almost dark. If the eye is defocused with respect to the camera (see Fig. 2.2B, for myopic and hyperopic eyes), the returning light is

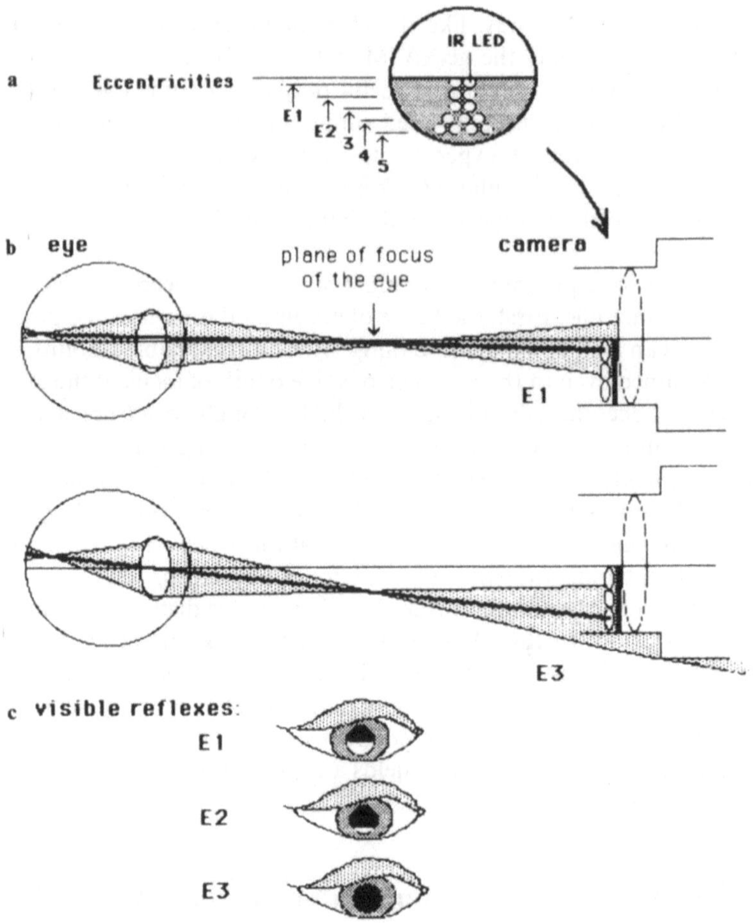

Fig. 2.2a–c. Optics and design of infrared photoretinoscopy. **a** The infrared photoretinoscope consists of rows of infrared light emitting diodes (*LED*) Positioned at different distances ("eccentricities", *E1–E5*) from the edge of a shield that covers about half of the camera lens aperture. **b** The infrared light emitted from the retinoscope enters the pupil and a small proportion returns to the camera upon reflection from the fundus. The vergence of rays returning from the eye provides information on the refractive state of the eye. In the case of a myopic eye the rays are brought into focus in front of the camera ("plane of focus of the eye"). Only the rays coming from the lower part of the pupil are detected by the camera (see paths of the rays), because the lower half of the camera lens aperture is occluded. Accordingly, the lower part of the pupil appears illuminated (**c**, *left column*, eccentricity *E1*). For an eye focused hyperopically with respect to the camera, the plane of focus lies behind the camera. Thus, rays do not cross between eye and camera and only those emerging from the upper part of the pupil can enter the camera aperture. As a result, the upper part of the pupil appears illuminated (**c**, *right column*, eccentricity *E1*). If LEDs from higher eccentricities are radiating, the light crescents seen in the pupil are smaller for a given amount of defocus, and disappear for even higher eccentricities (e.g. eccentricity "*E3*"). If an eye is perfectly focused on the camera, no light can be seen in the pupil at all because the light returning from the fundus is refocused on the light sources by the optics of the eye. **c** Schematic illustration of the appearance of the fundus reflexes seen in the pupil for three eccentricities, *E1* to *E3* (After Schaeffel et al. 1987)

refocused in a plane between the eye and the camera, and subsequently spreads out in a cone, the angle of which depends on the amount of defocus. The relative position of the reflex in the pupil indicates the sign of the refractive error. For an eye myopic relative to the camera, only rays coming from the lower part of the pupil (see Figs. 2.2B and 2.5B) can be detected by the camera due to the shield which covers the lower part of the aperture. For a hyperopic eye, only the upper part of the pupil appears illuminated (Figs. 2.2C and 2.5A). For a retinoscope arranged as shown in Fig. 2.2A, the amount of defocus of the eye in the vertical meridian can be calculated from the relative height of the light crescent in the pupil (Howland 1985). If light is projected into the eye from different eccentricities, the height of the light crescents seen in the pupil varies (Fig. 2.2C). The above method circumvents the problem of the entire pupil being filled with light for large amounts of defocus. Therefore, multiple eccentricities extend the range of measurement and also improve its precision (Schaeffel et al. 1987). If the LEDs are flashed in sequence, the fundus reflex appears to move in the pupil. In birds, the movement of the reflex in the pupil can always be seen very clearly due to the good quality of the eye's optics and because the intraocular media cause minimal scatter of infrared light.

2.3 Mechanisms of Accommodation in Terrestrial Birds

After a long lasting debate (e.g. Slonaker 1918; Gundlach et al. 1945; Sivak et al. 1986) it has finally been shown that the accommodation mechanism in birds includes a corneal component (Schaeffel and Howland 1987; Troilo and Wallman 1987). To measure corneal accommodation in alert birds, three video cameras were employed simultaneously. One camera was equipped with an infrared photoretinoscope and measured refractive state and the second camera measured the positions of eight corneal light reflexes created by eight infrared LEDs which were arranged in a circle of 150 mm diameter ("infrared photokeratometer"). A third camera controlled the distance of the bird's eye from the photokeratometer. The radius of curvature of the cornea was calculated from the distances between the corneal reflexes of the eight infrared LEDs. In the chicken, 50–70% of the total power of accommodation is accomplished by changes in the corneal radius of curvature (Fig. 2.3A), whereas the remainder is provided by changes in power of the crystalline lens.

In the pigeon, the corneal component is even more prominent, and can amount to close to 100%, at least in one individual bird studied (Fig. 2.3B). Corneal accommodation has not yet been shown in other birds. It is known that the changes in curvature of the crystalline lens are accomplished by a direct pressure of the ciliary body on the annular pad of the lens (Levy and Sivak 1980). However, the mechanism of corneal accommodation is still unclear. Schaeffel et al. (1988b) have developed a model of corneal accommodation based on a pressure-mediated mechanism, while a role of a circular portion of the ciliary muscle (Crampton's muscle) has also been suspected (Gundlach et al. 1945). In

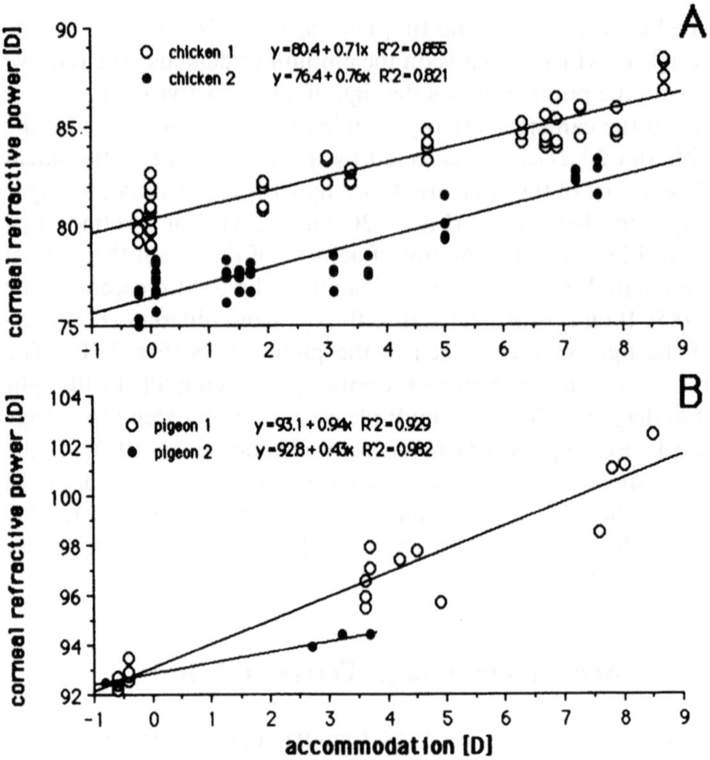

Fig. 2.3A, B. Corneal accommodation in chick and pigeon. Corneal refractive power (as determined by infrared photokeratometry from the anterior corneal surface, with an assumed corneal refractive index of 1.373) is plotted against accommodation. Both variables are in dioptres. **A** In the two chickens observed (ages: 90 days) the slopes were about 0.7, indicating that 70% of the total accommodative power was provided by the cornea. In younger chickens, the lenticular component is more prominent. **B** Corneal accommodation in two pigeons (ages: about 2 years). Only for one pigeon was enough data collected to allow the whole range of accommodation to be represented, demonstrating that the corneal contribution to accommodation was almost 100%. The other pigeon proved uncooperative, and only a small range of accommodation was tested. From the four data points collected, the slope appears flatter (After Schaeffel and Howland 1987, with additional data added)

addition, the presence of a ring of scleral ossicles in the bird eye (Walls 1942) may be important for a pressure-mediated mechanism of corneal accommodation (Schaeffel and Howland 1987). One advantage of corneal accommodation in comparison with lenticular accommodation is that it probably does not decline very much with age. It has been observed that even a 10-year-old pigeon was able to accommodate 6 D (dioptres) (Hodos, Howland, and Schaeffel 1989, unpubl. observ.). In humans, the accommodative range typically falls below 1 D at the age of 45, which is comparable to 10 years of age in a pigeon. Also, monkeys suffer a reduction with age in their ability to accommodate and so provide important models for the study of presbyopia.

2.3.1 Speed of Accommodation

A striking feature of accommodation in the chicken, barn owl and hawk owl (*Surnia ulula*), the only species on which data on speed are available, is that it is extremely fast. The high speed may result from the fact that the ciliary muscles are striated rather than smooth (Walls 1942; Oliphant et al. 1983). Also the pupillary reactions are very fast (Oliphant et al. 1983). In the chicken, which has a range of about 20 D of accommodation (Schaeffel and Howland 1991), a speed of more than 20 D/s is common (Fig. 2.4) and speeds up to 60 D/s are possible (Schaeffel and Howland pers. observ.).

Even in the barn owl, a speed of 15 D/s was measured (Wagner and Schaeffel 1991) despite the moderate range of accommodation (about 6 D) in the subspecies studied. Murphy and Howland (1983) even observed changes of 100 D/s in the hawk owl. The high speed of accommodation as compared to humans (about 8 D/s; Campbell and Westheimer 1960) could make accommodation particularly useful for distance estimation.

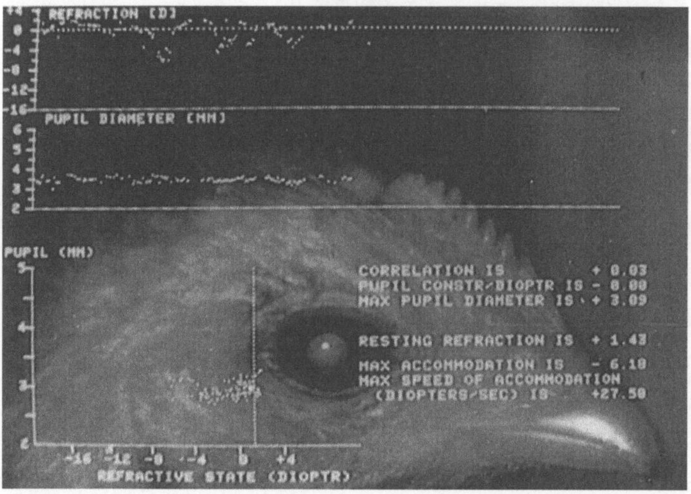

Fig. 2.4. On-line measurement of the dynamics of accommodation in an alert chicken. During automated infrared photoretinoscopy, an image processing computer programme finds the pupil of the chicken in the video image and performs all measurements automatically at a sampling rate of 8 Hz. Refractive state and pupil diameter are plotted against time during the measurements. Subsequently, the maximal speed of accommodation is calculated and plotted directly on the video monitor. Note that there is almost no near pupillary response (because of chick accommodated very little), and that the speed and accommodation was higher than 20 D/s

2.3.2 Coupled and Uncoupled Accommodation and the Convergence of Information

In contrast to mammals, which show a partial decussation (e.g. Polyak 1957), in birds most of the visual information crosses to the contralateral thalamic relay nucleus or to the contralateral optic tectum (Pettigrew 1979). Even so, functional binocular vision and stereopsis are achieved in some species like owls, because

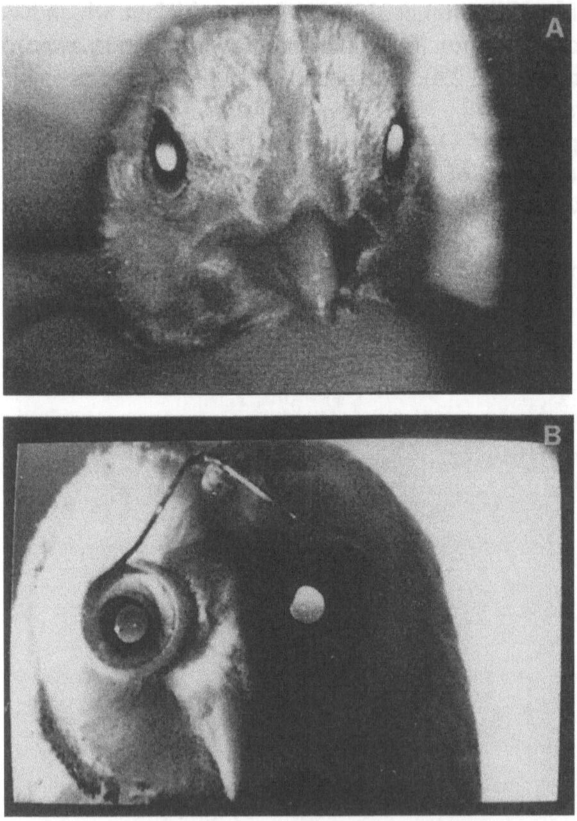

Fig. 2.5A, B. Coupling of accommodation in the chicken and in the barn owl. **A** Chickens can accommodate independently in both eyes. The left eye is focused at infinity (note *light crescent* in the top of the pupil), whereas the right eye accommodates to focus at a close target presented from the right side (note *light crescent* in the bottom of the pupil). **B** The barn owl displays different refractions in both eyes. Here, the differences are not the result of independent accommodation. Instead, the right eye has been artificially made myopic by a + 2 D lens. Because the owl cannot focus both eyes independently, one eye is always defocused by 2 D. Chickens with two different lenses in front of their eyes immediately refocus their retinal images independently (After Schaeffel and Wagner 1992)

the information from both eyes converges again in a higher visual centre, the visual wulst (Pettigrew 1989; see Sect. 3.3.1). It has also been demonstrated that owls *cannot* accommodate independently in both eyes (Schaeffel and Wagner 1992; Fig. 2.5B), confirming that information from the two eyes is not processed independently.

It may be that stereopsis and symmetrical accommodation are linked (Schaeffel and Wagner 1992; see Sect. 3.4). The occurrence of yoked accommodation in owls is in contrast to the chicken, where little interaction occurs between information processed from each eye Bell and Gibbs (1977) have shown that learned information from one eye cannot be transferred to the other. Also, accommodation can be completely independent for each eye in the chicken (Fig. 2.5A) and in the pigeon (Schaeffel, Nalbach and Howland pers. observ.). This does not imply that the refractions cannot be symmetrical in both eyes when the eyes converge (e.g. Martinoya et al. 1984a, b) to focus at a grain in front of the beak (Schaeffel et al. 1986). It also has been shown by means of behavioural experiments in pigeons (Watanabe et al. 1980; DiStefano et al. 1987; Martinoya et al. 1988) that visual performance improves under binocular viewing conditions. Although probability summation (statistical reduction of noise due to parallel copies of information which result in improved performance; see Sect. 1.6.2.3) is sufficient to explain the result in most cases, the findings also show that information from both eyes can converge even in birds with little binocular overlap. Maximum overlap in the pigeon is 30° (Nalbach et al. 1990; see Sect. 1.6.2) and in the chicken is about 25° (Burns and Wallman 1981, pers. observ.). The large extent of independence of the eyes in birds is further emphasized by the observation that pupillary reactions to light are independent in all cases, even in owls (Bishop and Stark 1965; Schaeffel and Wagner 1992).

2.4 Visual Guidance of Pecking Behaviour

Visual control of pecking behaviour has been extensively studied in the pigeon (Zeigler et al. 1980; Goodale 1983; Macko and Hodos 1985). Strikingly, pigeons show two points of fixation before the final descent of the beak occurs. The peck itself is mostly ballistic in nature (Martinoya et al. 1984b). If accommodative distance estimation is involved, then it has to occur at either the first or the second fixation point. The second fixation point is assumed to represent the near point of accommodation. Since this point is at a distance of about 64 mm from the target (Macko and Hodos 1985), the amplitude of accommodation in the pigeon would have to amount to about 16 D. The value is higher than the one reported by Schaeffel and Howland (about 9 D, 1987). The reason for this difference is unknown but experiments in barn owls show a similar discrepancy (see Sect. 2.6).

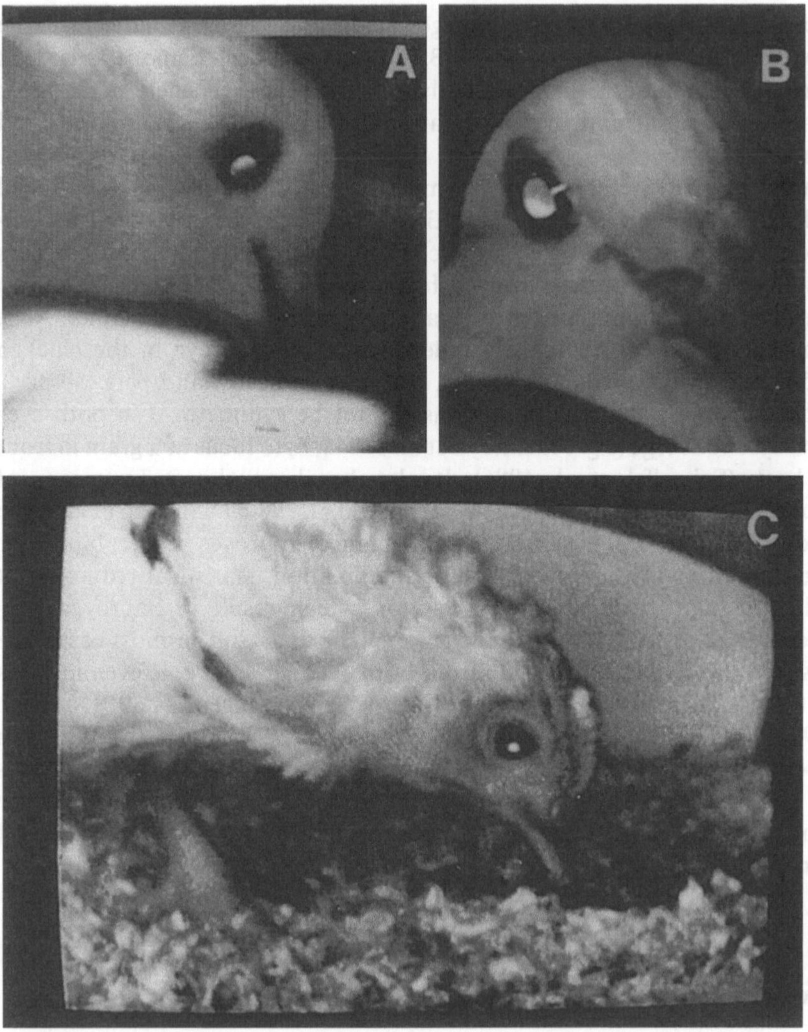

Fig. 2.6A–C. Accommodation during pecking behaviour in pigeons and chicks. **A** Pigeon during pecking with a hyperopic refraction in the lateral visual field. Because pigeons usually close their eyes during pecking (Macko and Hodos 1985), it is possible that the peck has just happened. **B** In pigeons, accommodation could be observed only during social interactions or if a target of interest to the bird was presented, but not during feeding. This pigeon observes a finger presented at a close range. Note the myopic refraction. **C** In contrast to pigeons, chickens never close their eyes during pecking but rather keep the ground in focus by accommodating up to about 20 D. Note the near pupillary response and the myopic crescent which almost fills the pupil

However, with reference to granivorous species, there are major differences in the pecking behaviour of the chicken and the pigeon. Pigeons appear to estimate their pecking distance from the second fixation point (Macko and Hodos 1985). After this fixation point the action must be ballistic in nature, as the eyes close, while in the chicken the eyes remain open until the beak touches the ground (Fig. 2.6C). Even more strikingly, no accommodation can be seen in pigeons at any phase of pecking behaviour, at least in the lateral visual field (Hodos, Howland and Schaeffel unpubl. observ. 1989; Fig. 2.6A). We have observed that pigeons accommodate most extensively during social interactions, or if an unfamiliar target is brought close to their eyes (Fig. 2.6B).

In contrast, chickens accommodate continuously up to 20 D during pecking (cf. Figs. 2.6A and 2.6C). It seems that pigeons rely on visual distance cues other than accommodation when they peck for food, although defocusing lenses can change the position of their fixation point (Martinoya et al. 1984b). Owls have been observed to behave in a similar way to pigeons during pecking (Wagner and Schaeffel 1991), as they often close their eyes during pecking, and because the lenses change the position of their fixation point. In contrast to pigeons, however, owls accommodate while they fixate their food. Consequently, there seem to be different strategies of visual control of pecking behaviour dependent upon the species under investigation. In particular, ocular accommodation varies in importance among the three species studied (chickens, pigeons and owls), even though all have large ranges of accommodation.

2.5 Lower Field Myopia: an Adaptation That "Keeps the Ground in Focus"?

A surprising observation has been described for the lower frontal visual field in pigeons (Millodot and Blough 1971; Rao and Erichsen 1984; Fitzke et al. 1985) and also for some other birds (Hodos and Erichsen 1990). The refractive state of the eye is claimed to vary across the visual field such that the ground is in focus without accommodation while the eye is emmetropic or hyperopic on-axis and in the upper visual field. The matching of refractive state to the average viewing distance across the visual field would be a plausible adaptation if one considers that visual acuity probably does not decline very much in the periphery (Ehrlich 1981). However, one has to be careful in deciding whether the variation in refractive state is a real adaptation or could also be explained by less precise emmetropization in the periphery. "Local emmetropization", the process of matching the location of the photoreceptor layer in the retina to the image shell all over the visual field (Miles and Wallman 1990), can only be accurate within the depth of focus. It is known that the refractive state off-axis varies considerably in humans, and that hyperopia is common (Millodot 1981). The same may be true in general for animals with one central fovea. Here, the variability in refractive state is not a problem because peripheral visual acuity is low.

If "local emmetropization" (Miles and Wallman 1990) accounts for the small amount of lower field myopia, the off-axis visual acuity must be high enough (and depth of focus must be small enough) to induce these small changes. Under extreme off-axis viewing conditions, the effective pupil aperture is very narrow in one meridian, and so the depth of focus is large in that meridian. Also diffraction may become critical at some point. Depth of focus remains unchanged, however, in the long meridian. The condition is similar to the one in animals with slit-shaped pupils (Howland et al. 1989; see Sect. 1.5). For lower field myopia to develop as a result of "local emmetropization", refractive development must be guided by a retinal image with a large amount of depth of focus in one meridian and with little depth of focus in the other, which would seem to present problems. It also seems unlikely that non-visual factors can control such subtle asymmetries in the optics of the eye as lower field myopia. It has to be kept in mind that 5 D of local myopia are produced in a chicken eye if the retina is moved backwards by only 0.3 mm, and this change must happen consistently to make lower field myopia a useful adaptation.

When refractive state is measured in chickens along the horizontal (Fig. 2.7A) and vertical meridians (Fig. 2.7B), a striking variability in the lower field refractive state is observed; it is more variable in the periphery than on-axis. A lack of precision of measurement can be ruled out, as demonstrated in Fig. 2.8. A similar amount of variability was observed in pigeons, at least in the two strains studied (Schaeffel and Howland 1987, unpubl. observ.). Some pigeons even failed to show any lower field myopia (Howland, Nalbach, and Schaeffel 1991 unpubl. observ.). The increased variability in refractive state off-axis could therefore be the result of lower acuity and, accordingly, less precise emmetropization. Finally, when anaesthetized birds are refracted sitting upright in a holder, the possibility must be considered that the ciliary body relaxes and the lens moves down. As a result, one could expect myopia selectively in the lower visual field because the distance of the lens from the upper retina increases.

However, there are also some arguments in favour of the assumption that lower field myopia is a real adaptation to the problem of "keeping the ground in focus" (Hodos and Erichsen 1990). First, the average refractive state in the lower visual field of the chicken is shifted to the myopic side (see Fig. 2.7B), and the upper visual field is more hyperopic. Therefore, although there is an increased variability in off-axis refractive state, the averages are in the right directions.

A second argument is that in contrast to humans, there is an obvious lack of off-axis astigmatism in chickens and pigeons. Therefore, the retinal image quality off-axis is still quite reasonable (Schaeffel et al. 1988a).

If distances are estimated from accommodation, the refractive state of the eye must be "known by the brain". The condition becomes more complex if refractive state is not constant but changes due to lower field myopia. If a pecking target is located within the myopic sector of the eye, the accommodative distance estimation must be calibrated for the more myopic refraction. To some extent, such a conclusion could explain why pigeons do not accommodate in the

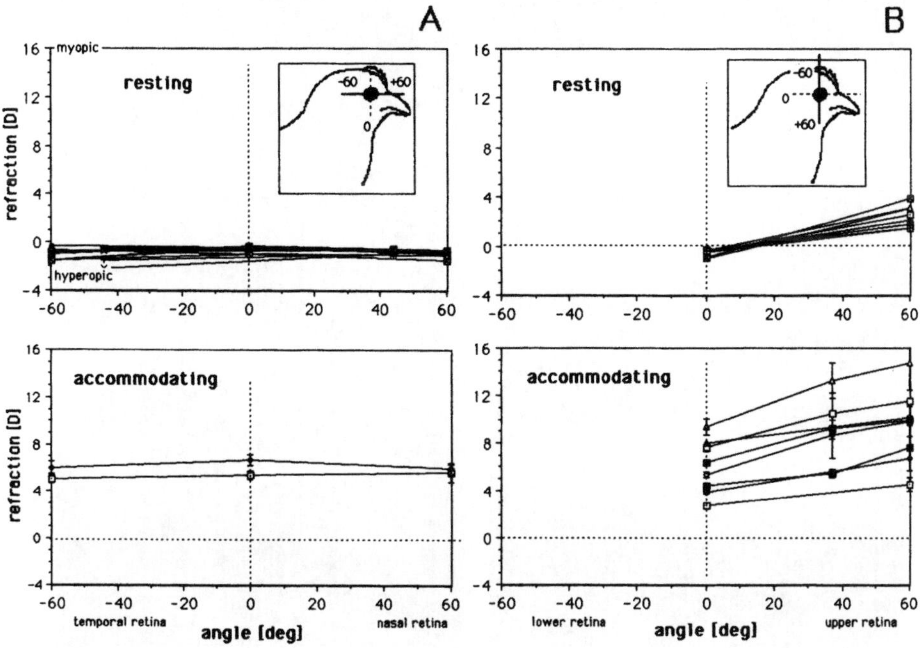

Fig. 2.7A, B. Refractive state across the visual field and radially symmetrical accommodation in the chicken. Measurements were performed simultaneously from two different locations (60° apart) with two identical photoretinoscopes. By this procedure, eventual fluctuations in accommodative tonus affected both measurements in the same way which makes comparisons reliable. **A** Refractive state did not vary across the horizontal meridian, either during resting accommodation (*upper panel*) or during accommodation (*lower panel*). **B** Refractive state changed across the vertical meridian and became more myopic in the lower visual field (the upper visual field was not measured). Note that the refractions were more variable 60° below the axis. The variability cannot be explained from variability of measurement, because the eyes were mostly free of astigmatism and the reflexes were clear (see Fig. 2.8). Note further that, during accommodation, refractive state changes to a similar extent at 0° as at 60° off-axis, indicating that accommodation does not focus selectively local retinal areas. *Error bars* are standard deviations from 3–6 measurements

lateral visual field when they peck (see Fig. 2.6A); the retinal image may be sufficiently focused due to the lower field myopia. It is clear, however, that 5 D of lower field myopia, as reported for the pigeon (Fitzke et al. 1985) are not sufficient to focus the retinal image at the second fixation point 64 mm away from the pecking target (Macko and Hodos 1985). The results could be reconciled if it could be shown that accommodation can selectively focus the frontal visual field only, while the lateral field remains hyperopic. We have measured natural accommodation in chickens with two photoretinoscopes simultaneously which were positioned at angles 60° apart (Fig. 2.7, lower panels). Chickens were presented with targets of interest in different parts of the visual field. The measurements showed that accommodation was never restricted to local areas but always changed refraction by the same amount in both

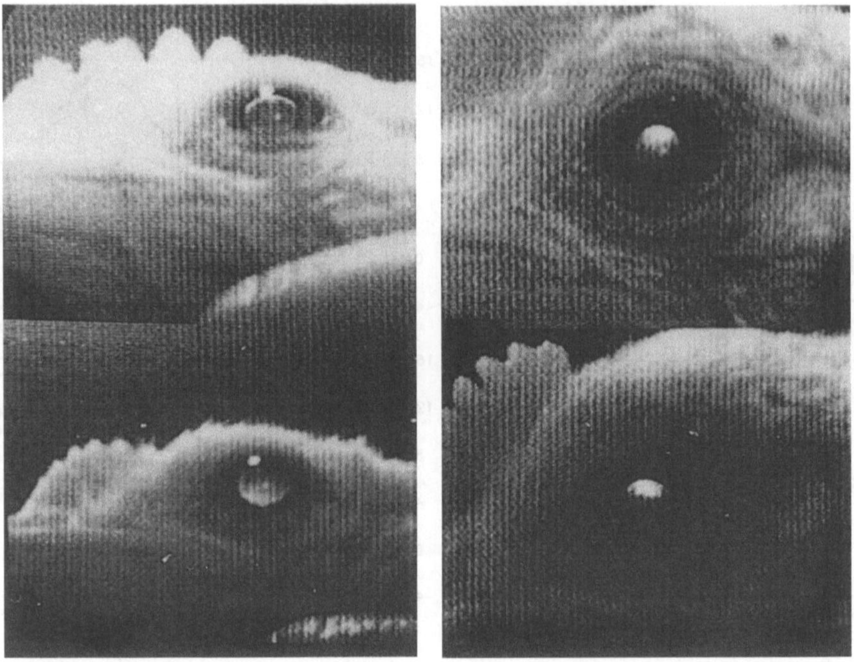

Fig. 2.8. Demonstration of the variability of lower field myopia in alert chickens. Chickens were refracted simultaneously on-axis (*panels on right*) and about 60° below the axis (*panels on left*). Both chicks are slightly hyperopic on-axis. The chicken shown in the *upper frames* does not show lower field myopia whereas the chick in the *lower frames* is clearly myopic. The chickens were 30 days old and originated from the same group

axes measured. Although similar data are not available from the pigeon, we assume that accommodation is also radially symmetrical. The results therefore suggest that distance information at the second fixation point in pigeons is not derived from accommodation but rather from other visual distance cues. It remains an unsolved question why the eyes are not properly focused at this point. One possible explanation is that the large amplitudes of rotation of the eye described for the pigeon during pecking (Nalbach et al. 1990) have an effect on refractive state in the frontal visual field, possibly by deforming the globe.

2.6 The Role of Accommodation in Judging Distances

The importance of accommodation during distance estimation can be most clearly demonstrated either by removing accommodation or by shifting the accommodative tonus by lenses. Removal of accommodation can be achieved by lesioning the parasympathetic motor output nucleus in the midbrain, the Edinger–Westphal nucleus. Such lesions have been performed bilaterally by

David Troilo in chickens (Schaeffel et al. 1990). The lesioned chickens learned to forage again quite rapidly, although it was obvious that their distance estimation was disturbed. In all cases, they pecked too short and could grasp the grain only after repeated attempts. Also young chickens in which accommodation was paralyzed by the use of vecuronium bromide (Pettigrew et al. 1990), showed a similar change in pecking behaviour (Fig. 2.9).

The observation that the chicks pecked too short after the motor side of the accommodative feedback loop was no longer functional is consistent with the idea that accommodation provides a distance cue. The chickens tried to focus their retinal images by accommodation. Because the ciliary muscle was not responding and the retinal image was not focused, the output command to the motor side of the accommodative feedback loop should be maximal. This condition, in turn, would normally correspond to a very close viewing distance and therefore explains why the chickens pecked too short. One has to keep in mind, however, that the oculomotor nuclei for the extraocular muscles are quite close to the Edinger–Westphal nucleus in the midbrain. The positioning of the lesions is open to error and thus there is some risk that they might have inadvertently affected eye movements. The same is true for pharmacological lesions of accommodation with vecuronium bromide. Therefore, since eye movements were not recorded during the experiment (Fig. 2.9) the possibility cannot be excluded that a defect in eye movements affected distance estimation.

Fig. 2.9. Effect of cycloplegia on pecking behaviour in young chicks. The chicks received 10 drops of vecuronium bromide into both eyes, applied over a period of 30 min which paralyzed accommodation and pupillary responses. The tip of the beak was traced (see *arrow*) in the digitized video frames with a temporal resolution of 25 Hz. Note that, during three pecking attempts, the ground was never actually touched. The observation is consistent with the hypothesis that accommodation provides distance information to the chicks

The role of accommodation can be tested more convincingly if ophthalmic lenses are used to shift the accommodative tonus by a defined amount. The pecking response should then be modified in a predictable way. Such experiments were first done by Harkness (1977) on the chameleon. Here, the errors in binocular distance estimation were in perfect agreement with the errors calculated from the power and sign of the lenses. Collett (1977) found that the effect of lenses on distance estimation in toads showed up predominantly under monocular viewing conditions, whereas, under binocular viewing conditions, stereopsis provided the major distance cue (see Sect. 16.2.3.1). Except for one study by Martinoya et al. (1984b) there are no comparable lens experiments available from birds. Recently, Wagner and Schaeffel (1991) tested the effects of lenses on pecking behaviour in the barn owl. A lens was applied monocularly while the other eye was covered to exclude disparity cues (Pettigrew and Konishi 1976). It was found that, during fixation of food items, the owls' accommodative tonus was consistently shifted by lenses of different powers

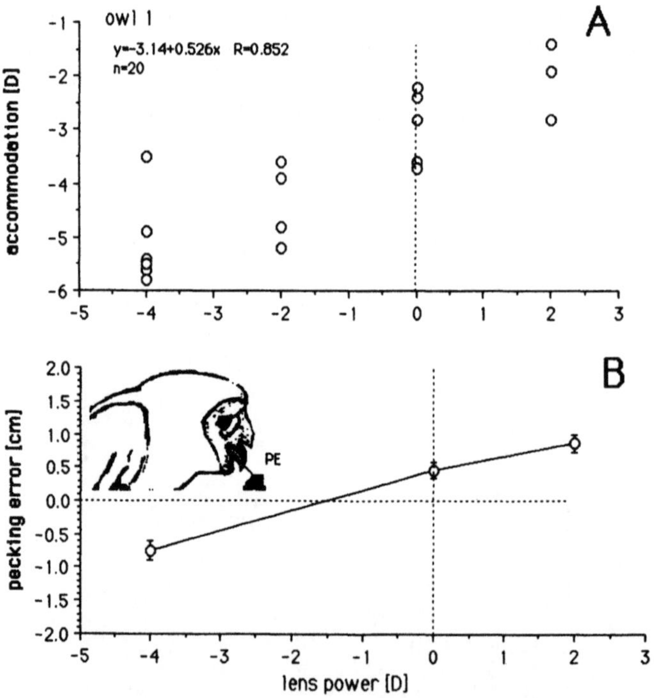

Fig. 2.10A, B. Accommodation and distance estimation in barn owls. **A** Tame, hand-raised barn owls were trained to tolerate spectacles with lenses (Wagner and Schaeffel 1991). They were then refracted while they were fixing their food, with one eye occluded and the other eye covered by a lens of variable power. The amount of accommodation was determined during fixation and plotted against the power of the lens. Note that a negative lens induces more accommodation than a positive lens. **B** Subsequent to a period of fixation, the owls pecked for the food. They were video taped from the side and the "pecking error" was determined. Each data point represents an average of 40–60 trials. *Error bars* are standard errors of the mean

(Fig. 2.10A). The maximal accommodative response elicited was about 6 D. During subsequent pecks, the owls undershot with negative lenses (negative pecking error, Fig. 2.10B) or overshot with positive lenses (positive pecking error, Fig. 2.10B). These results seem very clear and are certainly consistent with the idea that accommodative tonus provides a distance cue.

Even so, the interpretation of the above findings is complicated by a number of further observations. First, the effects of the lenses were less than predicted from their powers. Secondly, the owls always under-accommodated. In addition, when not wearing lenses, the owl found the accommodative range of 6 D apparently insufficient to focus the food appropriately for a fixation point 10 cm away from the food. A third point noted was that the fixation points were shifted depending on the power of the lenses, whereas pecking itself was largely ballistic with a constant "pecking length". Apparently, the owl initiated the peck from a fixed distance (equivalent to a fixed pre-programmed amount of accommodation or a pre-programmed retinal disparity). Thus, by employing lenses, the fixation distance was changed, with the final pecking error (Fig. 2.10B) arising from a shift in the fixation point *and* a shift in accommodative tonus (Fig. 2.10A). Some of these observations are similar to the ones by Martinoya et al. (1984b) in pigeons, although Wagner and Schaeffel (1991) measured accommodation directly for the first time. In conclusion, the results show that accommodation *is* involved in monocular distance estimation in the chicken, pigeon and barn owl, but in a more complex fashion than one might expect.

Acknowledgements. This work has been supported by the German Research Council (SFB 307, TP A7). The author thanks the editors for the invitation to contribute a chapter to their book, Gabi Hagel for help during measurements and data analysis, and Howard C. Howland and Hans-Ortwin Nalbach for stimulating discussions.

References

Bell GA, Gibbs ME (1977) Unilateral storage of monocular engram in day-old chick. Brain Res 124:263–270

Bishop LG, Stark L (1965) Pupillary responses of the screech owl, *Otus asio*. Science 148:1750–1752

Burns S, Wallman J (1981) Relation of single unit properties to the ocular motor functions of the basal optic root (accessory optic system) in chickens. Exp Brain Res 42:171–180

Campbell FW, Westheimer G (1960) Dynamics of accommodation responses of the human eye. J Physiol (Lond) 151:285–295

Collett TS (1977) Stereopsis in toads. Nature 267:349–351

DiStefano M, Kusmic C, Musumeci D (1987) Binocular interactions measured by choice reaction times in pigeons. Behav Brain Res 25:161–165

Ehrlich D (1981) Retinal specialization of the chick retina as revealed by the size and density of neurons in the ganglion cell layer. J Comp Neurol 195:643–657

Fitzke FW, Hayes BP, Hodos W, Holden AL, Low JC (1985) Refractive sectors in the visual field of the pigeon eye. J Physiol (Lond) 369:33–44

Foley JM (1978) Primary distance perception. In: Held R, Leibowitz HW, Teuber HL (eds) Perception. Handbook of sensory physiology, vol 8. Springer, Berlin Heidelberg New York, pp 181–213

Fox R, Lemkuhle SW, Bush R (1977) Stereopsis in the falcon. Science 197:79–81
Goodale MA (1983) Visually guided pecking behavior in the pigeon (*Columbia livia*). Brain Behav Evol 22:22–41
Green DG, Powers MK, Banks MS (1980) Depth of focus, eye size and visual acuity. Vision Res 20:827–835
Gundlach RN, Chard RD, Skahen JR (1945) The mechanisms of accommodation in the bird eye. J Comp Psychol 38:27–42
Harkness L (1977) Chameleons use accommodation cues of judge distance. Nature 267:346–349
Hodos W, Erichsen JT (1990) Lower-field myopia in birds: an adaptation that keeps the ground in focus. Vision Res 30:653:657
Howland HC (1985) Optics of photoretinoscopy: results from ray tracing. Am J Optom Physiol Opt 62:621–625
Howland HC, Murphy CJ, Howland B (1989) On the optical significance of pupillary shape. In: Erber J, Menzel R, Pflüger H-J, Todt D (eds) Neural mechanisms of behavior. Thieme, Stuttgart, p 184
Levy B, Sivak JG (1980) Mechanisms of accommodation in the bird eye. J. Comp Physiol 137:267–272
Macko KA, Hodos W (1985) Near point of accommodation in pigeons. Vision Res 25:1529–1530
Martinoya C, Le Houezec J, Bloch S (1984a) Pigeon's eyes converge during feeding: evidence for frontal binocular fixation in a lateral eyed bird. Neurosci Lett 45:335–339
Martinoya C, Palacios A, Bloch S (1984b) Participation and eye convergence and frontal accommodation in programming grain pecking in pigeons. Neurosci Lett (Suppl) 18:S233
Martinoya C, Le Houezec J. Bloch S (1988) Depth resolution in the pigeon. J Comp Physiol 163:33–42
McFadden SA (1987) The binocular depth stereo-acuity of the pigeon and its relation to the anatomical resolving power of the eye. Vision Res 27:1967–1987
McFadden SA, Wild JM (1986) Binocular depth perception in the pigeon. J Exp Anal Behav 45:149–160
Miles FA, Wallman J (1990) Local ocular compensation for imposed local refractive error. Vision Res 30:339–349
Millodot M (1981) Effect of ametropia on peripheral refraction. Am J Optom Physiol Opt 58:691–695
Millodot M, Blough P (1971) The refractive state of the pigeon eye. Vision Res 11:1019–1022
Murphy CJ, Howland HC (1983) Owl eyes: accommodation, corneal curvature and refractive state. J Comp Physiol 151:277–284
Murphy C, Howland HC (1986) On the gecko pupil and Scheiner's disc. Vision Res 26:815–817
Nalbach H-O, Wolf-Oberhollenzer F, Kirschfeld K (1990) The pigeon's eye viewed through an ophthalmoscopic microscope: orientation of retinal landmarks and significance of eye movements. Vision Res. 30:529–540
Oliphant LW, Johnson MR, Murphy CJ, Howland HC (1983) The musculature and pupillary responses of the great horned owl iris. Exp Eye Res 37:583–595
Pettigrew JD (1979) Comparison of the retinoscopic organization of the visual wulst in nocturnal and diurnal raptors, with a note on evolution of frontal vision. In: Cool SJ, Smith EL (eds) Frontiers of visual science. Springer, Berlin Heidelberg New York, pp 328–335
Pettigrew JD (1989) Is there a single, most-efficient algorithm for stereopsis? In: Blakemore C (ed) Vision: coding and efficiency. Cambridge University Press, Cambridge, UK, pp 283–290
Pettigrew JD, Konishi M (1976) Neurons selective for orientation and binocular disparity in the visual wulst of the barn owl (*Tyto alba*). Science 193:675–678
Pettigrew JD, Wallman J, Wildsoet CF (1990) Saccadic oscillations facilitate ocular perfusion from the avian pecten. Nature 343:362–363
Polyak S (1957) The vertebrate visual system. University of Chicago Press, Chicago
Rao VM, Erichsen JT (1984) The refractive state of the pigeon eye. Soc Neurosci Abstr 10:397
Schaeffel F, Howland HC (1987) Corneal accommodation in chick and pigeon. J Comp Physiol 160:375–384

Schaeffel F, Howland HC (1988) Visual optics in normal and ametropic chickens. Clin Vis Sci 3:83–93

Schaeffel F, Howland HC (1991) Properties of visual feedback loops controlling eye growth and refractive state in the chicken. Vision Res. 31:717–734

Schaeffel F, Wagner H (1992) Barn owls have symmetrical accommodation in both eyes but independent pupillary responses to light. Vision Res 32:1149–1152

Schaeffel F, Howland HC, Farkas L (1986) Natural accommodation in the growing chicken. Vision Res 26:1977–1993

Schaeffel F, Farkas L, Howland HC (1987) Infrared photoretinsocope. Appl Opt 26:1505–1509

Schaeffel F, Glasser A, Howland HC (1988a) Accommodation, refractive error and eye growth in chickens. Vision Res 28:639–657

Schaeffel F, Romano V, Howland HC (1988b) Accommodation in chicks: off-axis refraction and mechanical model. Opt Soc Am Tech Digest 11:MR33

Schaeffel F, Troilo D, Wallman J, Howland HC (1990) Developing eyes that lack accommodation grow to compensate for imposed defocus. Visual Nurosci 4: 177–183

Sivak JG, Hildebrant T, Lebert C, Myshiak L, Ryall L (1986) Ocular accommodation in chickens: corneal vs. Lenticular, effect of age. Vision Res 26:1865–1872

Slonakar JR (1918) A physiological study of the anatomy of the eye and its accessory parts of the English sparrow (*Passer domesticus*). J Morphol 39:351–459

Troilo D, Wallman J (1987) Changes in corneal curvature during accommodation in chicks. Vision Res 27:241–247

Wagner H, Schaeffel F (1991) Barn owls use accommodation as a distance cue. J Comp Physiol 169:515–521

Walls GL (1942) The vertebrate eye and its adaptive radiation. Cranbrook Inst Sci, Bloomfield Hills, Michigan

Watanabe S, Hodos W, Bessette BB (1980) Two eyes are better than one: superior discrimination learning in pigeons. Physiol Behav 32:847–850

Zeigler HP, Lewitt PW, Levine RR (1980) Eating in the pigeon (*Columba livia*): Movement patterns, stereotype and stimulus control. J Comp Physiol Psychol 94:783–794

3 Binocular Depth Perception

S.A. McFadden

3.1 Introduction

Successfully negotiating any spatial environment requires two distinct but related forms of visual information. The first is absolute distance information, which allows an animal to "know" its position with respect to other objects or points in space (e.g. *D* in Fig. 3.1). This is sometimes referred to as egocentric (self-centred) *distance* perception. It allows visual information to be used for precise motor action. The second form of spatial perception, *depth* perception, provides information regarding the relative position of two or more objects (e.g. *d* in Fig. 3.1). It is distinct from absolute distance perception in that it gives no information regarding the distance of the comparison points from the observer. What then is its use? It is argued in this chapter that such relative distance or depth perception is essential for complex object detection, but can only be used to guide reaching motor movements in the limiting case where one of the reference points (e.g. *F* or *P* in Fig. 3.1) is always at some predetermined absolute distance.

Absolute and relative distance perception can also interact in a more dynamic way, where the mechanisms underlying the former can be used to rescale the latter to enable veridical depth perception (McFadden 1992).

Visual systems have adapted multiple means for extracting depth and distance information. It is quite likely that different strategies dominate in different behavioural and perceptual contexts (see Chap. 16) and that the details of the neural mechanisms may differ in different species. However, the general strategies in vertebrates are similar, and reflect common optical and neural constraints of biological visual systems. These general strategies include the extraction of information through temporal channels, spatial channels (both binocular and monocular) and the oculomotor system.

In this chapter I will discuss only binocular strategies and their applicability to different motor actions in the avian world. Nearly all birds, and indeed nearly all vertebrates (Walls 1942; Hughes 1977), have a binocular field (see Sect. 1.6.2). This means that each eye can simultaneously sample the same position in space.

Department of Psychology, The University of Newcastle, NSW 2308, Australia

M.N.O. Davies and P.R. Green (Eds.)
Perception and Motor Control in Birds

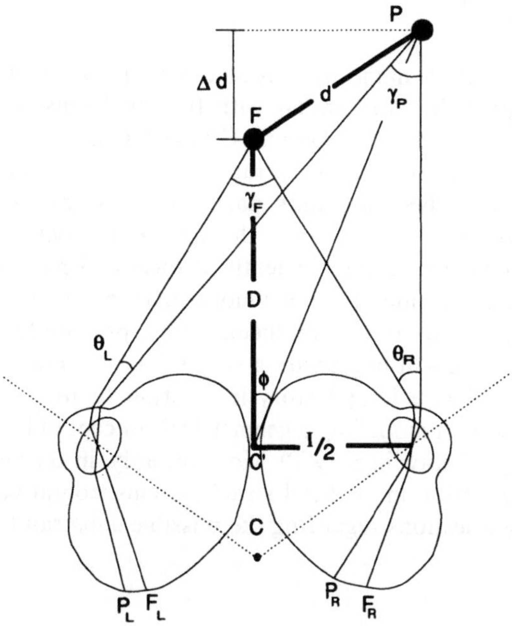

Fig. 3.1. Diagrammatic representation of retinal disparity. Variables are defined as follows: *F* fixation point; *P* arbitrary point; *d* depth; *D* absolute distance of *F*; ϕ visual direction of point *P*; γ_F & γ_P convergence angles of *F* and *P* respectively; θ_R & θ_L visual angles of *P* with respect to the fixation point for the right and left eyes respectively; *I/2* half the interocular separation; *C* and *C'* cyclopean centres ($C \approx C'$ when *I* is small relative to *D*); P_L & P_R matching retinal points for point *P* on the left and right retinas respectively; F_L & F_R corresponding retinal points for the fixation point *F* on the left and right retinas respectively; *dotted lines* represent the projection of the optic visual axes; $\Delta d \approx d$ for symmetrical convergence, i.e. when the visual direction $\phi = 0$. For most birds with laterally placed eyes and narrow binocular fields, ϕ can theoretically lie between 10° and 20°. Eq. (3.2) in the text relates only to symmetrical convergence conditions.

Provided the neural system can combine the information from each eye's sample (see Pettigrew and Konishi 1976), such a capacity generally leads to stereoscopic perception (cf. Sect. 1.6.2.3). However, stereopsis is not a unitary phenomenon, and can arise through many different mechanisms. All are dependent on the oculomotor system to work effectively, and some possibly drive and/or are driven by the vergence eye movement system. In some situations, this latter binocular convergence system may also act in isolation.

It will be seen in what follows that the particular neural strategy employed to achieve stereopsis varies between avian species and even within a single species, depending upon the different perceptual and motor demands of each different task. This in turn is dependent upon the optical and neural constraints associated with different niches (cf. Chap. 16).

3.2 What Exactly is Stereopsis?

The term "stereopsis" has been used to name numerous different (although associated) phenomena. We are probably all familiar with the vivid sense of depth that arises from looking at random dot stereograms (Julesz 1960), and the discovery of retinal disparity detectors in predatory raptors (Pettigrew and Konishi 1976; Pettigrew 1979a), but these are only parts of the story. As described below, the mechanisms of stereopsis arise through a number of separate channels of processing and operate independently at local and global levels. Some types of stereopsis require fusion of the two monocular images, but other types do not. Stereopsis has been studied mathematically, psychophysically, behaviourally and neurally. We will begin with a very brief overview of the mathematics, as this approach has a long history in relation to human vision, culminating in the obsession with modelling a geometrical concept of the horopter (for a review see Shipley and Rawlings 1970). More recently, in trying to model stereopsis for spatial perception in artificial visual systems, cognitive science has produced enlightening questions regarding the possible constraints and mechanisms of stereopsis.

3.2.1 Retinal Disparity and Stereopsis

Stereopsis requires that each eye must sample the same position in space (i.e. binocularly) from its different vantage position (Fig. 3.1). As the two eyes are separated horizontally, there arises a horizontal angular difference between the retinal positions of any two points. These disparate retinal images give rise to stereopsis (Wheatstone 1838). The associated horizontal (but not vertical) retinal disparity allows a direct angular measure of depth with respect to a fixation point (F in Fig. 3.1). Geometrically defined, horizontal retinal disparity directly measures the difference between the visual (θ) or convergence angles (γ) such that:

$$\theta_R - \theta_L = \gamma_F - \gamma_P \, . \tag{3.1}$$

The horizontal retinal disparity (δ) can be calculated as

$$\delta = 2\left[\tan^{-1}\frac{I/2}{D} - \tan^{-1}\frac{I/2}{D + \Delta d}\right]. \tag{3.2}$$

(See Fig. 3.1 for variable definitions.)

The horizontal retinal disparity predicts the relative distance between F and P dependent upon the interocular separation (I) and the absolute distance of the fixation point (D). A consequence of this relationship is that, for a fixed difference in depth and a fixed retinal disparity, as interocular separation decreases, the absolute distance of the fixation point must also decrease. Thus the smaller the bird and the eye (and hence the less the resolution capacity of the eye), the closer the operative range of stereopsis (cf. Sect. 1.4.1.2).

3.2.2 Types of Stereopsis

Stereopsis is not a unitary process, but involves at least two dissociable processes. These are known as local and global stereopsis. In humans, it has been found that global stereopsis can be impaired (as the result of temporal lobe excision), while local stereopsis remains intact (Ptito et al. 1991). This raises the question of whether local stereopsis acting alone may be a useful spatial cue for some birds under certain conditions. Its usefulness will depend upon whether birds encounter images in their natural environments that would not require a global solution. The local stereoscopic process itself can be divided into a number of different channels of processing, and there is evidence that the neural mechanism in some birds is quite selective in this respect. I will first briefly summarize the current understanding of the types of stereopsis, before addressing which mechanisms birds appear to have.

3.2.2.1 Local Stereopsis

Stereopsis is based on horizontal retinal disparity and requires that matching points on each retina [see (F_L, F_R) and (P_L, P_R) in Fig. 3.1] be binocularly combined. How this occurs is a complex question, but whatever the fusional mechanisms underlying binocular single vision, the neural mechanism of stereopsis must involve matching corresponding points in the images in the two eyes before disparity can be ascertained. Simple images with sparse contours contain edge information for which corresponding receptive fields can be unambiguously matched. Local stereopsis has been defined as the assignment of a retinal disparity value to such unambiguous matched corresponding points (Julesz 1978).

Spatial Channels in Local Stereopsis. The local matching process requires enough information so that corresponding receptive fields from each eye are assured. Changes in light intensity alone are not enough (Poggio and Poggio 1984), but higher level features such as orientated edges or some characteristic micropattern can provide the required substrate for matching. The required degree of similarity between the two corresponding images is dependent on the spatial frequency characteristics of the disparity detector (Bishop and Henry 1971) and in human stereopsis there are at least two spatial frequency channels (Julesz and Miller 1975). Thus for very small ($< 0.5°$) disparities (fine stereopsis), retinal rivalry and suppression occur if the two images are not sufficiently similar. In contrast, coarse stereopsis operates on images which can be quite dissimilar in form, luminance, contrast and spatial position. It may be the case for birds, therefore, that the need for oculomotor precision in fusion will depend upon the degree of stereoacuity required within a particular niche.

Apart from the possibility of channels varying in magnitude, human clinical evidence suggests that disparity neurones fall into three classes (Richards 1970). These classes are dependent upon whether the target is nearer than the fixation plane (coarse uncrossed disparity), further than the fixation plane (coarse crossed disparity; e.g. P behind F in Fig. 3.1) or at the same depth as it (zero disparity). Thus in sophisticated stereoscopic systems both the magnitude and the sign of the disparity are known.

3.2.2.2 Global Stereopsis

For all forms of local stereopsis, there is no doubt as to which two features correspond in each eye's image. However, for more complex images, there may be uncertainty in determining which pair of points or features correspond. This ambiguity is exemplified in random dot stereograms (RDSs; Julesz 1960), or in visual scenes which contain repetitive, finely grained textures. The so-called "correspondence problem" (see Julesz 1971; Poggio and Poggio 1984) can be resolved by selecting the most probable set of matching pairs based on comparison of the disparity values assigned to neighbouring pairs from the local process. Whatever the criterion employed for selecting the correct set of corresponding pairs, the process requires a global solution and hence is referred to as global stereopsis.

Computational and Neural Models of Global Stereopsis. The neurophysiological basis of global stereopsis is unknown. However, in the rhesus monkey, 20% of depth sensitive binocular cells in A17 were found to respond to dynamic RDSs, which consist of a series of different static RDSs presented in fast temporal succession (Poggio and Poggio 1984). These were generally complex cells with little orientation specificity.

Computational models of global stereopsis have provided insight into its possible mechanisms (recently reviewed by Blake and Wilson 1991). Resolving the correspondence problem commonly involves making assumptions about the image, such as surface continuity (Nelson 1975), although this approach has problems with discontinuities in disparity such as those occurring at borders. The notion of the avian neural system evoking inbuilt assumptions is not without foundation, since in chicks intrinsic assumptions regarding 3 D surface shading are known to control the perception of the concavity of a surface (Hess 1950; Hershberger 1970). Obviously, image constraints are not so critical if the false matches arising from ambiguity can be reduced. Some models avoid false matches by using progressive spatial filtering (Poggio and Poggio 1984; Blake and Wilson 1991) or by including the vertical disparities of a small number of corresponding points (Longuet-Higgins 1981; Mayhew and Longuet-Higgins 1982), although vertical disparities are very small and may be difficult for a biological system to monitor.

3.3 Stereopsis in Birds

The evidence for stereopsis in birds stems from two sources, neurophysiology and behavioural demonstrations. The results of these experiments are discussed below in the context of the different levels of stereopsis described above.

3.3.1 Neural Mechanisms for Local Stereopsis in Birds

In the avian visual system there is almost complete crossing of the optic fibres to the contralateral first order nuclei. However, this initial segregation does not prevent binocular interaction since subsequent decussations potentially allow superimposition of monocular visual information. In particular, it is well established that the visual Wulst, the major termination of the thalamofugal visual pathway is one such site (Karten et al. 1973; Bagnoli et al. 1990).

The discovery of disparity sensitive neurones in the visual Wulst of both the barn owl (Pettigrew and Konishi 1976; Pettigrew 1979b) and the kestrel (Pettigrew 1979a) led to the first proposals that stereopsis may be present in these raptors. In the owl, 10% of cells in the Wulst were purely binocularly driven in that they only responded to simultaneous stimulation of both eyes The preferred disparity of these cells was 3°–5°, and generally orientation sensitive. Similarly, the Wulst of the kestrel has an expanded representation of the zone surrounding the binocular fovea and disparity sensitive neurons of similar stimulation requirements to those found in the owl. Since the acuity of the kestrel has been reported to be at least as high as 60 cycles deg^{-1} (Fox et al. 1976; Hirsch 1982), the sensitivity of these disparity cells is much coarser than the resolution capacity of the kestrel. Thus, these disparity neurons could support a local stereoscopic mechanism that is orientation sensitive, but at a relatively coarse level.

Both the owl and the kestrel are predators which use their talons to catch prey. Binocular cells have also been found in the Wulst of non-predators such as pigeons (Perisic et al. 1971; Micheli and Reperant 1982) and chickens (Wilson 1980), but they are relatively scarce. A deliberate search for disparity sensitivity in the pigeon Wulst yielded negative results (Frost et al. 1983), although full anaesthesia may have limited the response properties of binocular cells. In the studies of barn owls and kestrels by Pettigrew, in contrast, anaesthesia was very light and eye movements were allowed and monitored.

The neurophysiological evidence suggests that the visual Wulst, at least in predatory birds, has cells which may mediate local stereopsis at a coarse resolution level. There is as yet no neurophysiological evidence of global stereopsis in birds.

3.3.2 Behavioural Tests of Stereopsis in Birds

Behavioural studies have used operant conditioning techniques to examine the role of retinal disparity in local and global stereoscopic tasks in both the kestrel and the pigeon. These studies have obtained different results depending in part upon whether the visual stimuli were context enriched or impoverished. Depth acuity appears to depend upon the complexity of the stimulus.

3.3.2.1 Global Stereopsis

Random dot stereograms (RDS) are theoretically the best way to isolate retinal disparity as the only cue for global stereopsis. Dynamic RDSs were presented to each eye of an American kestrel (Fox et al. 1977) using a red or green filter placed before each eye. The bird was rewarded by flying to the stereoscopic display at a distance of 1.68 m, and was able to discriminate the presence of a vertical rectangle with 8°–12° of uncrossed disparity.

In the kestrel study, the pattern of random dots was replaced every 16 ms on a colour monitor. The flicker fusion frequency of the kestrel is unknown but in the pigeon has been measured up to 150 cycles s^{-1} (Powell and Smith 1968). Although perhaps unlikely, it is therefore possible that the onset delay with which the pattern was generated may have been detected, without any discrimination based on global stereopsis. The reported decline in performance at disparities about the training disparity could be attributed to a generalization gradient, particularly given inequalities in training at the different data points. Despite these criticisms, the experiments support the view that global stereopsis is present in this predatory raptor.

The discrimination of the stereoscopic form was also found to be orientation sensitive, which is interesting since the complex cells which respond to RDSs in the monkey are orientation insensitive (Poggio and Poggio 1984).

3.3.2.2 Local Stereopsis

Global stereopsis from an RDS is dependent upon some process operating to resolve the ambiguities arising from two patterns of identical dots to determine which dots in each eye's image correspond. In contrast, to test for local stereopsis, the images in each eye must contain sufficient information so that little or no ambiguity arises in this matching process. One way is to provide features such as orientated lines to reduce false matches. The discrimination of depth in the Frisby stimuli (Fig. 3.2A) primarily relies on such local stereopsis, and has been used to examine these processes in the pigeon (McFadden and Wild 1986). Here, pigeons were trained to peck at a stimulus (less than 10 cm from the eye) with a central circle displaced in depth (1° of retinal disparity). The discrimination of depth was sensitive to the sign of the disparity with discrimination between crossed and uncrossed disparity easily obtained. Performance consistently fell to chance when the binocular field was reversibly blocked. Interestingly, adult birds that have had monocular visual deprivation for the first 30 days of life are unable to learn even the easiest of our depth discriminations, and fail on the visual cliff (McFadden unpubl. observ.). Thus it would appear that the pigeon has at least a local stereoscopic capacity which may require balanced binocular input during development.

3.3.2.3 Stereoacuity in Context Rich Stimuli

In man, the threshold for local stereopsis is a hyperacuity (2–4 s of arc compared to a minimum angle of resolution of 25 sec of arc; Westheimer 1975), and this

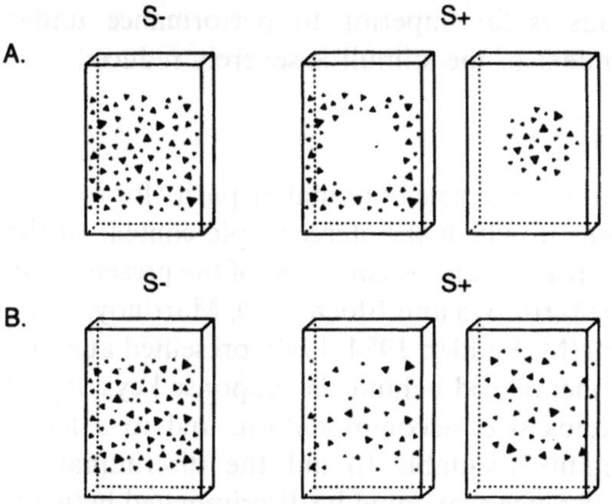

Fig. 3.2A, B. Frisby stimuli used for testing local stereopsis in the pigeon. Components within an array of random sized triangles were presented either in the same plane (negative stimulus $S-$) or in two different planes (positive stimulus $S+$). The two planes in $S+$ are shown separately for clarity. **A** In $S+$, a displaced circle was presented behind a background. **B** No coherent displaced form was present in $S+$; half the triangles were in the foreground and half were in the background. The density of the triangle array was equal in the two conditions. The stimuli are presented on optically clear glass and illuminated from behind. In both cases, when viewed binocularly, the perception of two separate planes is clearly appreciated

capacity is thought to occur through co-operative neural interactions (Westheimer 1979). In contrast, the threshold for local stereopsis in the pigeon does not appear to be a hyperacuity, as the value of about 1 min of arc is closely matched to the anatomical resolving power of the pigeon eye (McFadden 1987). It is unknown whether local stereopsis is a hyperacuity in other birds, such as birds of prey. Even the spatial acuity of the eagle, although much greater than in many other species, is theoretically matched to the limits imposed by physiological optics and the spacing of photoreceptors (Reymond 1985).

The stereoacuity of the pigeon is influenced by the configuration of the stereoscopic display. In experiments using Frisby stimuli with no emergent stereoscopic form present, so that the depth impression is one of two lacy transparent planes separated in depth (Fig. 3.2B), the smallest depth discriminable was ten times greater (9.5 min of arc, SE = 1.5). This task is also twice as difficult to learn, requiring more than 2000 trials. It is possible that without the emergence of a solid shape with clear borders, greater ambiguity is present in the stimulus and a global stereoscopic process is required to resolve the presence of retinal disparity. If this is true, then local and global stereopsis have different acuities in the pigeon, with some resolution being lost at the global level. In the kestrel (Fox et al. 1977) best performance on the RDSs occurred in a disparity band between 8–12 min of arc, with chance performance at 2 min of arc, a value well above the spatial resolution capabilities of this bird.

Both RDSs and the Frisby stereoacuity tests are context rich, with multiple nearest neighbour comparisons possible to promote stereoperception. The stereoacuity under these conditions is far superior to performance under conditions in which the pattern context of the stimuli is severely reduced.

3.3.2.4 Context-Restricted Displays

Two studies in the pigeon have examined the role of retinal disparity for relative distance perception under conditions in which the stereoscopic context of the stimuli was impoverished. The stimulus conditions consisted of the presentation of two red lights (1–3°) in the dark (Martinoya and Bloch 1980; Martinoya et al. 1988) or brightly lit white disks (15°) (McFadden 1984, 1990) presented at equal or unequal depths down a long tunnel placed behind the response keys. Apart from binocular cues, relative stimulus size, accommodation, and to a lesser extent luminous flux were potentially available to aid the discrimination. Despite this, the smallest relative difference that could be discriminated between the two disks at a distance of 10–13 cm from the eye was equivalent to a retinal disparity of 2.95° (Martinoya et al. 1988) and 3.05° (McFadden, unpubl. observ.). When retinal disparity cues are removed by blocking the binocular field of one eye, performance is significantly poorer (binocular threshold is increased by 45%), but discrimination is still possible. This is not surprising given the availability of the monocular cues mentioned above (cf. Chap. 16).

These studies probably fail to provide enough context for stereoscopic perception, but demonstrate that some binocular cue aids performance under impoverished conditions, allowing a depth threshold of only about 3°. This binocular depth cue could be convergence, as the threshold is equal to the step size of these eye movements (see Sect. 3.4.4).

3.4 Binocular Vision and the Oculomotor System in Birds

The detection of retinal disparity and its use in mediating the various forms of stereopsis require not only a binocular field but also the appropriate eye movements and retinal specializations required for binocular fixation and/or fusion. These aspects of binocular vision effectively govern where and in what situation stereopsis can be effectively used.

3.4.1 The Position of the Binocular Field

Binocular spatial perception will be limited by the position of the binocular field defined as that area of space that can be seen simultaneously by the retinas of both eyes (see Chap. 1). The binocular field increases in size as the eyes become more frontally placed, as in the owls, but this is traded off against a loss of monocular panoramic vision. The evolution of frontal vision may be related to a reduced need for panoramic vision coupled with a niche that requires binocular enhancements over a greater range of space, rather than simply some general "urge towards binocularity" as thought by Walls (1942). As discussed in Sect. 1.6, the binocular enhancement may have to do with visual sensitivity rather than binocular depth perception. Whatever the size of the binocular field, the symmetrical positions of the eyes in the head mean that if the mandibles lie within the binocular field (see Sect. 1.6.2.1), they will be centred horizontally within it. However, this co-localization may not be causal. As detailed in Sect. 3.5.1, activities in which objects are manipulated with a beak are not necessarily guided by binocular perception.

One implication of the above observations is that where the position of the eyes is relatively lateral, the binocular field will be small and binocular perception will be restricted to objects located centrally in the field of view. The height of binocular fields is also generally much greater then their width; for example, 144° high × 24° wide in the pigeon (McFadden and Reymond 1985). The significance of this is unclear, but it may indicate a role for vertical disparities, or be related to the distance of the ground plane.

The angular size of the binocular field is usually defined with reference to a fixation point and does not increase in size as the eyes converge, as is often stated. Rather, the fixation point and thus the relevance of the binocular field, becomes closer to the eyes. Binocular fixation requires each eye to observe simultaneously the same point in space, and the accuracy with which this can be done depends in turn upon eye movement control and specializations in that part of the retina which subserves the binocular visual field.

3.4.2 The Visual Trident in Birds

Many birds are bifoveate (e.g. hawks, eagles, swallows, terns, kingfishers, bitterns, hummingbirds and some wing-feeding passerines), thus allowing a dual mode of vision which was described by Rochon-Duvigneaud (cited in Walls 1942) as the "visual trident" (Fig. 3.3). This involves a central fovea specialized for the monocular lateral field of view, and a temporal fovea which projects into the binocular visual field. In birds with a single central fovea, a temporal area of increased cell density and synaptic complexity is present (Galifret 1968; Yazulla 1974; Hayes and Holden 1983); in pigeons, this is termed the area dorsalis.

Prior to the peck response of pigeons, it has been calculated that the head is positioned so that the grain is imaged on the area dorsalis (Goodale 1983; McFadden and Reymond 1985). Similarly, boundaries present in the Frisby stimuli also appeared to be fixated with the area dorsalis during binocular threshold discriminations (McFadden 1984). Thus it is quite likely that the temporal binocular area or fovea is used in binocular fixation. If this is true, then the more developed this retinal specialization, the greater the accuracy of fixation. Lateral shift of the image across a foveal pit may be required for the finest fixation control.

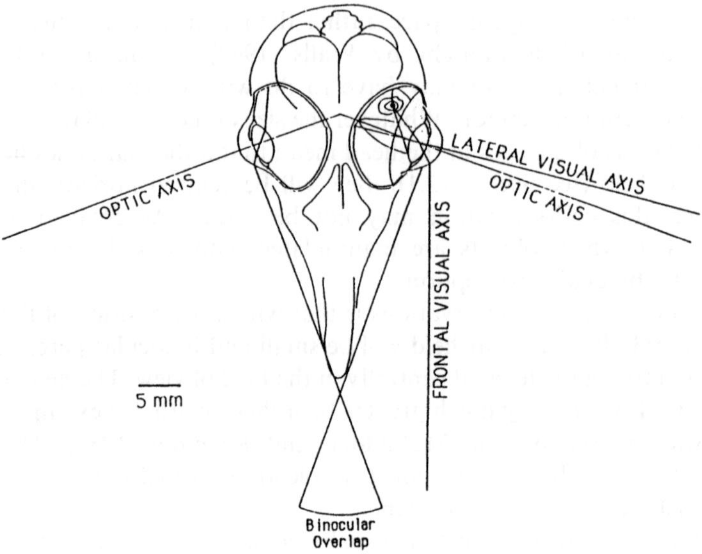

Fig. 3.3. The dual viewing system of the pigeon, represented from a dorsal view

3.4.3 Binocular Fixation and Fusion

Binocular fixation differs from binocular fusion. In both cases, vergence eye movements are essential if fixation and/or fusion is to occur for a range of viewing distances. However, only in the latter case is neural integration of the two monocular images required. Binocular fixation without fusion may seem an odd concept. Even so, despite the presence of double images, it can be used to support a small range of patent stereopsis in man (Ogle 1950) which occurs outside Panum's fusion area (see Mitchell 1966 for a review). Furthermore, qualitative stereopsis is unaffected; here, an object can be detected as nearer or further than the fixation point but the precise disparity/depth relationship of patent stereopsis is lost.

Binocular fixation without fusion may be precisely what species without accommodation (such as invertebrates) use to guide prey reaching movements. Here, the focal plane is fixed for any particular visual direction. Thus the cue to strike at prey targets in these species may be the time at which the prey object falls on this fixation locus, and fusion may not be necessary. Indeed, it has been elegantly shown in the praying mantis (Rossel 1983) that only a narrow range of convergence angles elicits strike activity. Unless it can be shown that the endpoint position of the body part being extended varies, and not just the timing of the strike, it is likely that the action is based on absolute perception (using the convergence angle) rather than relative distance perception (using retinal disparity). The confusion between these two measures has arisen because retinal disparity has been defined as equivalent to the convergence angle with geometric respect to forward looking foveae, that is, a hypothetical point at infinity, rather than the usual definition with respect to some fixation point (toads: Collett 1977; praying mantis: Rossel 1983). The situation in amphibians is interesting, as frogs and toads can estimate accurately the absolute distance of prey objects using both accommodation and convergence angle (Ingle 1976; Collett 1977). However, without a vergence system and a binocular fixation mechanism (Collett 1977), it is difficult to envisage the precise role of convergence in such distance computations.

Birds, on the other hand, can vary their angle of eye convergence and align the binocular visual axes despite differing degrees of frontality (pigeon: Bloch et al. 1987; little eagle and tawny frogmouth: Wallman and Pettigrew 1985; starling: Martin 1986; kingfishers: Moroney and Pettigrew 1987). Such binocular fixation relies on coordinated vergence eye movements.

3.4.4 Vergence Eye Movements

In general, avian eye movements are more independent and more weakly yoked than those of mammals. However, saccades in the two eyes do tend to be initiated simultaneously; within 8 ms in the chicken (Wallman and McPhun, cited in Wallman and Pettigrew 1985) and in less than 0.2 s in the pigeon (Bloch et al. 1984). Small saccades (less than 5°) can be conjugate or disjunctive (either convergent or divergent) in the eagle and the frogmouth (Wallman and Pettigrew 1985) and in the pigeon (Martinoya et al. 1984a; Bloch et al. 1984).

As a pigeon's head descends during pecking, convergent saccades are simultaneously initiated in each eye and occur in discrete steps of approximately 3° (Fig. 3.4). Whether there is a neural signal from proprioceptive sources proportional to the vergence state of the eyes is unknown. It is also difficult to imagine how such a measure could provide enough accuracy to enable calculation of the eye movement trajectory required to ensure sufficiently precise alignment of the temporal foveal or area for binocular fusion. Retinal disparity may be a better signal to guide the saccade size required for fusion. This possibility is corroborated by examination of convergent eye movements under

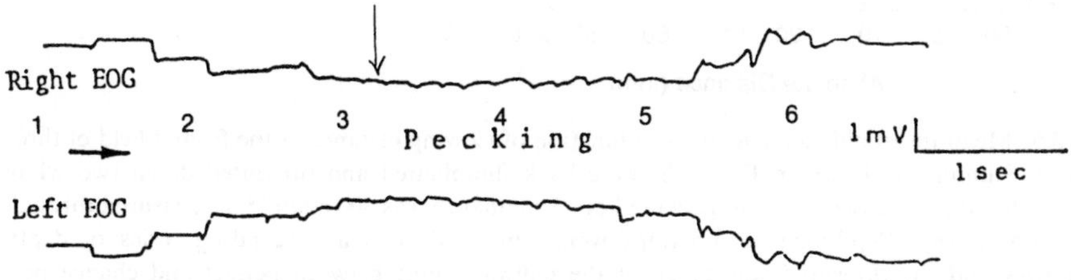

Fig. 3.4. *EOG* traces of convergent eye movements in the pigeon during pecking. The *arrow* indicates the point at which the peck occurs. (McFadden et al. 1986)

open loop conditions in which retinal disparity is removed either by occlusion of the binocular field of one eye or by binocular neural dissociation (split brain pigeon) (McFadden 1990). Under both of these conditions, the eyes overconverge, suggesting that retinal disparity may control the halt position of the convergence eye movement. The initiation of the movement may be under accommodative control because convergence in the pigeon still occurs when one eye is covered (Bloch et al. 1984; see Sect. 2.3.2).

3.4.5 Stereoscopic Limits Imposed Through the Oculomotor System

The limiting range of stereopsis is the difference between the nearest fixation plane (near point) and the furthermost distance (far point) that can support stereoscopic perception. This range is limited by accommodative and convergence capacity. In the pigeon, threshold stereoacuity using the stimuli shown in Fig. 3.2B varies as a function of the absolute distance of the stimulus array from the eye (Fig. 3.5).

The near point for threshold stereopsis was found to lie at a distance of 90 mm from the eye, suggesting that the tip of the beak lies outside the threshold stereoscopic range. This value is likely to reflect the nearest point which can be

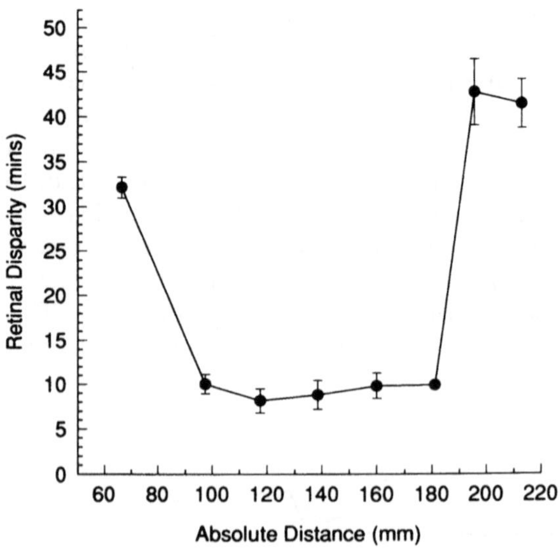

Fig. 3.5. Mean threshold stereoacuity as a function of viewing distance in the frontal field of three pigeons. Stimuli (as shown in Fig. 3.2B) were back illuminated and presented down two white tunnels placed behind two adjacent pecking keys. Luminance was 500 cd m^{-2} as measured through the response key. Psychometric functions were generated with a descending series of depth differences and the threshold was taken at the halfway point between perfect and chance performance. The depth threshold was converted to retinal disparity using Eq. (3.2) in the text and distance values obtained from video images of the head fixation position during threshold testing

accurately focused through accommodation, rather than to be a function of binocular fixation limitations, as the near point for threshold spatial acuity measured with high contrast gratings viewed with one eye occurs at a similar position (100 mm; Rounsley et al. 1993). The near point for precise accommodation is thus somewhat greater than the 50 mm predicted on the basis of the stationary head positions that occur during key pecking in the pigeon (Hodos et al. 1985).

The far point of stereopsis is normally defined in terms of the greatest distance at which an object can be just detected as nearer than an object at infinity (Ogle 1962). Based on a threshold stereoacuity of 9.5 min of arc in the pigeon (see Sect. 3.3.2.3), it can be calculated that the limiting far point would be about 8 m (McFadden 1993). Since the experimentally measured far point is only 190 mm (Fig. 3.5) additional constraints are at work. It is known that the pigeon eye varies in its refractive state as a function of elevation (Fitzke et al. 1985; see Sect. 2.5) and is also myopic in the frontal binocular field (Millodot and Blough 1971). In the anaesthetized resting state, we have measured the refractive state within the binocular field 10° below the eye beak axis to be 2 D (SE = 0.4 D) myopic. This is equivalent to an accommodative resting state of 500 mm. Thus the limited stereoscopic range is not easily explained in terms of accommodative limitations. It is possible that fine fixation and fusion capacity are the determining limitation in a species which does not have a binocular fovea. This could mean that the presence of a binocular fovea increases stereoscopic range through an enhancement of fixation accuracy.

3.5 Role of Binocular Vision in the Guidance of Avian Behaviour

The different forms of binocular vision appear to have different levels of precision. Local stereopsis can detect both the sign and the magnitude of disparity with the highest level of precision. However, it is restricted to determining relative distance between local surfaces and their backgrounds. On its own, it cannot give absolute distance information. Behaviours involving relative distance information with simplistic images may arise when the distance between a predetermined fixed absolute position in space and an object needs to be estimated. This may occur if the reaching body part is extended into some automatic and constant position, and it can be fixated within the binocular visual field. It is also possible that the output from coarse local retinal disparity detectors could drive the fine eye movements required for binocular fusion without diplopia.

Global stereopsis, with a possibly coarser resolution level, may be most useful in the detection of visual patterns rather than in directly guiding motor movements. Camouflaged images can be decoded on the basis of surface groupings of differing retinal disparity values. This is obviously important in identifying and locating objects relative to their surroundings (cf. the use of relative motion to break camouflage; Sect. 12.1.1).

3.5.1 Guidance of the Peck Movement

The peck response is a good example of a reaching movement, in which the egocentric position of the beak is relatively fixed. The peck movement of the pigeon is made up of a number of fast saccade-like movements, in which head velocity is up to 165 cm s^{-1} (Smith 1974; Hodos et al. 1976), interspersed with longer periods (100–200 ms) in which the head remains stationary (Goodale 1983). After the final fixation, visual feedback is limited as the eyes begin to close and are fully closed at the point of contact (see Sect. 2.4). Accuracy in pecking at individual grains of food (3 mm) improves with practice (up to a 30% residual error rate in the most highly trained birds). Coincident with this improvement in movement accuracy is a change in the positions of the fixation points (McFadden 1990).

Binocular vision does not appear to guide directly these fixation positions (McFadden 1990), but is partially involved in peck accuracy. For example, if a pigeon is deprived of binocular vision by occluding the binocular field of one eye (so there is no blind spot), pecking accuracy (at small single grains) is significantly retarded, although with practice birds can learn to partially overcome the deficit. The lateral position of the final peck is also not really affected by monocular conditions (Jager and Zeigler 1991). Prisms placed before the eyes can significantly affect peck performance, but adaptation occurs (Hess 1956; Martinoya et al. 1984b).

In contrast, accommodation clearly plays a role, as disrupting the refractive state of the eye with lenses of varying powers not only retards peck accuracy (Martinoya et al. 1984b) but has also been shown to affect partially the fixation position in owls (Wagner and Schaeffel 1991; see Sect. 2.6). Use of accommodation to guide absolute distance has also been found in other species, such as the chameleon (Harkness 1977) and toad (Collett 1977).

What then is the role of binocular vision in guiding movement? Theoretically, binocular fixation without fusion can be used to estimate absolute distance, by a transformation of the convergence signal of the eyes. Stereopsis, on the other hand, can only provide relative distance information. In the studies mentioned above, the conditions usually involved pecking at an obvious target in a simple context. Complex pattern decoding was not required. It is likely that binocular vision is not primarily used to guide such absolute distance perception, but rather is used in pattern decoding and object detection. Thus, for example, the range of maximal stereoscopic acuity correlates precisely with the position of the head up to, and finishing precisely with, the first fixation position (McFadden unpubl. observ.). This suggests that stereopsis is used prior to the actual peck movement, and could be important for textured images in which the position of the grain is partially camouflaged.

The peck movement therefore relies on multiple cues (cf. Chap. 16), but the most prominent is accommodation linked into a ballistic movement pattern (cf. Sect. 2.4). Convergence on its own plays a very minor role in determining the movement associated with this absolute distance task. Since convergence in each eye still occurs when one eye is covered (Bloch et al. 1984), the role of convergence may be primarily limited to that driven by accommodation (Mulhearn and McFadden 1993; cf. Sect. 2.3.2). On the other hand, the role of binocular vision in peck accuracy suggests that vergence eye movements underpin binocular fixation and fusion, and thus play a prominent role in object perception rather than localization.

3.5.2 Dependence of Behaviour on the Frame of Reference

An egocentric frame of reference is important to guide reaching motor movements. A free floating angular depth measurement is of no use except in the limiting case where one of the reference points is in a fixed position relative to the observer. This may occur in the extension of the feet during prey capture in raptors or in estimating landing position (cf. Sect. 16.2.2), provided the reference point can be clearly fixated binocularly and has a predetermined end position. In non-limiting cases in which the reference point is variable, retinal disparity would require a more sophisticated transformation if it is to aid reaching movements.

For simplistic images, local stereopsis would be sufficient, the output of which is effectively automatically recalibrated. In other words, because γ_F (see Fig. 3.1) is effectively a constant, and given that from Eq. (3.1) the following relation holds:

$$\gamma_P = \gamma_F - \delta \tag{3.3}$$

then the absolute distance of a target at position *P* is directly related to the retinal disparity δ (for example between the talons and the target). As the limbs move, the binocular field would also need to be large enough to encompass activity within a binocular field without recourse to excessive head and/or eye movement. Narrow binocular fields imply that it would be desirable to keep the fixated body part fairly central. If the visual direction is not central ($\phi > 0$ in Fig. 3.1), then the perceived depth computation has to include this additional variable. Visual direction could be derived from conjugate saccade amplitude. Since conjugate saccades have been found to be limited to less than 6° (Wallman and Pettigrew 1985) it would be predicted that behaviours based on local stereopsis may fail if the limbs and the object are laterally separated by more than 6°.

Few studies (e.g. Davies and Green 1991) have specifically examined the role of binocular vision in reaching movements, although it is known that optic flow variables are involved in some avian species (Davies and Green 1990; see also Chap. 13), suggesting that multiple cues may be available to guide these behaviours (cf. Chap. 16).

Relative distance information derived from stereopsis can, however, be used directly for object detection. Raptors can discriminate figure from ground from a great height, and it has commonly been assumed that this is achieved through stereopsis. Global stereopsis would aid such object detection if underpinned by hyperacute local stereopsis and fine oculomotor control, neither of which has yet been demonstrated. Global stereopsis would also be useful in many behaviours which rely on object detection in camouflaged environments.

3.6 Conclusions

Binocular depth perception and stereopsis are not unitary phenomena, nor a high level capacity bestowed on the élite (see Fox 1978) but may occur in various forms, and vary with both the ecological niches and behavioural needs of birds.

Binocular spatial perception arises from a number of processes, some of which are interdependent, and some of which are dissociable. In summary, there are three main components, which have complementary purposes. Some signal associated with convergence eye movements in isolation (possibly without fusion) can support absolute distance perception, although it has a relatively coarse level of resolution. This mechanism normally works in conjuction with accommodation, and is useful for behaviours with a fixed frame of reference. Horizontal retinal disparity appears to guide at least the halt position of these convergence eye movements, and provides the basis for two levels of stereopsis.

At the first level, local stereopsis can be used in relative distance perception and may guide external limb movements in contacting or avoiding objects. This first level may exist independently of the second, at which global stereopsis uses input from local retinal disparity neurons to enhance object detection and location with respect to a patterned background, but is unlikely to guide motor behaviour directly.

References

Bagnoli P, Fontanesi G, Casini G, Porciatti V (1990) Binocularity in the little owl, *Athene noctua*. Brain Behav Evol 35:31–39

Bishop PO, Henry GH (1971) Spatial vision. Annu Rev Psychol 22:119–160

Blake R, Wilson HR (1991) Neural models of stereoscopic vision. Trends Neurosci 14:445–452

Bloch S, Rivaud S, Martinoya C (1984) Comparing frontal and lateral viewing in the pigeon. III Different patterns of eye movements for binocular and monocular fixation. Behav Brain Res 12:173–182

Bloch S, Lemeignan M, Martinoya C (1987) Coordinated vergence for frontal fixation, but independent eye movements for lateral viewing, in the pigeon. In: O'Regan JK, Levy–Schoen A (eds) Eye movements: from physiology to cognition. Elsevier, Amsterdam, pp 47–56

Collett TS (1977) Stereopsis in toads. Nature 267:349–351

Davies MNO, Green PR (1990) Optic flow-field variables trigger landing in hawk but not in pigeons. Naturwissenschaften 77:142–144

Davies MNO, Green PR (1991) The adaptability of visuomotor control in the pigeon during landing flight. Zool Jahrb Physiol 95:331–338

Fitzke FW, Hayes BP, Hodos W, Holden AL, Low J (1985) Refractive sectors in the visual field of the pigeon eye. J Physiol 369:33–44

Fox R (1978) Binocularity and stereopsis in the evolution of vertebrate vision. In: Cool SJ, Smith EL (eds) Frontiers in visual science, vol 3. Springer Berlin Heidelberg New York, pp 316–327
Fox R, Lehmkuhle SW, Westendorf DH (1976) Falcon visual acuity. Science 192:263–265
Fox R, Lehmkuhle SW, Bush RC (1977) Stereopsis in the falcon. Science 197:79–81
Frost BJ, Goodale MA, Pettigrew JD (1983) A search for functional binocularity in the pigeon. Proc Soc Neurosci 9:823
Galifret Y (1968) Les diverses aires fonctionelles de la rétine du pigeon. Z Zellforsch 86:535–545
Goodale MA (1983) Visually guided pecking in the pigeon (*Columba livia*). Brain Behav Evol 22:22–41
Harkness L (1977) Chameleons use accommodation cues to judge distance. Nature 267:346–349
Hayes BP, Holden AL (1983) The distribution of displaced ganglion cells in the retina of the pigeon. Exp Brain Res 49:181–188
Hershberger W (1970) Attached shadow orientation perceived as depth by chickens reared in an environment illuminated from below. J Comp Physiol Psychol 73:407–411
Hess EH (1950) Development of chick's responses to light and shade cues of depth. J Comp Physiol Psychol 43:112–122
Hess EH (1956) Space perception in the chick. Sci Am 195:71–80
Hirsch J (1982) Falcon visual sensitivity to grating contrasts. Nature 300:57–58
Hodos W, Leibowitz RW, Bonbright JC Jr (1976) Near-field visual acuity of pigeons: Effects of head location and stimulus luminance. J Exp Anal Behav 25:129–141
Hodos W, Bessette BB, Macko KA, Weiss SRB (1985) Normative data for pigeon vision. Vision Res 25:1525–1527
Hughes A (1977) The topography of vision in mammals of contrasting lifestyle: Comparative optics and retinal organisation. In: Cresticelli F (ed) Handbook of sensory physiology, vol VII/5. The visual system in vertebrates. Springer, Berlin Heidelberg New York, pp 614–642
Ingle D (1976) Spatial vision in anurans. In: Fite K (ed) The amphibian visual system: a multidisciplinary approach. Academic Press, New York, pp 119–140
Jager R, Zeigler HP (1991) Visual field organization and peck localization in the pigeon (*Columba livia*). Behav Brain Res 45:65–69
Julesz B (1960) Binocular depth perception of computer generated patterns. Bell System Tech J 39:1125
Julesz B (1971) Foundations of cyclopean perception. University of Chicago Press, Chicago
Julesz B (1978) Global stereopsis: Cooperative phenomena in stereoscopic depth perception. In: Held R, Leibowitz HW, Teuber H-L (eds) Handbook of sensory physiology, vol 8. Perception. Springer, Berlin Heidelberg New York, pp 215–256
Julesz B, Miller JE (1975) Independent spatial-frequency-tuned channels in binocular fusion and rivalry. Perception 4:125–143
Karten HJ, Hodos W, Nauta WJH, Revzin AM (1973) Neural connections of the 'visual Wulst' of the avian telencephalon: Experimental studies in the pigeon (*Columba livia*) and owl (*Speotyto cunicularia*). J Comp Neurol 150:253–278
Longuet-Higgins C (1981) A computer algorithm for reconstructing a scene from two projections. Nature 293:4133–4135
Martin GR (1986) The eye of a passeriform bird, the European starling (*Sturnus vulgaris*): eye movement amplitude, visual fields and schematic optics. J Comp Physiol 159:545–557
Martinoya C, Bloch S (1980) Depth perception in the pigeon: looking for the participation of binocular cues. In: Grastyan E, Molnar P (eds) Sensory functions: advances in physiological sciences, vol 16. Pergamon Press, Oxford, pp 477–482
Martinoya C, Le Houezec J. Bloch S (1984a) Pigeons' eyes converge during feeding: Evidence for frontal binocular fixation in a lateral eyed bird. Neurosci Lett 45:335–339
Martinoya C, Palacios A, Bloch S (1984b) Participation of eye convergence and frontal accommodation in programming grain pecking in pigeons. Abstr 8th Eur Neuroscience Congr, The Hague, The Netherlands, Sept 11–15. Neurosci Lett Suppl 18
Martinoya C, Le Houezec J, Bloch S (1988) Depth resolution in the pigeon. J Comp Physiol 163:33–42

Mayhew JEW, Longuet-Higgins C (1982) A computational model of binocular depth perception. Nature 297:376–378
McFadden SA (1984) Depth perception in the pigeon. Thesis, The Australian National University, Canberra, Australia
McFadden SA (1987) The binocular stereoacuity of the pigeon and its relation to the anatomical resolving power of the eye. Vision Res 27:1741–1746
McFadden SA (1990) Eye design for depth and distance perception in birds: an observer orientated perspective. J Comp Psychol: Comp Stud Perception Cognition 3:1–31
McFadden SA (1992) Depth constancy in the pigeon. Proc 6th Biennial Meet Int Soc Comp Psychol, 20–24 July, Brussels, S1
McFadden SA (1993) Constructing the 3 D image. In: Zeigler HP, Bischof H-J (eds) Vision, brain and behaviour in birds. MIT Press, Cambridge, Mass (in press)
McFadden SA, Reymond L (1985) A further look at the binocular visual field of the pigeon (*Columba livia*). Vision Res 25:1741–1746
McFadden SA, Wild JM (1986) Binocular depth perception in the pigeon (*Columba livia*). J Exp Anal Behav 45:149–160
McFadden SA, Lemeignan M, Martinoya C, Bloch S (1986) Effect of commissurotomy on pecking and eye convergence in the pigeon. Neurosci Lett 26:572
Micheli D, Reperant J (1982) Thalmo-hyperstriatal projections in the pigeon (*Columba livia*) as demonstrated with retrograde double-labelling with fluorescent tracers. Brain Res 245:365–371
Millodot M, Blough P (1971) The refractive state of the pigeon eye. Vision Res 11:1019–1022
Mitchell DE (1966) A review of the concept of "Panum's fusional areas". Am J Optom Physiol Opt 43:387
Moroney MK, Pettigrew JD (1987) Some observations on the visual optics of kingfishers (Aves, Coraciformes, Alcedinidae). J Comp Physiol 160:137–149
Mulhearn JT, McFadden SA (1993) Comparison of accommodation during coordinated and independent eye movements in the pigeon. Proc Aust Soc Neurosci 4:130
Nelson JI (1975) Globality and stereoscopic fusion in binocular vision. J Theor Biol 49:1–88
Ogle KN (1950) Disparity limits of stereopsis. Arch Ophthalmol 48:50
Ogle KN (1962) Spatial localisation through binocular vision. In: Davson H (ed) The eye, vol 4. Academic Press, New York, pp 271–320
Perisic M, Mihailovic J, Cuenod M (1971) Electrophysiology of contralateral and ipsilateral projections to the Wulst in the pigeon (*Columba livia*). Int J Neurosci 2:7–14
Pettigrew JD (1979a) Comparison of the retinotopic organisation of the visual Wulst in nocturnal and diurnal raptors, with a note on the evolution of frontal vision. In: Cool SJ, Smith EL (eds) Frontiers of visual science, vol III. Springer, Berlin Heidelberg New York, pp 328–335
Pettigrew JD (1979b) Binocular visual processing in the owl's telencephalon. Proc R Soc Lond B 204:435–454
Pettigrew JD, Konishi M (1976) Neurones selective for orientation and binocular disparity in the visual Wulst of the barn owl (*Tyto alba*). Science 193:675–678
Poggio GF, Poggio T (1984) The analysis of stereopsis. Annu Rev Neurosci 7:379–412
Powell RW, Smith JC (1968) Critical-flicker-fusion thresholds as a function of very small pulse-to-cycle fractions. Psychol Rec 18:35–40
Ptito A, Zatorre RJ, Larson WL, Tosoni C (1991) Stereopsis after unilateral anterior temporal lobectomy. Brain 114:1323–1333
Reymond E (1985) Spatial visual acuity of the eagle *Aquila audax*: A behavioural, optical and anatomical investigation. Vision Res 25:1477–1491
Richards W (1970) Stereopsis and stereoblindness. Exp Brain Res 10:380–388
Rossel S (1983) Binocular stereopsis in an insect. Nature 302:821–822
Rounsley R, Watson T, McFadden SA (1993) Variation in visual acuity as a function of absolute distance in the frontal field of the pigeon eye. Proc Aust Neurosci Soc 4:130
Shipley T, Rawlings SC (1970) The nonius horopter. I. History and theory. Vision Res 10:1225
Smith RF (1974) Topography of food-reinforced key peck and the source of the 30-millisecond interresponse time. J Exp Anal Behav 21:541–551

Wagner H, Schaeffel F (1991) Barn owls (*Tyto alba*) use accommodation as a distance cue. J Comp Physiol 169:515–521

Wallman J, Pettigrew JD (1985) Conjugate and disjunctive saccades in two avian species with contrasting oculomotor strategies. J Neurosci 5:1418–1428

Walls GL (1942) The vertebrate eye and its adaptive radiation. Hanfer, New York

Westheimer G (1975) Visual acuity and hyperacuity. Invest Ophthalmol Visual Sci 14:570

Westheimer G (1979) Cooperative neural processes involved in stereoscopic acuity. Exp Brain Res 36:585–597

Wheatstone C (1838) Contributions to the physiology of vision. I. On some remarkable, and hitherto unobserved, phenomena of binocular vision. Philos Trans R Soc Lond 128:371–394

Wilson P (1980) The organisation of the visual hyperstriatum in the domestic chick: Topology and topography of the visual projection. Brain Res 188:319–332

Yazulla S (1974) Intraretinal differentiation in the synaptic organisation of the inner plexiform layer of the pigeon retina. J Comp Neurol 153:309–324

4 Sound Cues to Distance: The Perception of Range

P.K. McGregor

4.1 Introduction

A major theme of this book is that the perceptual abilities of a species have evolved in relation to the specific demands of information gathering in the species' environment. It follows, therefore, that we must identify the nature of the information that is important and salient to the study species when exploring its perceptual and sensory abilities. Without such information it will be difficult to design appropriate experiments and to interpret the capabilities which they reveal.

The sense of hearing is commonly considered to be one of birds' major senses, ranking equal in importance with vision. A diverse array of perceptual abilities based on sound have been documented for birds. From the detection of prey by sound (see Chap. 14) to the identification of particular individuals such as parents, offspring, mates and territorial neighbours (reviewed by Falls 1982), birds have demonstrated an ability to gather important information from the sounds in their environment. In this chapter I will discuss how information on distance is obtained from vocalizations used in intraspecific communication (songs and calls), although the same principles will apply to any sounds made by birds. The distance information which will be discussed is referred to as *range* (Morton 1982); it is the distance between the signal receiver (the listening bird) and the signaller (the vocalizing bird). Therefore, we are largely concerned with birds as receivers and interpreters of sound signals, rather than as signallers, and in principle we are interested in both males and females, calls and songs.

4.2 Why Range?

Separation distance plays an important role in bird social behaviour. Birds in cohesive feeding or migrating flocks must maintain relatively small inter-individual distances, whereas most territorial individuals must maintain larger distances. In some situations, visual inspection will give the necessary distance

Behaviour and Ecology Research Group, Department of Life Science, University of Nottingham, Nottingham NG7 2RD, UK

M.N.O. Davies and P.R. Green (Eds.)
Perception and Motor Control in Birds

information, but in many others vision will be limited by habitat or diurnal variation in light levels. Bird vocalizations are often long range signals that transmit effectively in habitats and over distances where visual signalling is of limited usefulness; these are the contexts where an ability to range using sound cues has evolved. For example, flocking is generally considered to have some anti-predator role (e.g. Gill 1990; Brooke and Birkhead 1991) and, if it is to be effective, individual birds in flocks need to maintain close spacing. Visual cues will be restricted in dense vegetation, such as the deciduous temperate woodland habitat of family parties of long-tailed tits (*Aegithalos caudatus*); the dense aggregations of some species, such as flocks of geese, will have a similar effect; and darkness will cause obvious problems for flocks of waders feeding on mudflats. In all these cases we might expect birds to use sound cues to range and in all these situations vocalizations are common.

Similar advantages to ranging with sound can be identified in the context of male territorial songbirds; an ability to decide whether a singer is within the receiver's territory without expending energy in approaching would be of advantage to males. A singer's position is important because a bird inside the receiver's boundary poses a threat to territory tenure and the receiver should respond aggressively, initially by approaching the intruder. However, if the singer is outside the territory then it poses much less of a threat and the receiver should respond weakly, if at all. Territorial receivers are surrounded by singing males, by far the most common being their territorial neighbours, and an ability to range will increase the efficiency of territory defence by minimizing the number of "false" responses, that is, approaches to neighbours singing some distance outside the receiver's boundaries.

4.3 Ranging Cues

Sound is affected by the sound propagation characteristics of the environment during transmission. Early studies of these effects dealt with influences on the evolution of signal structure and signaller behaviour (e.g. Morton 1970, 1975, 1980; Chappius 1971; Nottebohm 1975; Marten and Marler 1977; Marten et al. 1977; Hunter and Krebs 1979; Richards and Wiley 1980; Gish and Morton 1981; Michelsen and Larsen 1983; Cosens and Falls 1984; Handford 1988). Although perception of habitat-induced changes in signals was not considered until later (Michelsen 1978; Hansen 1979; Richards 1981), they are potential cues to range and will be considered in this section.

The influence of habitat transmission acoustics can be thought of as affecting both the quantity and quality of sound. The quantity of sound energy in the signal falls as distance increases because energy is absorbed by the environment and because sound radiates out from a source. In this chapter, the quantity effect will be referred to as a loss of overall amplitude. The quality of the signal also falls with increasing distance because the signal is distorted by many influences including reverberations (echoes from the ground and objects such as tree trunks and vegetation) and irregular fluctuations in amplitude (often caused by variation in the wind speed). Such qualitative effects are referred to in a variety of ways; the terms "smearing" (Michelsen 1992) and "blurring" (Dabelsteen et al. 1993) have recently been used, but the commonest term in the bird song literature is "degradation" (e.g. Morton 1970, 1975, 1986; Richards and Wiley 1980; Wiley and Richards 1982; McGregor 1991a). Confusingly, degradation is also used as an overall term for all changes to a signal during transmission, including overall amplitude effects (Dabelsteen et al. 1993). This is common in the literature on other animal groups (e.g. Michelsen 1992; but cf. Römer 1992) and it has been proposed that this meaning should be adopted in the songbird literature (Dabelsteen et al. 1993; Pedersen et al. submitted). While acknowledging the advantages of such a proposal, I will use the term degradation in the more restrictive sense of changes in signal quality (i.e. excluding overall amplitude effects) in order to facilitate comparisons between source material and the ideas developed here.

In principle, effects on either overall amplitude or degradation, or both, could be used by a receiver to extract information on the distance of a vocalizing individual. One of the reasons for distinguishing between the effects of overall amplitude and of degradation on signals is to clarify the a priori argument that birds should use degradation cues to range. The argument is that signal amplitude can be controlled by the signaller (by turning the head, opening the bill or increasing the rate of syringeal airflow), and therefore a low amplitude signal could indicate either a distant signaller or one signalling quietly from close at hand (Richards 1981; Morton 1982; Wiley and Richards 1982). A signaller cannot affect degradation in such a way; for example, the height at which a blackbird (*Turdus merula*) perches has little influence on blurring (Dabelsteen et al. 1993). Therefore degradation is potentially a more reliable cue to range than decrease in overall amplitude. It has been argued that degradation changes with distance in a more predictable way than does overall amplitude (Wiley and Richards 1978, 1982; Richards and Wiley 1980), although recent measurements only found a clear effect when the receiver was placed high above the ground (Fig. 5B in Dabelsteen et al. 1993).

The effects of degradation on the song of a great tit (*Parus major*) can be seen clearly by comparing recordings of the same song made from different distances (Fig. 4.1). To ensure that the differences illustrated in the figure were due to distance rather than changes in the signal or signaller behaviour, the same song was re-recorded from a loudspeaker. The wide-band (Fig. 4.1a) and narrow-band (Fig. 4.1b) spectrogram displays show the same loss of detail in the component elements (or notes) in the distant recording (righthand section) when compared with the close recording (lefthand section). The details of this loss of clarity are apparent in a wide-band spectrogram of the first two elements (Fig. 4.1c). In the distant recording, rapid frequency modulations of the first element are less clear and this element is lengthened by echo (reverberation). The amplitude of the second element is very low in the distant recording, and is virtually invisible in the figure, as a result of high frequency filtering. An

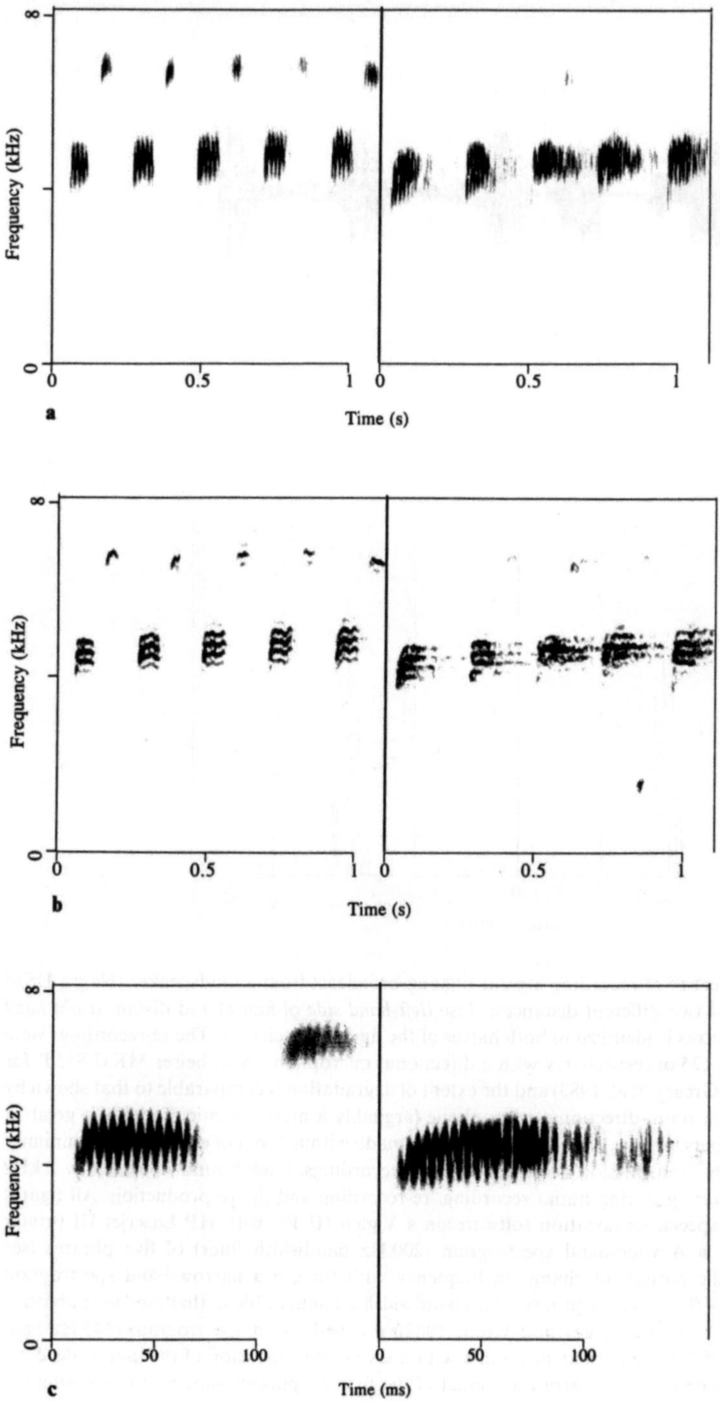
Frequency (kHz)
8
0
0
0.5
1
0
0.5
1
a
Time (s)
Frequency (kHz)
8
0
0
0.5
1
0
0.5
1
b
Time (s)
Frequency (kHz)
8
0
0
50
100
0
50
100
c
Time (ms)

Fig. 4.1a–c.

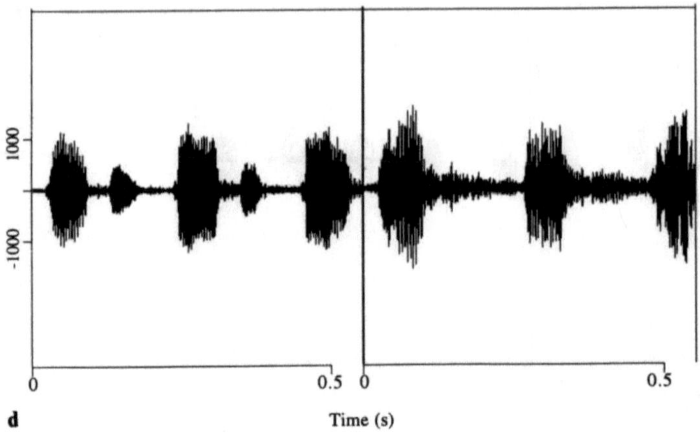

d

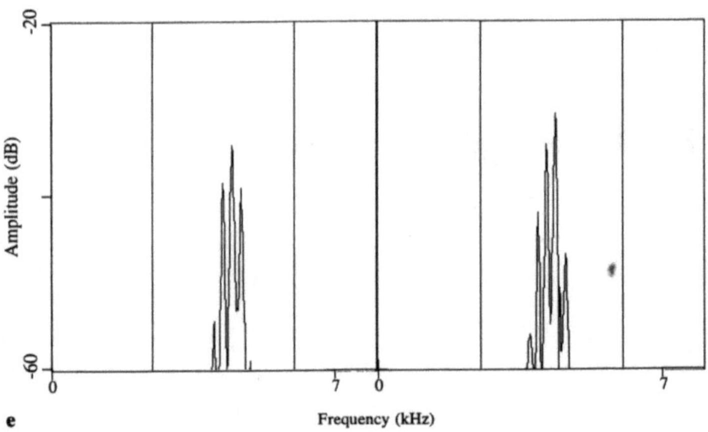

e

Fig. 4.1a–e. The effect of re-recording a great tit song broadcast from a loudspeaker (Nagra DSM speaker/amplifier) at two different distances; close (*left-hand side* of figure) and distant (*right hand side*). The scaling of axes is identical in both halves of the figure in each case. The re-recordings were made at 5 m and at 125 m respectively with a directional microphone (Sennheiser MKH 815T; for other details see McGregor et al. 1983) and the extent of degradation is comparable to that shown by re-recordings with an omni-directional microphone (arguably a more realistic model of a great tit ear) at 3 and 35 m (pers. observ.). The recordings were made within 5 min of each other to minimize differences in habitat transmission acoustics between recordings. Background noise below 2 kHz was removed by filtering during initial recording, re-recording and figure production. All figures were made by LSI speech workstation software on a Viglen III PC with HP Laserjet III printer (McGregor 1991b). **a** A wide-band spectrogram (200 Hz bandwidth filter) of five phrases (see Fig. 4.2), showing the pattern of change in frequency with time; **b** a narrow-band spectrogram (40 Hz bandwidth) of the same five phrases. This bandwidth is comparable to that used for published great tit sonagrams (e.g. McGregor and Krebs 1982); **c** a wide-band spectrogram (340 Hz bandwidth) of the first of the five phrases in **a** and **b** with a seven fold expansion of the time scale; **d** an oscillogram (amplitude on *y-axis*, arbitrary units) of the first 2.5 phrases shown in **a**, showing the pattern of change in amplitude with time; **e** A spectral slice display (amplitude vs frequency, bandwidth 32 Hz) of the first low frequency element

oscillogram (Fig. 4.1d) illustrates this effect of distance on the high frequency element; two high frequency elements are visible as the two smaller pulses between the three low frequency elements in the close recording, but in the distant recording the high frequency elements cannot be distinguished from background noise and the echoes ("tails") of the main elements. The distance effects on the distribution of energy in relation to frequency are shown by the difference in the relative height and position of the peaks in Fig. 4.1e. All, or any, of these effects provide features of sound degradation that birds could use to perceive range.

4.4 The Experimental Evidence for Ranging Ability

The capabilities of birds as receivers and interpreters of signals have been investigated using two rather different approaches. The first is by way of laboratory psychophysical experiments where the usual object is to characterize the auditory system of birds. The second approach uses playback of recorded vocalizations to subjects in the field to document the response to signals in virtually natural contexts. The rapid aggressive response of males of most territorial species of songbirds, which can be elicited by playback in the field, is ideally suited to experiments on ranging ability. Indeed, I am aware of no studies of ranging abilities using other approaches or in different contexts.

It was argued above that territorial males should respond less strongly to song outside their territory than inside, and this effect provides a way of investigating whether males use cues in the degradation of song as a source of range information. The experimental design is to play back the same song, or group of songs, with different extents of degradation; playback stimuli are "less degraded" or "more degraded". The stimuli for playback are produced by re-recording broadcast songs at different distances, taking care to minimize the amount of background noise recorded. The resulting recordings show clear signs of different extents of degradation (see Fig. 4.1 and also Fig. 1 in McGregor et al. 1983) and little difference in the amount of background noise in the range of frequency covered by the song (Fig. 4.1a–c). The songs are played back at the same overall amplitude and from the same position inside the territory boundary. Such a design presents the birds with conflicting signal properties with respect to distance, because the overall amplitude is the same in both stimuli, and the test of ranging abilities is therefore conservative (T. Dabelsteen pers. comm.). A conflicting properties design has recently been used specifically to investigate the perception of alarm calls (Brémond and Aubin 1992).

If the extent of degradation provides males with information on distance to a singer, then the more degraded songs should be perceived as coming from outside the territory and the less degraded from inside, resulting in a weak territorial response to more degraded songs and a strong response to less degraded songs.

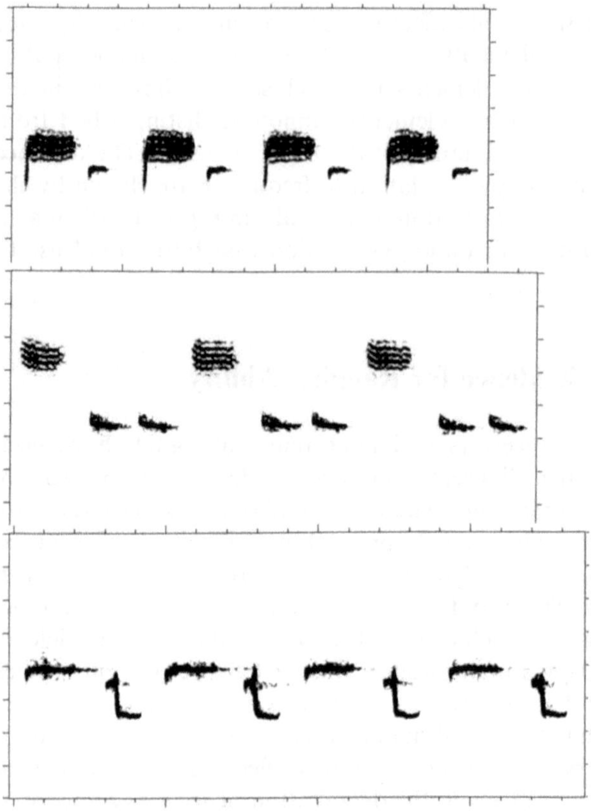

Fig. 4.2a. Caption see p. 82

This experimental paradigm has been used with five bird species. The first experiment, with Carolina wrens (*Thryothorus ludovicianus*), found a difference in response in the direction expected if degradation is used in ranging (Richards 1981). When presented with less degraded songs, the experimental males did not sing, but closely approached the sound source; the same response is elicited by naturally occurring intrusions by singing males. However, males responded to playback of more degraded songs by singing, without approaching the playback loudspeaker; the same response occurs to a male singing outside the territory. While this result is consistent with a role of song degradation in ranging, alternative interpretations are that more degraded song elicits a different (weaker) response because it lacks some species-specific releasing stimuli, or because it is less detectable (sensu information theory). The fact that the qualitatively different responses of Carolina wrens (approach, no song versus no approach, song) are the same as the responses elicited naturally by singing intruders inside and outside the territory boundary (Richards 1981) makes the alternative interpretations unlikely, but cannot rule them out. However, subsequent experiments with two other species have ruled out an explanation in

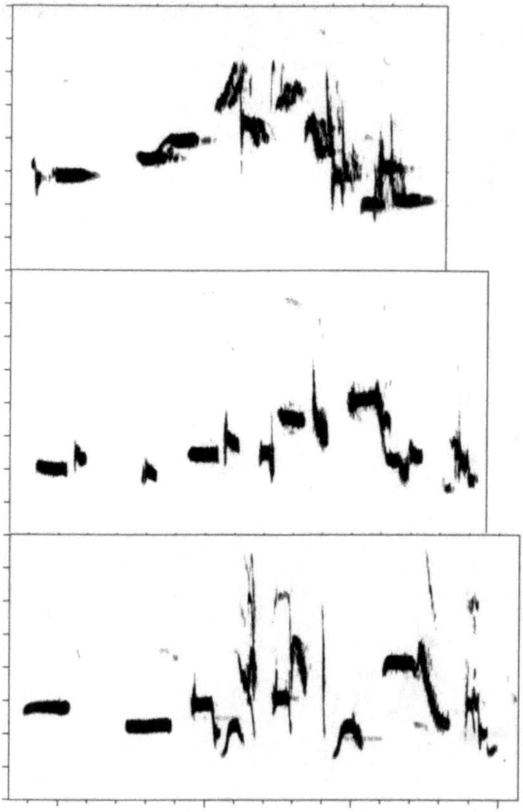

Fig. 4.2bi. Caption see p. 82

terms of a lack of releasing stimuli, by showing that the difference in response to more and less degraded songs only occurs for some songs. In these experiments with great tits and western meadowlarks (*Sturnella neglecta*), two kinds of song were played back in more and in less degraded versions; some songs were familiar to the test male and some were unfamiliar. In order to explain the design and interpretation of these experiments it is necessary to establish the meaning of the terms familiarity and similarity, and also to give a little background to song variation in these species to distinguish between songs and song types.

Great tits and western meadowlarks are examples of species in which each male sings a number of discrete songs, making up a song repertoire (Fig. 4.2). The constituent songs of each male's repertoire can be categorized into *song types* on the basis of a number of criteria (great tits: McGregor and Krebs 1982; western meadowlarks: Horn and Falls 1988). Songs sung by different males that are classified into the same song type are therefore more similar to one another than to a song of a different song type. Similarity can be quantified in a number of ways (e.g. appendix of Falls et al. 1982) as well as the more qualitative approach of visual inspection of sonagrams which uses an unknown, but

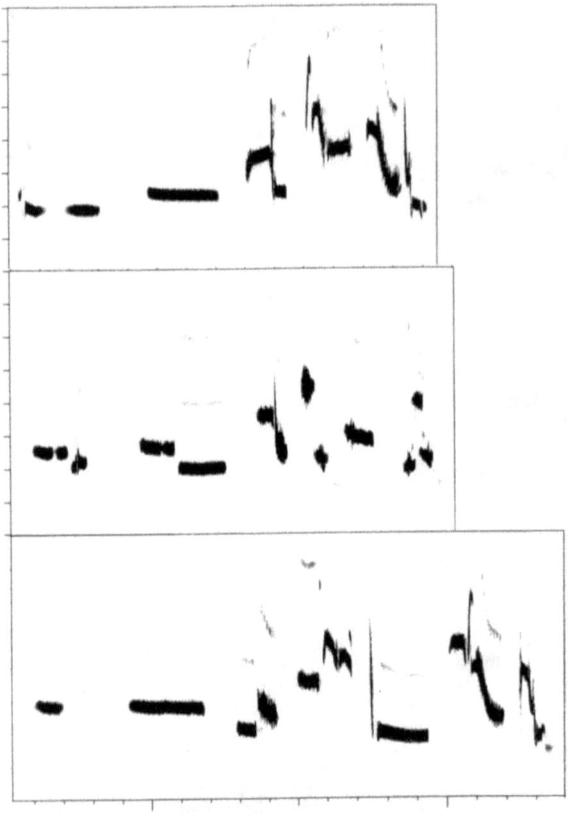

Fig. 4.2a. A representative great tit repertoire of three song types, categorized as F, D and X (*top to bottom*) according to the criteria of McGregor and Krebs (1982). Songs are composed of elements (also termed notes) which are discrete spectrogram traces (there are 8, 9 and 8 elements in F, D and X respectively). Elements are grouped into a repetitive phrase (there are 4, 3 and 4 phrases in F, D and X respectively). Phrases of different song types can have different numbers of elements per phrase (numbers are 2, 3 and 2 in F, D and X respectively). The *x-axis* divisions are at 0.1-s intervals, those of the *y-axis* at 1 kHz intervals. The sonagrams were made with a 16 kHz sampling rate and 40 Hz bandwidth. **b** A representative western meadowlark repertoire of six song types. Closely associated elements are referred to as syllables, particularly common in the middle and end of each song. The *x-axis* divisions are at 0.1-s intervals, the *y-axis* at 1 kHz intervals. The sonagrams were made with a 16 kHz sampling rate and 200 Hz bandwidth (see Fig. 4.1 for further details)

probably large, number of features of the whole song. Familiarity is used in its everyday sense and the term can be applied both to songs and to song types. Therefore if a neighbour of a male sings a song of song type A, the male will be familiar both with that neighbour's *song* (a version of song type A) and also with the category of *song type* A. Males will be most familiar with both the songs and the song types of contiguous neighbours, because they will hear them frequently. Males will be unfamiliar with songs of males that are out of earshot, but they could be familiar with their song types, provided that these song types are also sung by neighbours or by themselves.

Both the similarity and the familiarity of songs are known to affect the response to playback (great tits: Falls et al. 1982; western meadowlarks: A.G. Horn, pers. comm.). To control for these two effects in the degradation experiments, the songs used for playback were all recorded from males out of earshot of the test males (McGregor et al. 1983; McGregor and Falls 1984; McGregor and Krebs 1984); all playback *songs* were therefore of the same level of unfamiliarity. However the *song types* used in the experiments were chosen to have different degrees of familiarity. An "unfamiliar" playback stimulus song was of a song type that was sung by neither the test male, nor any of its neighbours, nor any males within earshot of the test male. A "familiar" playback song was of a song type that was sung by either the test male, or a contiguous neighbour, or both.

The results of the great tit and western meadowlark experiments showed a significant difference in the strength of the aggressive response to playback of familiar stimuli; weaker responses were elicited by more degraded than by less degraded familiar song types. However, there were no significant differences in the response to unfamiliar stimuli (Fig. 4.3; McGregor et al. 1983; McGregor and Falls 1984; McGregor and Krebs 1984).

The spatial distribution of song types in the study areas used for these experiments enabled most of the playback stimuli to be used as familiar songs in

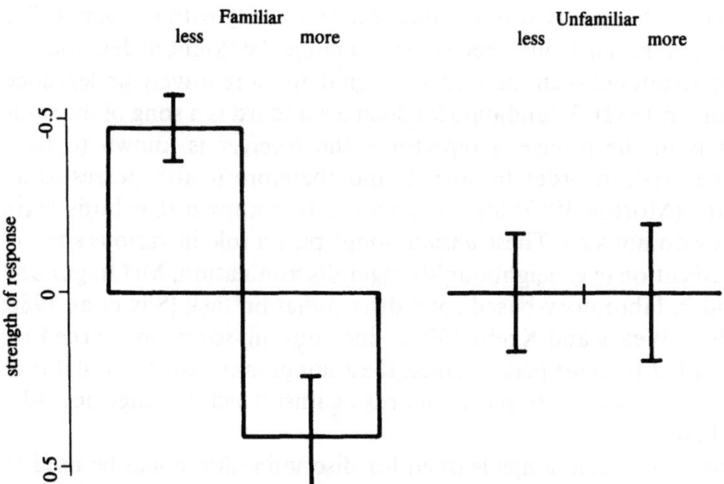

Fig. 4.3. The effect of the extent of degradation (less or more, see text) on the mean response ($\pm$ 1 SE) to playback of familiar and unfamiliar song types by 32 male great tits. The strength of response is a single variable derived from eight original variables by principal component analysis; note that a large negative value indicates a strong aggressive response to playback (see McGregor 1992, for details using the same data set). Two-tailed paired t-tests: Familiar $p < 0.0001$; Unfamiliar $p \approx 0.43$

one part of the area and as unfamiliar songs in another. In other words, there was no consistent difference between familiar and unfamiliar stimuli that could have produced the less degraded/more degraded difference in response demonstrated by these experiments. Therefore these experiments are clear evidence for a role of degradation in ranging. Recent experiments with Kentucky warblers (*Oporornis formosus*) have found comparable effects (Wiley and Godard 1992).

Additionally, the western meadowlark results suggest that an ability to range using cues from sound degradation can evolve even when visual ranging is possible. This species lives in open grassland habitat and has striking breast plumage, so that singing males are visually conspicuous if they are facing a receiver. Of course, these experiments (McGregor and Falls 1984) do not exclude the possibility that visual cues to range are used, but they do show that ranging is possible in the absence of visual cues. However, these sound ranging abilities may prove to be an example of the use of multiple redundant cues, which can provide a degree of insurance against unpredictable circumstances. This theme is explored more fully in the context of bird navigation in Chapter 5 and of depth perception in Chap. 16.

4.5 Mechanisms of Degradation Perception

Theoretically, degradation cues would permit a crude estimate of the range of a sound which was unfamiliar to the receiver (Wiley and Richards 1982). However, the finer resolution needed to decide whether a singer is just inside, or just outside, the territory boundary will require some familiarity with the signal. The most likely role for familiarity in a mechanism to judge the extent of degradation involves a comparison between the received signal and a relatively undegraded "standard" (Morton 1982). A candidate for such a standard is a song of the same song type that is in the receiver's repertoire; the receiver is known to have learned the song type, in order to sing it, and therefore it also possesses an internal standard (Morton 1982) for comparison. It is known that birds learn songs which they do not sing. These *unsung* songs play a role in various sorts of individual identification (e.g. neighbour/stranger discrimination; McGregor and Avery 1986) and in laboratory-based song discrimination task (Shy et al. 1986; Weary 1988, 1992; Weary and Krebs 1992). Since unsung songs are learned for discrimination rather than for performance, these songs may also be available to act as relatively undegraded internal standards against which to judge degradation of received songs.

The question of whether songs learned for discrimination could be used to assess degradation was addressed by playing great tits different sorts of familiar stimuli (McGregor and Krebs 1984). One set of males were played song types that were only sung by the test male and not by any contiguous neighbours or males within earshot; therefore only the test male's own song types were available as internal standards. A second group of males were played song types not sung by the test male but sung by at least one contiguous neighbour. For these birds, the test male's own song types were not available as internal standards for the familiar stimuli (because they were of different song types), but songs learned for neighbour identification could act as internal standards. The final group of males were played song types sung both by the test male and by at least one contiguous neighbour, and therefore two sorts of internal standard were available.

As Fig. 4.4 shows, there were significantly stronger aggressive responses to less degraded than to more degraded stimuli for all three sorts of familiar songs; there were no significant differences between unfamiliar stimuli. The result demonstrates that songs learned for both discrimination and for performance can be used as internal standards to judge degradation. Although there were no significant differences between the three sorts of familiar stimulus (one-way analysis of variance on less degraded values minus more degraded values; $F_{2,29} = 1.66$, not significant), it is tempting to speculate on the tendencies apparent in Fig. 4.4. The clearest difference in response to less and to more degraded stimuli occurs when the only internal standard available is a song in the test male's repertoire. This could indicate a difference in quality between an internal standard of own song and of a neighbour's song learned for discrimination; certainly the neighbour's song will be heard from further away than the bird's own song. It is surprising that the smallest difference in response to less and to more degraded song is seen when the test male has two possible internal standards (own song and neighbour's song). Interference between the two standards is a possible explanation of these results, as memory interference has

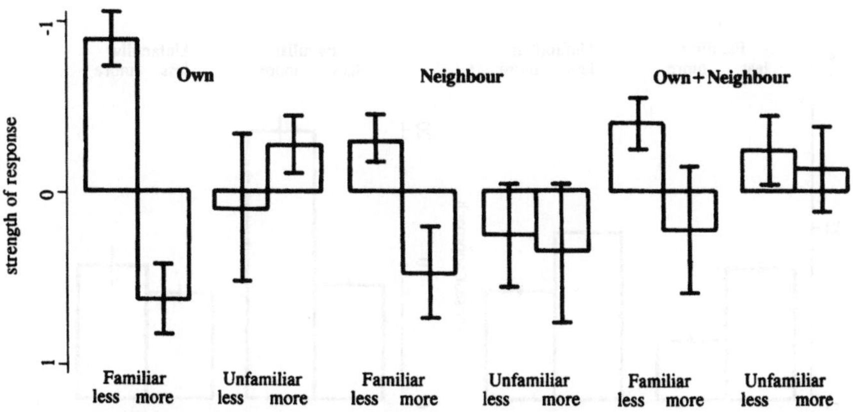

Fig. 4.4. The effect of the extent of degradation (less or more, see text) on the mean response (± 1 SE) to playback of unfamiliar song types and to three different sorts of familiar song types by 32 male great tits. *Own* only the test male sings the song type, $n = 7$; *Neighbour* only a neighbour sings the song type, $n = 13$; *Own + Neighbour* both the test male and a neighbour sing the song type, $n = 12$. The strength of response is measured in the same way as in Fig. 4.3. There are significant differences between less and more degraded forms for all three familiar categories (2-tailed paired t-tests), but no significant differences for any of the unfamiliar categories

been demonstrated in great tit neighbour/stranger discrimination field experiments (McGregor and Avery 1986; McGregor 1989).

As well as being consistent with Morton's (1982) internal standard mechanism of degradation assessment, these results demonstrate that internal standards can be songs learned both for performance and for other discrimination tasks.

Similar experiments on species with different patterns of song variation will give further insight into the mechanisms of degradation assessment. For example, species with highly variable songs generally have little overall similarity between successive songs, but each song is constructed from the same repertoire of elements or syllables. The appropriate level of familiarity is therefore at the level of element or syllable, rather than that of the song. In species with repertoires of many hundreds of syllables, most of which appear to be shared by all birds (regardless of whether they are neighbours), it may be impossible to produce the familiarity effect (e.g. Fig. 4.3) since all syllables and therefore songs are familiar to a large extent. Experiments with the European robin (*Erithacus rubecula*) provide some support for this idea. The robin has a very large syllable repertoire, it sings highly variable songs and there is extensive sharing of syllables between individuals in a population even when they are not immediate neighbours (Brémond 1968; Hoelzel 1986; Brindley 1992). Each of seven measures of strength of response to playback showed a significantly stronger response to playback of less degraded than more degraded song for both familiar and unfamiliar songs (Brindley 1992, Fig. 6.2; two of the measures are shown in Fig. 4.5).

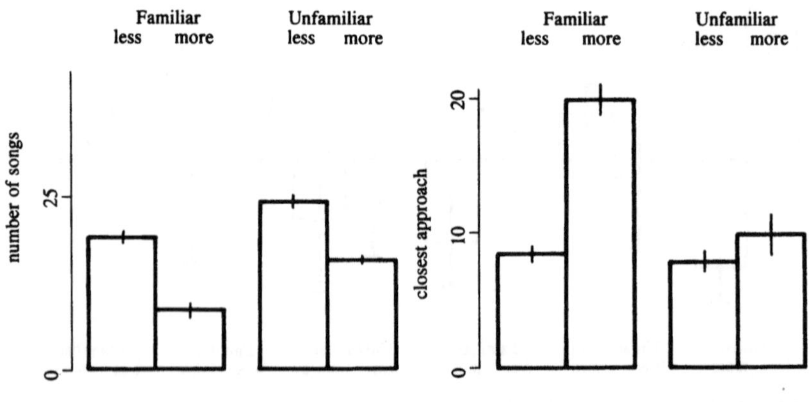

Fig. 4.5. The effect of the extent of degradation (less or more, see text) on the mean response ($\pm$ 1 SE) to playback of familiar and unfamiliar song types by eight male robins. Familiar song is recorded from a neighbour, and unfamiliar song is recorded from a non-neighbour at least 250 m from the test male. Two response variables are shown; the number of songs during playback (large value = strong aggressive response), and the closest approach during playback (a small value = strong aggressive response). There are significant effects of both degradation (more vs less) and familiarity (familiar vs unfamiliar) (non-parametric two-way analysis of variance; Meddis 1984)

4.6 Ranging and Honesty

Evidence for the importance of ranging information, especially in a territorial context, has led to the development of two quite different arguments about the evolution of ranging ability. The first considers how the efficiency of transfer of range information could be maximized, whereas the second considers the withholding of range as a territorial defence tactic. Each of these suggestions will be examined for further insights into the selection pressures affecting the perception of range information.

The basic underlying nature of communication has been discussed from a variety of theoretical viewpoints; as a cooperative exchange of mutually beneficial information (Smith 1977), as the manipulation of receivers by senders (Krebs and Dawkins 1984), or in terms of distinctive features such as reliability and memorability (Dawkins and Guilford 1991; Guilford and Dawkins 1991). Dawkins (1986) provides a good review of the issues. One instance where there is a mutual benefit in communicating range is between two singing, neighbouring territory holders, each inside their respective boundaries. Since neither constitutes a serious threat to the other and the status quo is maintained, the mutual benefit in ranging is a reduction in effort expended in territory defence (see Sect. 4.2). A common feature of singing behaviour in this situation – matched counter-singing – may be explicable in terms of a ranging mechanism based on comparison of the extent of degradation against an internal standard.

Matched counter-singing is a common feature of singing interactions in a number of species (e.g. Falls and Krebs 1975; Krebs and Kroodsma 1980; Krebs et al. 1981; Schroeder and Wiley, 1983; Horn and Falls 1988); counter-singing males reply to one another with the same song type, picked from the repertoire of possible alternatives each possesses. If degradation is judged by reference to an internal standard (see Sect. 4.5; Morton 1982), then a male (the replier) that chooses to counter-sing with the same song type as that already being sung (by the initiator) is doing two things (Krebs et al. 1981; Falls et al. 1982; Morton 1982). First, he is allowing the initiator to judge distance by singing a song that he, the replier, knows can be ranged; since the initiator is singing the same song type, the initiator must have an appropriate internal standard against which to judge the degradation of the song being sung by the replier. Second, the replier is announcing that he can also range the initiator, by showing the initiator that he, the replier, has an appropriate internal standard. In this way, matched counter-singing could be seen as an exchange of information on range and also on ability to range. Thus in all cases where a replying individual begins, or changes song types, to match counter-sing, it is actively choosing to transmit information on range and ranging ability.

In contrast, it has been suggested that males may withhold information on distance by singing a song which cannot be ranged by the receiving male (Morton 1982), and the possibility of such "anti-distance assessment mechanisms" is discussed in more detail by Morton (1986). The effect of singing such "unrangeable" songs is that the receiver must spend time and energy further investigating the range of the singer (e.g. by approaching) to determine whether or not he constitutes a threat.

Clearly the feasibility of this idea rests on an ability to sing unrangeable songs, or songs that give incorrect range information (e.g. Morton 1986). Experiments on great tits (McGregor et al. 1983; McGregor and Krebs 1984; see also Figs. 4.3 and 4.4), western meadowlarks (McGregor and Falls 1984) and robins (Brindley 1992) have clearly established the importance of similarity and familiarity in the ability to perceive degradation. This conclusion is in agreement with Morton's (1982) suggested mechanism of assessing the degree of degradation by comparison of the received song with a less degraded internal standard (see also Morton 1986). Songs learned for discrimination (e.g. neighbour identification) are familiar and can be used to range (McGregor and Krebs 1984; see Fig. 4.4). Many species are known to be able to identify neighbours on the basis of song even if they do not sing the neighbour's song type (Falls 1982). A consequence of neighbour recognition is that males cannot sing unrangeable songs to their neighbours because appropriate internal standards against which to judge degradation of all neighbours' song types exist – these are songs learned in order to identify neighbours.

It was argued above that the robin playback results supported the idea that species with large repertoires, a high proportion of song or syllable sharing, or both, could range all songs in a population because of the extent of familiarity with the syllables making up the songs in the whole population. In this case too, there is little prospect of intruders singing unrangeable songs.

A territory holder could be faced with a male singing unfamiliar songs if the intruder was an immigrant from a population singing very different songs. This is the situation simulated by the unfamiliar playback stimuli in the great tit and western meadowlark experiments (but not in those with robins), and also in an experiment with Carolina wrens (Shy and Morton 1986), although in the wren experiment the stimuli were the test male's own song and a song from a distant population. In none of these cases is there evidence of an ability to range, and it is possible that the intruder gains an initial, disruptive, advantage when trying to establish a territory. However, both males will suffer a disruptive effect and, as argued above, this advantage will only last as long as it takes for songs to be learned for identification.

Therefore, the suggested role in disruption of foraging of Morton's (1982) ranging hypothesis (see also Morton 1986) is untenable for all species with the ability to recognize neighbours and for species with extensive sharing of syllables; in both cases, birds cannot sing unrangeable songs. If disruption occurs, it can only be for a brief period during territory establishment.

Although there is no information on ranging in contexts other than territory defence, benefits of flocking can be mutual or reciprocal (e.g. Krebs and Davies 1987) and many flocks contain relatives (e.g. family parties of long-tailed tits). It is under such conditions of a high degree of genetic relatedness and/or mutually beneficial behaviour that we might expect signals to evolve to contain unambiguous information on range. It would therefore be interesting to examine the pattern of degradation of contact calls of family parties and the perceptual abilities of flock members.

4.7 Some Developments of Ranging Studies

To date, studies of ranging and associated ideas have been limited to the context of two territorial males, invariably neighbours, signalling and gathering information on position in relation to their territory boundaries. This is understandable in that some of the situations in which ranging could be advantageous (see Sect. 4.2) are not amenable to study. In this section I discuss two areas where there is potential for further ranging studies.

4.7.1 Ranging as a Component of Other Signalling Behaviour

This first area still concerns the territory defence context but considers the use of range information in interactive exchanges of signals rather than the use of range information per se. A new piece of equipment, the digital sound emitter, has allowed playback experiments that are interactive; the experimenter can alter playback song by song in response to the test bird (Dabelsteen and Pedersen 1991; Dabelsteen 1992), in contrast to the more usual fixed duration tape loop playback. Recent research with interactive playback has led to the proposal that the various forms of song matching constitute a series of signals which give progressively more specific information on the intended receiver of the signal (McGregor et al. 1992). Briefly, the hypothesis is that bout singers (e.g. the great tit) which sing with eventual variety (repeating each song of the same song type a number of times before switching to a different song type, i.e. aaaaaaaaaaa aaabbbbbb) can give only limited information on the intended receiver by matching song types in counter-singing. This is because changes between song types occur relatively infrequently. However, the number of phrases in a song, which provides a measure of song (or strophe) length (Lambrechts and Dhondt 1987), can be changed song by song and therefore matching strophe length is a more specific indication of the intended receiver in a song duel than just matching song type.

A finer level of specificity still involves matching the delay between songs, or even overlapping songs (Dabelsteen et al. submitted). It has been suggested that overlapping song during duels is an aggressive signal (e.g. Brindley 1991; Dabelsteen et al. submitted), but there is a confounding effect with distance. Distant singers probably constitute less of a threat (and therefore elicit less overlapping) than close singers, but the speed of sound transmission imposes a constraint (i.e. a delay) that means that overlapping is less easy to achieve with increasing separation distance (Dabelsteen 1992). Although males hear the timing of the interaction only at their own locations, this does not prevent them from knowing what happens at the location of a singing rival. Ranging using song degradation cues would, in theory, allow the time the opponent produces and receives songs to be deduced. Theoretically, males could compensate for the delay imposed by the speed of sound transmission constraint during the process of co-ordinating singing with a rival and alter the singing rhythm to achieve or avoid overlap. This argument may seem far fetched, but a similar sort of prediction to overcome reaction time constraints has been proposed for the intricate coordination of wheeling in bird flocks (Potts 1984).

It would be possible to study this idea experimentally using the interactive playback equipment described above (Dabelsteen 1992). One playback design would use a loudspeaker at a fixed distance from the bird, and the stimuli broadcast interactively would vary in the extent of degradation. Therefore the actual delay would be constant (because distance to the bird is constant), but if the bird is using ranging information from degradation cues in the playback it would have a variable assessment of the distance of the playback and hence the delay. A related design could use variable loudspeaker positions (i.e. variable actual delays) and varying extents of stimulus degradation to counter the changes in speaker position so that playback is perceived by the bird as coming from the same distance and therefore having a constant delay. An alternative approach would be to play back songs which could not be ranged (such as unfamiliar, dissimilar song; see Sect. 4.5) and to compare the timing of the resulting interactions with that during playback of rangeable songs.

4.7.2 Resolution of Ranging

The second idea concerns the correspondence between the level of resolution of range information and the context of ranging. For example, in the context of relating the position of an intruder to territory boundaries it may be acceptable to have a resolution of $\pm$ 5 m over a distance of 100 m. In contrast, to maintain tight grouping in a foraging flock may require a resolution of $\pm$ 0.5 m over a few metres.

There are a number of ways in which questions of resolution could be tackled. In the laboratory, operant techniques could be used to document minimum resolvable differences in degradation of stimuli. In the field, establishing the form of the relationship of the curve of change in degradation with distance will give insights into the optimum distances for range information. Either type of investigation will require a method of quantifying degradation. Such a method has now been developed (Pedersen et al. submitted; Dabelsteen et al. 1993). The first results, with blackbird song, have shown effects both of signal structure and of receiver behaviour. The "twitter" and "motif" sounds making up blackbird song (see Fig. 1 in Dabelsteen 1992; Fig. 1 in Dabelsteen et al. 1993) degrade at different rates, with twitter sounds degrading more rapidly than motifs, suggesting that twitters are optimized for shorter transmission distances than motifs. The experiments also found that signaller perch height had little effect on the rate of degradation, but a high receiver perch dramatically decreased the extent of degradation experienced relative to low perches at the same distance. Also, the shape of the relationship between degradation and distance suggested higher resolution at shorter distances, and the relationship between degradation and perch height suggests high resolution at high receiver heights. For distance, the log-linear relationship found over the 10–100 m range (Dabelsteen et al. 1993; Fig. 13A) suggests best resolution at a few metres, although this was not established directly.

Studies relating the physical structure of signals, the shape of the degradation – distance curve and the tuning of perceptual abilities are now possible, and it will be fascinating to see the results of research in this area.

4.8 Conclusions

This chapter began by arguing that a knowledge of the type of information that was important to a species was central to an understanding of the perceptual and sensory abilities of birds. I have argued that an ability to range is important in many contexts and that such information is often only available from habitat-induced changes in vocalizations. So far, there is evidence to support these arguments in one context, territorial defence, and as the arguments developed above make clear, there are a number of facets to this perceptual ability. Ranging in different contexts, such as flocking, different types of vocalization (e.g. calls) and the abilities of females have yet to be investigated. To date, only the field playback approach has been used in ranging, and laboratory studies of degradation would provide interesting complimentary information. Much remains to be done!

Acknowledgements. I thank John Krebs for permission to use joint data. Some of the ideas resulted from discussions with Torben Dabelsteen and it was particularly helpful to have access to his current manuscripts. Haven Wiley, Gene Morton and the editors made several suggestions that improved the manuscript.

References

Brémond JC (1968) Recherches sur la sémantique et les elements vecteurs d'information dans les signaux acoustiques du rougegorge (*Erithacus rubecula* L.). Terre Vie 2: 109–220

Brémond JC, Aubin T (1992) The role of amplitude modulation in distress-call recognition by the blackheaded gull *Larus ridibundus*. Ethol Ecol Evol 4: 187–191

Brindley EL (1991) Response of European robins to playback of song: neighbour recognition and overlapping. Anim Behav 41: 503–512

Brindley EL (1992) Variation in Robin (*Erithacus rubecula*) song: Effects of season, sex and habitat acoustics. Thesis, University of Nottingham, Nottingham

Brooke M, Birkhead TR (1991) The Cambridge encyclopedia of ornithology. Cambridge University Press, Cambridge

Chappius C (1971) Un example de l'influence du milieu sur les émissions vocales des oiseaux: l'évolution des chants on forét équatoriale. Terre Vie 25: 183–202

Cosens SE, Falls JB (1984) A comparison of sound propagation and song frequency in temperate marsh and grassland habitats. Behav Ecol Sociobiol 15:161–170
Dabelsteen T (1992) Interactive playback: a finely tuned response. In: McGregor PK (ed) Playback and studies of animal communication. Plenum Press, New York, pp 97–110
Dabelsteen T, Pedersen SB (1991) A portable digital sound emitter for interactive playback of animal vocalizations. Bioacoustics 3:193–206
Dabelsteen T, Larsen ON, Pedersen SB (1993) Habitat induced degradation of sound signals: quantifying the effect of communication sounds and bird location on blurring, excess attenuation and signal-to-noise ratio in blackbird song. J Acoust Soc Am 93:2206–2220
Dabelsteen T, McGregor PK, Shepherd M, Whittaker X, Pedersen SB (1993) Overlapping and signalling in interactive playback experiments (submitted)
Dawkins MS (1986) Unravelling animal behaviour. Longman, Harlow, Essex
Dawkins MS, Guilford T (1991) The corruption of honest signalling. Anim Behav 41:865–873
Falls JB (1982) Individual recognition by sounds. In: Kroodsma DE, Miller EH (eds) Evolution and ecology of acoustic communication in birds, vol II. Academic Press, New York, pp 237–278
Falls JB, Krebs JR (1975) Sequences of songs in repertoires of western meadowlarks (*Sturnella neglecta*). Can J Zool 53:1165–1178
Falls JB, Krebs JR, McGregor PK (1982) Song matching in the Great tit (*Parus major*): the effect of similarity and familiarity. Anim Behav 30:997–1009
Gill FB (1990) Ornithology, Freeman, New York
Gish SL, Morton ES (1981) Structural adaptations to local habitat acoustics in Carolina Wren songs. Z Tierpsychol 56:74–84
Guilford T, Dawkins MS (1991) Receiver psychology and the evolution of animal signals. Anim Behav 42:1–14
Handford P (1988) Trill rate dialects in the rufous-collared sparrow, *Zonotrichia capensis*, in northwestern Argentina. Can J Zool 66:2658–2670
Hansen P (1979) Vocal learning: its role in adapting sound structures to long-distance propagation and a hypothesis on its evolution. Anim Behav 27:1270–1271
Hoelzel AR (1986) Song characteristics and response to playback of male and female robins, *Erithacus rubecula*. Ibis 128:115–127
Horn AG, Falls JB (1988) Repertoires and countersinging in western meadowlarks (*Sturnella neglecta*). Ethology 77:337–343
Hunter ML, Krebs JR (1979) Geographical variation in the song of the great tit (*Parus major*) in relation to ecological factors. J Anim Ecol 48:759–785
Krebs JR, Davies NB (1987) An introduction to behavioural ecology, 2nd edn. Blackwell Scientific, Oxford, pp 111–133
Krebs JR, Dawkins R (1984) Animal signals: Mind-reading and manipulation. In: Krebs JR, Davies NB (eds) Behavioural ecology: An evolutionary approach, 2nd edn. Blackwell Scientific, Oxford, pp 380–402
Krebs JR, Kroodsma DE (1980) Repertoires and geographical variation in bird song. In: Rosenblatt JS, Hinde RA, Beer C, Busnell M-C (eds) Advances in the study of behaviour, vol 11. Academic Press, New York, pp 143–177
Krebs JR, Ashcroft R, van Orsdol K (1981) Song matching in the great tit (*Parus major* L.). Anim Behav 29:918–923
Lambrechts M, Dhondt AA (1987) Differences in singing performance between male great tits. Ardea 75:43–52
Marten K, Marler P (1977) Sound transmission and its significance for animal vocalization. I. Temperate habitats. Behav Ecol Sociobiol 2:271–290
Marten K, Quine D, Marler P (1977) Sound transmission and its significance for animal vocalization. II. Tropical forest habitats. Behav Ecol Sociobiol 2:291–302
McGregor PK (1989) Pro-active memory interference in neighbour recognition by a song bird. Int Ornithol Congr XIX:1391–1397
McGregor PK (1991a) The singer and the song: on the receiving end of bird song. Biol Rev 66: 57–81

McGregor PK (1991b) The Loughborough Sound Images speech workstation: spectrum analysis for a PC. Bioacoustics 3:223–234
McGregor PK (1992) Quantifying responses to playback: one, many or multivariate composite measures? In: McGregor PK (ed) Playback and studies of animal communication. Plenum Press, New York, pp 79–96
McGregor PK, Avery MI (1986) The unsung songs of great tits (*Parus major*): learning neighbours' songs for discrimination. Behav Ecol Sociobiol 18:311–331
McGregor PK, Falls JB (1984) The response of Western Meadowlarks (*Sturnella neglecta*) to the playback of degraded and undegraded songs. Can J Zool 62:2125–2128
McGregor PK, Krebs JR (1982) Song types in a population of great tits: their distribution abundance and acquisition by individuals. Behaviour 52:126–152
McGregor PK, Krebs JR (1984) Sound degradation as a distance cue in great tit (*Parus major*) song. Behav Ecol Sociobiol 16:49–56
McGregor PK, Krebs JR, Ratcliffe LM (1983) The response of great tits (*Parus major*) to the playback of degraded and undegraded songs: the effect of familiarity with the stimulus song type. Auk 100:898–906
McGregor PK, Dabelsteen T, Shepherd M, Pedersen SB (1992) The signal value of matched singing in great tits: evidence from interactive playback experiments. Anim Behav 43:987–998
Meddis R (1984) Statistics using ranks. Blackwell Scientific, Oxford
Michelsen A (1978) Sound perception in different environments. In: Ali MA (ed) Perspectives in sensory ecology. Plenum Press, New York, pp 345–373
Michelsen A (1992) Hearing and sound communication in small animals: Evolutionary adaptations to the laws of physics. In: Webster DB, Fay RR, Popper AN (eds) The evolutionary biology of hearing. Springer, Berlin Heidelberg New York, pp 61–78
Michelsen A, Larsen ON (1983) Strategies for acoustic communication in complex environments. In: Huber F, Markl H (eds) Neuroethology and behavioural physiology. Roots and growing points. Springer, Berlin Heidelberg New York, pp 321–331
Morton ES (1970) Ecological sources of selection on avian sounds. Thesis, Yale University, Newhaven
Morton ES (1975) Ecological sources of selection on avian sounds. Am Nat 109:17–34
Morton ES (1980) The ecological background for the evolution of vocal sounds used at close range. Int Ornithol Congr XVII:737–741
Morton ES (1982) Grading, discreteness, redundancy and motivational-structural rules. In: Kroodsma DE, Miller EH (eds) Evolution and ecology of acoustic communication in birds, vol. I. Academic Press, New York, pp 183–212
Morton ES (1986) Predictions from the ranging hypothesis for the evolution of long distance signals in birds. Behaviour 99:65–86
Nottebohm F (1975) Continental patterns of song variability in *Zonotrichia capensis*: some possible ecological correlates. Am Nat 109:605–624
Pedersen SB, Dabelsteen T, Larsen ON (1993) A digital technique for noise-compensated quantification of the degradation of transmitted sounds signals. (submitted)
Potts WK (1984) The chorus-line hypothesis of manoeuvre coordination in avian flocks. Nature 309:344–345
Richards DG (1981) Estimation of distance of singing conspecifics by the Carolina wren. Auk 98:127–133
Richards DG, Wiley RH (1980) Reverberations and amplitude fluctuations in the propagation of sound in a forest: implications for animal communication. Am Nat 115:381–399
Römer H (1992) Ecological constraints for the evolution of hearing and sound communication in insects. In: Webster DB, Fay RR, Popper AN (eds) The evolutionary biology of hearing. Springer, Berlin Heidelberg New York, pp 79–94
Schroeder DJ, Wiley RH (1983) Communication with shared themes in tufted titmice. Auk 100:414–424
Shy E, Morton ES (1986) The role of distance, familiarity, and time of day in Carolina wrens responses to conspecific songs. Behav Ecol Sociobiol 19:393–400

Shy E, McGregor PK, Krebs JR (1986) Discrimination of song types by male great tits. Behav Proc 13:1–1
Smith WJ (1977) The behavior of communicating: an ethological approach. Harvard University Press, Cambridge, Mass
Weary DM (1988) Experimental studies on the song of the great tit. Thesis, University of Oxford, Oxford
Weary DM (1992) Bird song and operant experiments: a new tool to investigate song perception. In: McGregor PK (ed) Playback and studies of animal communication. Plenum Press, New York, pp 201–210
Weary DM, Krebs JR (1992) Great tits classify songs by individual voice characteristics. Anim Behav 43:282–287
Wiley RH, Godard R (1992) Ranging of conspecific songs by Kentucky warblers *Oporornis formosus* reduces the possibilities for interference in territorial interactions. Abstr IVth Int Behav Ecol Congr T54c. Univ Press, Princeton
Wiley RH, Richards DG (1978) Physical constraints on acoustic communication in the atmosphere: Implications for the evolution of animal vocalizations. Behav Ecol Sociobiol 3:69–94
Wiley RH, Richards DG (1982) Adaptations for acoustic communication in birds: sound transmission and signal detection. In: Kroodsma DE, Miller EH (eds) Evolution and ecology of acoustic communication in birds, vol I. Academic Press, New York, pp 131–181

5 Avian Orientation: Multiple Sensory Cues and the Advantage of Redundancy

R. Wiltschko and W. Wiltschko

5.1 Theoretical Considerations

Orientation, in the broadest sense, designates two basically different phenomena: the control of an animal's position and stability in space, and the control of an animal's path through space. The present chapter is concerned with the latter – the problem of getting from point A to a goal at point B. We also restrict our considerations to long distance orientation, where the animal has no direct visual, auditory or olfactory contact with the goal.

In birds, long distance orientation occurs in two behavioural contexts, namely *homing* and *migration*. Homing means that the bird, after foraging or displacement, whether natural or artificial, returns to its "home", the place where the excursion began. Depending on the species, the distances involved in homing vary considerably. Under natural conditions these distances correlate with the size of the bird's home range. For example, the home range varies from less than a kilometre up to a few kilometres in most small passerines, or up to 20 km in birds like rock doves or gulls. Distances of hundreds or thousands of kilometres, as covered by some pelagic seabirds, are exceptional. Migration, in contrast, means that birds transfer their home range to a distant region. The distances involved are generally much larger, and even small passerines like willow warblers (*Phylloscopus trochilus*) may cover more than 10 000 km in one migration. The record is held by sea birds like the Arctic tern (*Sterna paradisea*) a species breeding in the tundra at the coast of the Arctic Sea and wintering at the ice edge around the Antarctic continent, thus flying around the world once a year.

Long distance flights in homing and migration are initiated by specific motivational states. In homing, this may be the urge to return to the nest after foraging, to return to a specific place to rest at night or to return to the home region after experimental displacement. Migration, on the other hand, is a seasonal phenomenon. The urge to move reflects a certain phase in the bird's annual cycle associated with a characteristic physiological state (for a summary, see Berthold 1988). Having no direct contact with the destination makes it

Zoologisches Institut, J.W. Goethe-Universität, Siesmayerstr 70, D-60323 Frankfurt am Main 11, Germany

M.N.O. Davies and P.R. Green (Eds.)
Perception and Motor Control in Birds

necessary for birds to establish contact with the distant goal in some way, in order to head in the correct direction. This is done via an *external reference*; the direction to the goal is first determined as a compass course, and then a compass is used to locate this course, transforming it into an actual flight direction. Bird orientation thus generally represents a two-step process (see Kramer 1959).

The current view of the avian orientation system is summarized in Fig. 5.1. The nature of the first step, the determination of a direction as a compass course, varies according to the behavioural context of the orientation task. In migration, the course depends on the geographical relationship of the breeding ground and the wintering area. Birds possess innate information on this course, which is genetically transmitted from one generation to the next. The birds face the task of transforming this genetically encoded directional information into an actual migratory course. In homing, however, the correct course varies according to the bird's current position relative to its home. Here the bird must be able to interpret environmental information in order to determine the course, which is then located by compass mechanisms.

The above considerations alone indicate that orientation, at least in the case of homing, cannot depend on a single sensory modality. Factors with fundamentally different characteristics are required for use as a compass and for navigation, in the sense of establishing the position relative to home. In migration, the situation is different insofar as the same factor might, at least in theory, serve as a reference for the genetically encoded information and for the actual flight direction. Experimental evidence, however, clearly shows that in migration as well as in homing a multitude of factors are involved. These factors interact in a complex way, supplementing as well as replacing each other. Birds appear to have more than just one option for compass orientation as well as for

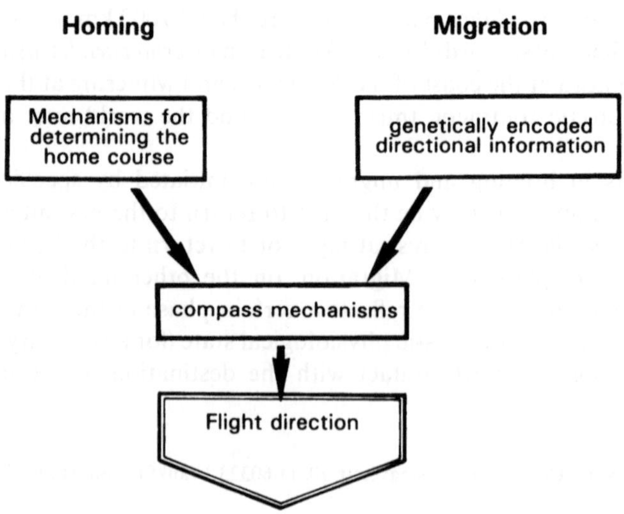

Fig. 5.1. Model describing bird orientation as a two-step process (for more information, see text)

determining the home course. Even the genetically encoded migratory direction seems to be represented more than once, using different external references.

The present chapter will summarize current knowledge about the avian orientation system and will emphasize the interaction of cues of different nature and their relative importance in various orientation tasks. First we will describe the compass mechanisms, as they also form the basis of the mechanisms used to determine the desired course.

5.2 Compass Mechanisms and Their Interrelation

Compass mechanisms allow the location of directions relative to an external reference, such as geographic or magnetic north. Several environmental factors provide the required type of information, the most prominent ones being the *sun*, the *stars* and the *geomagnetic field*. All three of them are known to be used by birds. The two celestial compasses are based on visual information in the same way as humans use the sun or the stars in direction finding. In sun compass orientation, birds compensate for the sun's apparent movement with the help of their internal clock (for a summary, see Schmidt-Koenig 1960), while in star compass orientation, they seem to use the constant spatial relationship between stars just as we are able to derive north from the configuration of the Great Dipper (*Ursa major*) regardless of its position (Emlen 1967). The magnetic compass, on the other hand, involves a sensory modality which is available to humans only through technical means. Birds' use of the magnetic field differs fundamentally from the technical compass used by sailors and surveyors. Therefore, we will briefly summarize how the magnetic compass of birds functions (for a more detailed review, see Wiltschko and Wiltschko 1988).

5.2.1 The Magnetic Compass of Birds

A magnetic compass was first described for European robins (*Erithacus rubecula*), a night-migrating passerine. During the migration season, captive nocturnal migrants are restless at night. Their activity is concentrated in that part of the cage which points to the migratory direction of their free-flying conspecifics. When the magnetic field was experimentally altered around the test cage, the birds reacted to a shift in magnetic north with a corresponding deflection of their directional tendencies. This indicated that they used magnetic information in direction finding (Wiltschko 1968). Subsequent experiments on the functional characteristics of the magnetic compass using the same methods and the same species yielded the following important findings:

1. The magnetic compass of birds does not use the polarity of the field lines, but depends on their axial course and their inclination in space. Therefore it does not distinguish between "north" and "south", but between "poleward",

the direction where the axial field lines are inclined to the ground, and "equatorward", the direction where they point upwards (see Fig. 5.2). At the magnetic equator itself, the field lines run horizontally and the magnetic compass becomes ambiguous.

2. The functional range of the magnetic compass is closely tuned to the intensity of the local magnetic field. In European robins, *either* an increase *or* a decrease of about 25% in the local total intensity of 46 000 nT caused disorientation. The birds were, however, able to adjust to lower and higher intensities when they remained in the altered fields for 3 days or more.

Meanwhile, a magnetic compass has been demonstrated in ten other bird species (for a summary, see Wiltschko and Wiltschko 1988). It seems to be a rather widespread mechanism, among birds as well as other animals (Wiltschko and Wiltschko 1991a).

Our knowledge of the perception and processing of magnetic compass information is still rather limited. As primary processes, optical pumping (Leask

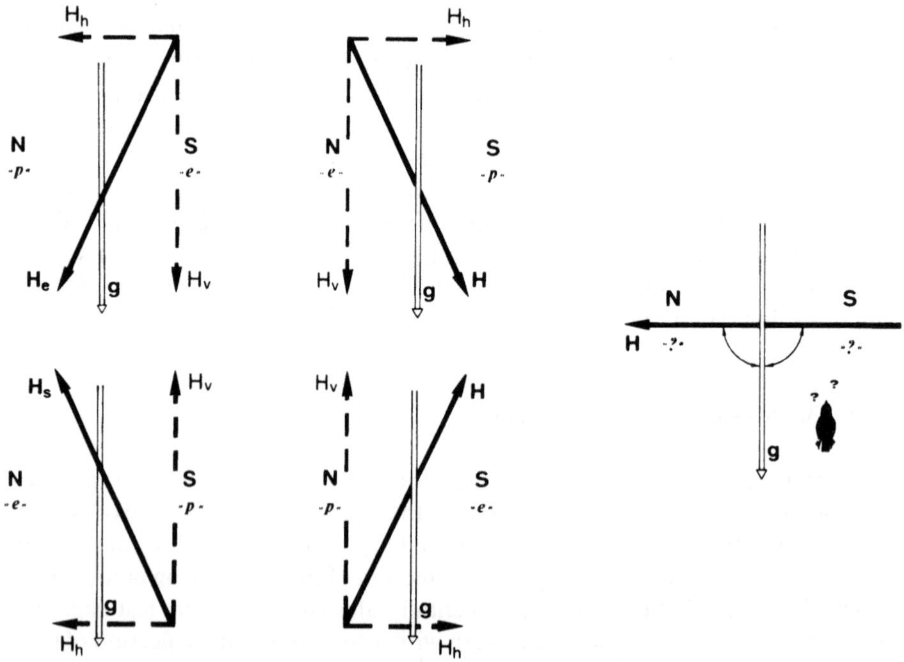

Fig. 5.2. The magnetic compass of birds is based on the axial course of the magnetic field lines and their inclination. It does not distinguish between north and south, but between "poleward" and "equatorward". *Upper left* geomagnetic field in the northern hemisphere; *Upper right* horizontal component reversed; *Lower left* vertical component reversed (= geomagnetic field in the southern hemisphere); *Lower right* both components reversed; *Far right* situation at the magnetic equator where the field lines run horizontally; *H* magnetic vector; H_e vector of the local magnetic field at the test site; H_h, H_v horizontal and vertical component of the magnetic field; *g* gravity vector; *N*, *S* geographic north and south; »*p*«, »*e*«: poleward and equatorward in term of the birds

1977) and processes involving magnetite (Walcott et al. 1979) have been discussed. Electrophysiological recordings by Semm et al. (1984) suggest that the visual system, and particularly the accessory optical system, is involved in the perception of magnetic compass information. Reaction to changes in the direction of the magnetic field have been recorded in the nBOR and the optic tectum, parts of the brain where spatial information is processed (Semm and Demaine 1986; see Chap. 12). Semm and Beason (1990) found magnetically sensitive neurons also in the trigeminal system.

5.2.2 The Interrelation Between Magnetic Compass and Sun Compass

Experiments demonstrating sun compass orientation make use of the fact that the internal clock is involved in the mechanism compensating for the sun's movement (Schmidt-Koenig 1960). In the classic clock-shift tests, the internal clock is reset by subjecting the birds for a few days to an artificial light regime, with the light/dark period corresponding in duration to the natural day, but with the beginning and end of the light period shifted, usually by 6 h. This results in the birds misjudging time of day. When pigeons are displaced and released after such treatment, they show as a consequence a characteristic deviation from the direction taken by untreated controls (Fig. 5.3).

Manipulating the birds' internal clock interferes only with the sun compass, leaving intact the magnetic compass. The fact that pigeons rely on their sun compass, although their magnetic compass could have given them correct directional information, clearly demonstrates the dominant role of the sun

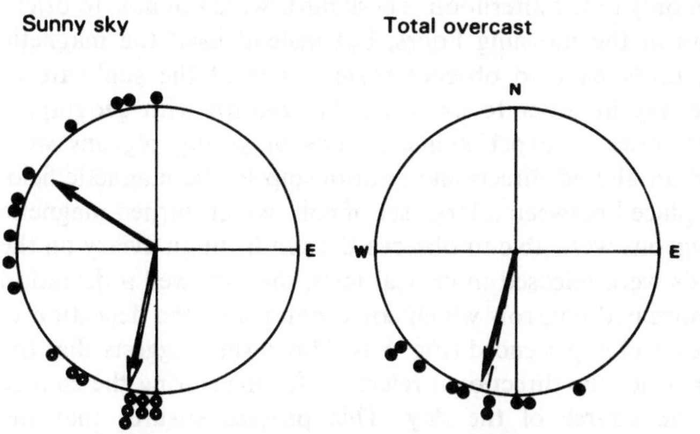

Fig. 5.3. Compass orientation in homing pigeons. Under a sunny sky, shifting the internal clock by 6 h results in a characteristic deflection demonstrating the use of the sun compass (see text for further explanation). When the sky is totally overcast, the pigeons are equally well oriented. The tests were performed at the same site; the home direction (192°) is indicated by a *dashed radius*. The *symbols at the periphery of the circle* mark the vanishing bearings of individual pigeons; the *arrows* represent the mean vector the length of which is proportional to the radius of the circle. *Left diagram: open symbols* indicate untreated controls and the *solid symbols* pigeons whose internal clock was shifted 6 h slow

compass when the sun is available. Yet on overcast days, the pigeons' orientation is equally good (Keeton 1969; see Fig. 5.3). This leads to the conclusion that mechanisms of directional orientation in birds are partially redundant; the sun compass is preferentially used, but it can be replaced without loss when the sun is not visible.

The findings mentioned above seem to suggest that the magnetic compass represents just a subsidiary mechanism for overcast days. The magnetic compass, however, also plays a very important role during the development of the orientation system. The behaviour of very young homing pigeons suggests that when birds begin to fly, the sun compass is not yet available to them, so they initially rely on the magnetic compass for locating direction. This is clearly indicated by the fact that during this period their orientation can be disturbed with magnets even when the sun is visible (Keeton 1971). Also when clock-shifted, they do not show the typical deviation, which indicates that the birds do not use the sun compass (Wiltschko and Wiltschko 1981). This is not surprising, since the magnetic compass is a simpler mechanism than the sun compass; perception of the field lines per se provides directional information, whereas the use of the sun as a compass requires compensation for the sun's movement. Since the changes in the sun azimuth depend on the geographic latitude and on the season, the sun compass has to be adjusted to the local situation. The ability to make this adjustment is ensured by a learning process, in which the magnetic compass is used to calibrate the sun compass. This process takes place as soon as the young birds gain some flying experience.

A series of experiments analyzing the development of the sun compass indicates that the magnetic compass is involved in the process (for a summary, see Wiltschko 1983). A group of young pigeons was raised in conditions where they could see the sun only in the afternoon. These birds were not able to orient with the sun compass in the morning hours, but instead used the magnetic compass. Obviously, birds have to observe major parts of the sun's arc at different times of the day in order to associate its azimuth with geographic direction. In another series of experiments, groups of young pigeons were exposed to the sun in an altered directional relationship to the magnetic field; here a small loft was placed between a large set of coils which turned magnetic north by 120°. The pigeons were able to observe the sun from an aviary on the roof. When these birds were released in critical tests, they showed a deviation compared with the untreated controls which corresponded to the deflection of the magnetic field they had experienced (Fig. 5.4). This result suggests that the magnetic compass provides the directional reference for measuring the change of sun azimuth in the course of the day. This process ensures that the mechanisms compensating for the sun's movement are perfectly adapted to the sun's arc in the bird's home region.

Thus, during the development of the navigational system, the magnetic compass seems to control the sun compass, while later the sun compass dominates. There are indications, however, that the magnetic compass continues to serve as a back-up system, not only for overcast days, but also in

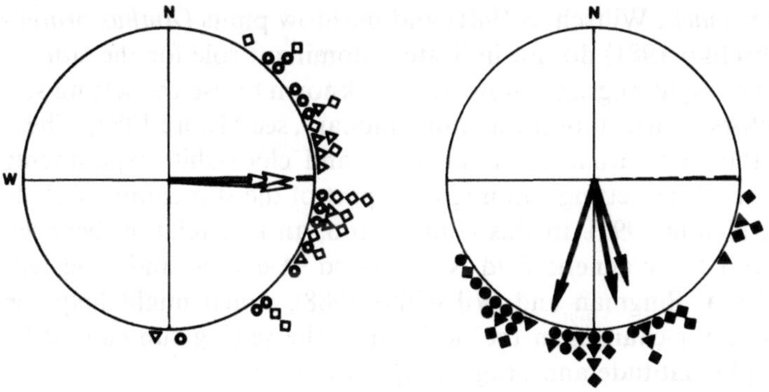

Fig. 5.4. Staying in an altered magnetic field during the period in which the sun compass is established affects the orientation of young pigeons in later releases. Magnetic north had been shifted by 120° to ESE. *Open symbols* controls; *solid symbols* altered magnetic field. The results of three separate releases are given in *different symbols* (Data from Wiltschko and Wiltschko 1990)

situations when the sun compass is impaired. After clock-shift tests, many experimental pigeons return on the day of their release, before their internal clocks could be normalized through natural sunset and sunrise. This suggests that they had realized that their sun compass was leading them astray and had given it up in favour of a non-time-compensating mechanism. Further, the sun compass of homing pigeons can be dramatically altered by subjecting the birds for a long time to a shifted photoperiod. When pigeons living in a 6-h slow shifted day had access to free flight during the overlap time between natural and artificial day, their sun compass adjusted to the experimental situation. For these pigeons, the sun in their (subjective) "morning" was in the south and it reached west at "noon", while they spent their "afternoon" under artificial light (Wiltschko et al. 1984). Adapting the sun compass to the experimental situation means that an entirely new relationship between sun azimuth, time and geographic direction has to be formed. This process is equivalent to establishing a new sun compass, which suggests that similar mechanisms are involved. The magnetic compass may again provide a directional reference system, thus controlling the calibration of the sun compass. In the same way, the birds could also adapt their sun compass to the changes which occur in the sun's arc in the course of seasons, which are especially significant in the tropics.

The findings presented so far pertain to homing pigeons, and reflect the use of the sun compass within a limited area. It is not yet clear whether the sun compass is also used during migration, when birds cover considerably larger distances. Theoretical considerations show that in this case the immense changes in geographic latitude might pose a severe problem, because they result in dramatic changes in the sun's arc, which would require permanent readjustments. So far, there is little experimental evidence on the question whether day migrants use the sun compass to locate their migratory direction. Data on starlings (*Sturnus vulgaris*, Wiltschko 1981) and meadow pipits (*Anthus pratensis*, Orth and Wiltschko 1981) do not indicate a dominant role for the sun.

Several species of night migrants, however, are known to use the setting sun and associated factors as orientation cues (for summary, see Moore 1987). These birds often start their nocturnal flights at dusk, and clock-shift experiments reveal that their use of the setting sun might be part of the sun compass (Able and Cherry 1986; Helbig 1991). In this context, too, an interrelation between sun-related cues and the magnetic field is suggested (Alerstam and Högstedt 1983; Bingman 1983a; Bingman and Wiltschko 1988), which might help the birds to compensate for changes in the position of the setting sun caused by changes in geographic latitude and progressing season.

5.2.3 Directional Orientation at Night

Night-migrating birds also have more than one means of directional orientation. Besides the magnetic field, they may use the stars as a celestial compass on clear nights. For several species, among them the European robin, the garden warbler (*Sylvia borin*), the indigo bunting (*Passerina cyanea*) and the pied flycatcher (*Ficedula hypoleuca*), a star compass *and* a magnetic compass have been demonstrated in cage experiments (for summary, see Wiltschko and Wiltschko 1991b).

The two compasses are not independent from each other. This has been shown in several studies in which the stars and the magnetic field were arranged to give conflicting information. The first reaction of the birds varied between the species tested. Whereas some species seemed to follow the stars, others, mainly long distance migrants, followed the magnetic field. After a few nights of exposure, however, all birds oriented according to magnetic north (Fig. 5.5, upper diagrams), which indicated that the magnetic field was the dominant cue (e.g. Wiltschko and Wiltschko 1975a, b; Beason 1987; Bingman 1987). When the magnetic field was partly compensated and did not provide meaningful information, the birds oriented by the stars alone. The ones that had previously been exposed to an altered magnetic field now continued to prefer the new magnetically derived direction (Fig. 5.5, lower diagrams). This result suggests that the birds had recalibrated the stars according to the experimental magnetic field (Wiltschko and Wiltschko 1975a, b; Bingman 1987).

These findings demonstrate a rather complex interrelationship between the star compass and the magnetic compass. Each system alone may provide sufficient information for specifying the direction. When both types of cue are present, the magnetic compass proves in the long run to be dominant over the stars and to control their directional significance. However, the frequency with which the stars are checked against the magnetic field seems to vary greatly. Some species follow the stars for one or two days despite the conflicting information from the magnetic field, while others react immediately to the

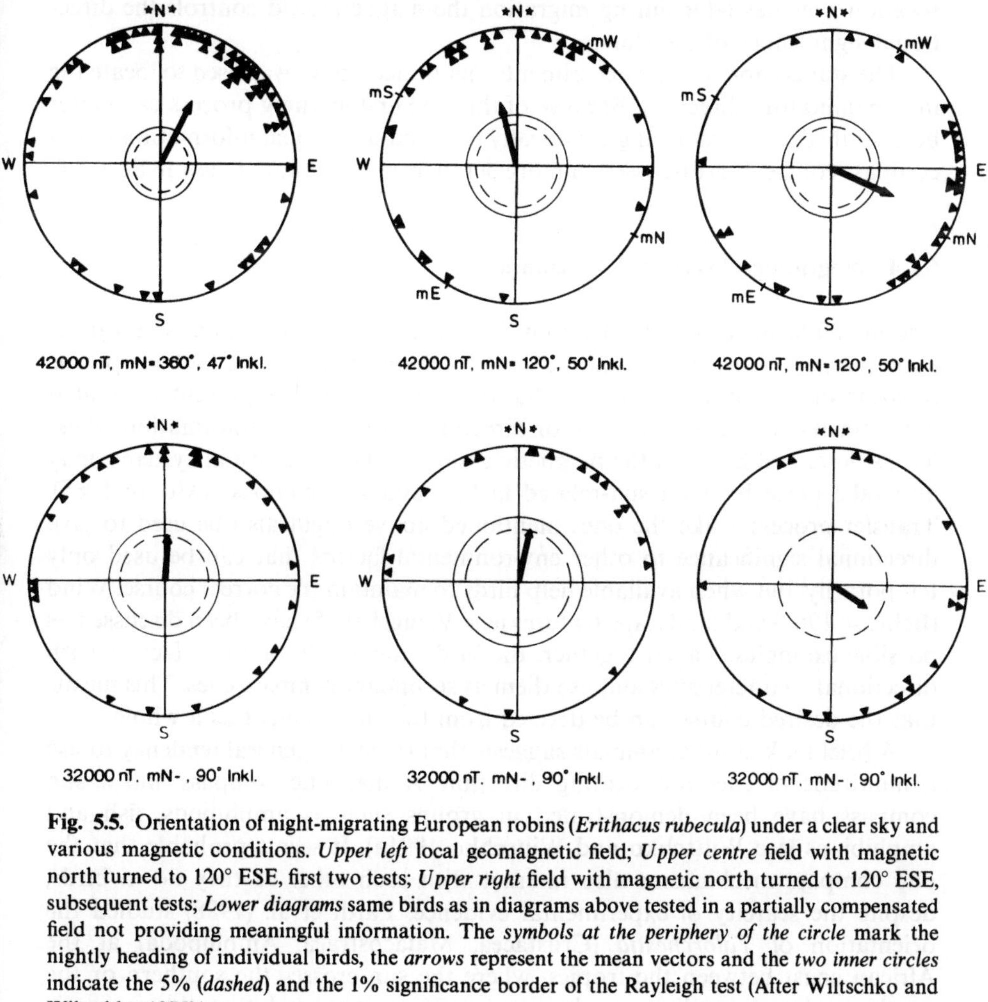

Fig. 5.5. Orientation of night-migrating European robins (*Erithacus rubecula*) under a clear sky and various magnetic conditions. *Upper left* local geomagnetic field; *Upper centre* field with magnetic north turned to 120° ESE, first two tests; *Upper right* field with magnetic north turned to 120° ESE, subsequent tests; *Lower diagrams* same birds as in diagrams above tested in a partially compensated field not providing meaningful information. The *symbols at the periphery of the circle* mark the nightly heading of individual birds, the *arrows* represent the mean vectors and the *two inner circles* indicate the 5% (*dashed*) and the 1% significance border of the Rayleigh test (After Wiltschko and Wiltschko 1975b)

change of magnetic north. In bobolinks (*Dolichonyx oryzivorus*), Beason (1987) found considerable differences even between individual birds.

The interrelationship between the star compass and the magnetic compass appears to be even more complex when the ontogeny of the star compass is considered. Birds do not possess an innate knowledge of the stars. In experiments the natural sky can easily be replaced by a planetarium sky (Emlen 1967; Beason 1987) or even by a simpler pattern of artificial "stars" (Wiltschko et al. 1987a; Able and Able 1990). Experiments with hand-raised birds showed that birds must have the opportunity to observe the stellar sky during the first summer of life, prior to their first migration, in order to be able to use a star compass (Emlen 1972). The reference for these first learning processes is celestial rotation, whereas later during migration the magnetic field controls the directional significance of the stars.

The star compass in the experiments mentioned above was used to locate the innate migratory direction. Because of this, the first learning process can better be described as transforming genetically encoded directional information into a compass course. It is discussed in more detail from this point of view in Sect. 5.4.

5.2.4 Integrating Directional Orientation

The manifold interrelations between the sun compass, star compass and magnetic compass clearly show that the three mechanisms cannot be looked upon as separate or substitute systems for the birds. Rather, they represent integrated components of a general system for directional orientation. Information transfer, as described between the magnetic compass and the celestial systems, may also take place between sun-related factors and star compass (Moore 1985). Transfer processes like the ones mentioned above might also be used to give directional significance to other environmental factors that can be used only temporarily, but when available help birds to maintain the correct course. Wind (Bellrose 1967) and landscape features (e.g. Vleugel 1955) have been discussed as possible examples. Taken together, the birds might calibrate any factors with directional characteristics and use them as secondary compass cues. This means that the desired course can be derived from the environment as a whole.

A brief look at other animals suggests that there is a general tendency to use a multitude of cues for locating direction. A magnetic compass and a sun compass have been demonstrated in groups such as amphipods, fish and amphibians (see Wiltschko and Wiltschko 1991a). It seems probable that the two compass mechanisms also interact with each other as they do in birds, despite the scarcity of experimental evidence. Pardi et al. (1988) studied the orientation of *Talorchestia* (Crustacea, Malacostraca: Amphipoda) at the African coast between the tropics, where the sun crosses the southern or the northern sky, depending on the season. They suggested that the magnetic compass might be used to adapt the sun compass to these dramatic seasonal changes, which would be a parallel to the findings in birds.

5.3 Mechanism for Determining the Home Direction

The ability to return home from any location requires information on the position of the respective site in relation to the home site. To obtain this information, two basic strategies can be used: Animals can rely on information obtained en route during the outward journey, or they can use local information obtained at the starting point of the return flight (in displacement experiments, this is the release site). Experimental evidence suggests that birds make use of both strategies.

5.3.1 Navigation by Route-Specific Information

Route reversal appears to be a fairly simple homing strategy because the birds only have to obtain information on the route of the outward journey and process it in an adequate way. The fact that birds return by more or less direct routes, even when the outward journey was a detour, excludes the possibility that birds just follow, in reverse order, the sequence of landmarks which they passed on the outward journey. In addition, we know from experiments that the necessary information can also be obtained when birds are passively displaced, in conditions where they do not actively control the course of the outward journey and are deprived of any view of the sun and landmarks.

Different mechanisms based on route-specific information have been discussed. At the turn of the century several authors suggested some kind of inertial navigation, proposing that birds record and double-integrate all accelerations experienced during an outward journey. However, the deviation observed in clock-shift experiments (cf. Fig. 5.3) clearly shows that the home course is first determined with respect to an external reference system, and argues against inertial navigation.

Another mechanism, however, is supported by experimental evidence. Homing pigeons became disoriented when they were displaced in a distorted magnetic field, although the same magnetic field did not interfere with orientation when it was presented to them at the release site (Fig. 5.6). Apparently, the birds recorded the direction of the outward journey using their magnetic compass. If necessary, integrating detours would tell them the net direction of displacement and, simply by reversing it, they could obtain the homeward course (Wiltschko and Wiltschko 1978).

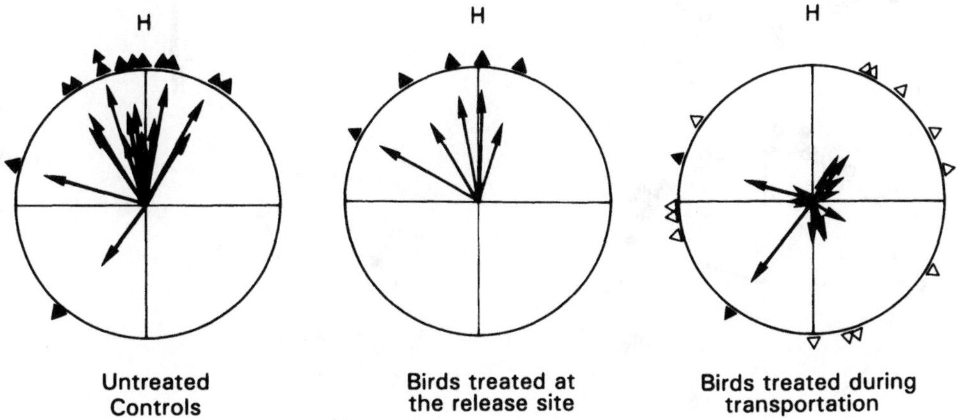

Fig. 5.6. Effect of transportation in a distorted magnetic field on the initial orientation of very young, untrained pigeons. The mean vectors of various releases are given as *arrows* with respect to the home direction (*H* upward); the mean directions are marked by *triangles at the periphery of the circle*, *open symbols* indicating non-significant samples (After Wiltschko and Wiltschko 1978, 1985)

Experiments with pigeons of different age and flying experience showed that this strategy is only used by young, inexperienced birds during their very first flights. As soon as the pigeons gain more experience, the effect of the treatment decreases rapidly. In experiments with old, experienced pigeons, the effect of transporting the birds in a distorted magnetic field was very small or non-existent (e.g. Kiepenheuer 1978; Wallraff et al. 1980; Wiltschko and Wiltschko 1985). This suggests that experienced pigeons normally use other strategies.

Another type of route-specific information has been discussed by Papi et al. (1984). When they released experienced pigeons at considerable distances from the loft, they observed that initial orientation was affected by the route of the outward journey. A detour in the initial leg of the route to the left induced a deviation to the left and vice versa (Fig. 5.7). The authors attribute this effect to olfaction; they assume that the pigeons, as long as they were still in the familiar vicinity of their loft, recorded the direction in which they were transported by the typical combinations of odours of the locations which they passed. Even if attempts to replicate these findings were not always successful (see Keeton 1980), it appears possible that information on the route during the outward journey

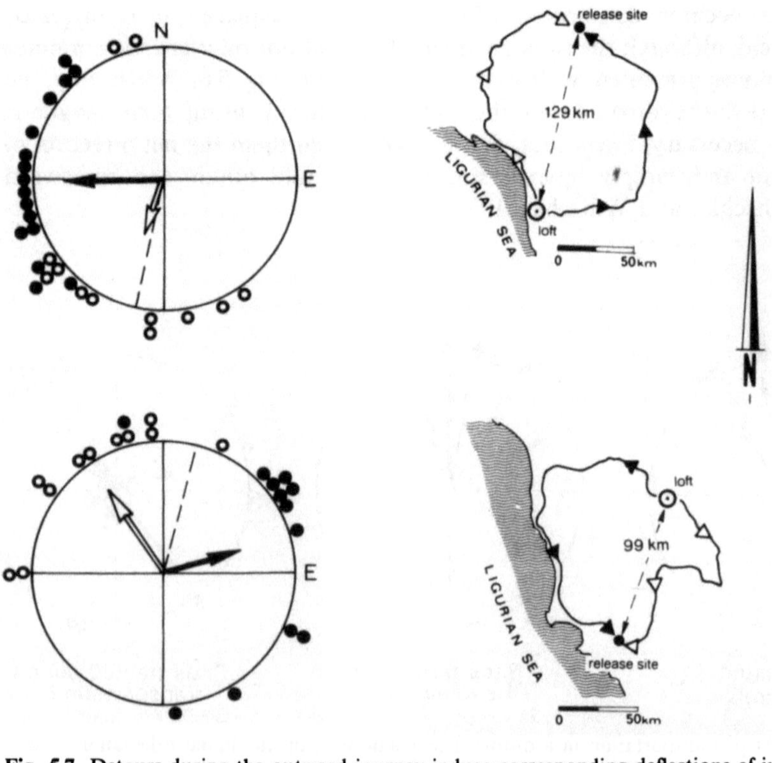

Fig. 5.7. Detours during the outward journey induce corresponding deflections of initial orientation. *Open symbols* refer to the group transported on the routes marked with *open arrows* and vice versa (After Papi et al. 1984)

can be incorporated into the homing process of experienced pigeons (Tögel and Wiltschko 1992). The nature of this information, however, is still open and recent findings (Ganzhorn and Burkhardt 1991) argue against a solely olfactory interpretation.

5.3.2 Site-Specific Information – the Navigational "Map"

Orientation based on site-specific information means that birds must be able to derive their home direction by comparing local values of certain environmental variables with the home values. In contrast to route-specific information, which is collected during the outward journey and processed shortly thereafter, the use of site-specific information requires knowledge of the spatial distribution of navigational factors which must be memorized on a long-term basis. The model describing our concept of navigation by local information dates back to the last century (Viguier 1882). It is based on the assumption that birds use at least two environmental gradients which intersect at an angle. The birds must know the direction of the gradients. For example, a bird may know that factor A increases towards the east, while it also must remember the home values very accurately. Encountering values of A that are smaller than the home values would then tell the bird that it is west from home and therefore has to fly east (for a detailed discussion, see Wallraff 1974).

Birds' knowledge of the distribution of navigational factors, the navigational "map", is believed to represent a directionally oriented mental picture of their spatial array. It is established by learning when the young birds begin to undertake their first extended flights. As birds obviously can also use local information at unfamiliar sites, they must be able to extrapolate the course of gradients beyond the area of direct experience. This means the "map" is not limited to the familiar area, but extends considerably beyond it (Wallraff 1974; Wiltschko and Wiltschko 1987).

Large-scale, even world-wide, gradients seemed at first to be prime candidates as potential components of the "map". The first hypotheses were based on magnetic parameters (Viguier 1882) or the coriolis force (Yeagley 1947), both of which show more or less north–south gradients, or on features of the sun's arc (Matthews 1953). Gravity was also considered (Larkin and Keeton 1978). But since the "map" is a learned system, factors of a more local nature may also be involved. This led to the consideration of cues like odours, which may form locally characteristic combinations and gradients (Papi 1976, 1986), or infrasound, which is propagated over very large distances with little attenuation (Quine 1982). In the latter case, the birds could locate several sound sources and use the fact that their bearings differ from different locations.

The research efforts of recent years clearly indicate that the birds' "map" represents a multiple cue system. For example, the use of magnetic cues is indicated by the following findings: The orientation of pigeons was found to be affected by magnetic storms, which induced certain deviations from the bearings on magnetically quiet days (Keeton et al. 1974). Further, pigeons become disoriented when they are released at strong magnetic anomalies, and the size of effect is correlated with the steepness of local magnetic gradients (see Fig. 5.8 and Walcott 1978). Wiltschko and Wiltschko (1988) give a summary of these magnetic effects.

Experimental evidence also suggests the involvement of olfactory input. Depriving pigeons of olfactory information often affected initial orientation, deflecting the bearings from those of untreated controls or leading to disorientation (see Fig. 5.9), and also decreased the homing performance (e.g. Benvenuti et al. 1973; for summary, see Papi 1986). Manipulations at the home site designed to affect olfactory components of the "map" did not yield a clear-cut picture. The effects did not always agree with the predictions of the olfactory hypothesis (e.g. Ioale 1982; Papi 1986). In general, the olfactory effects proved highly variable. The largest effects of olfactory deprivation were observed in Italy, whereas in upstate New York, USA, and in Frankfurt, FRG, effects were almost negligible (Wiltschko et al. 1987b).

Recent tests suggest the involvement of infrasound as a third factor. At greater distances from their loft, pigeons that had been partially deprived of infrasound by piercing their *tympanum* with a small needle showed differences

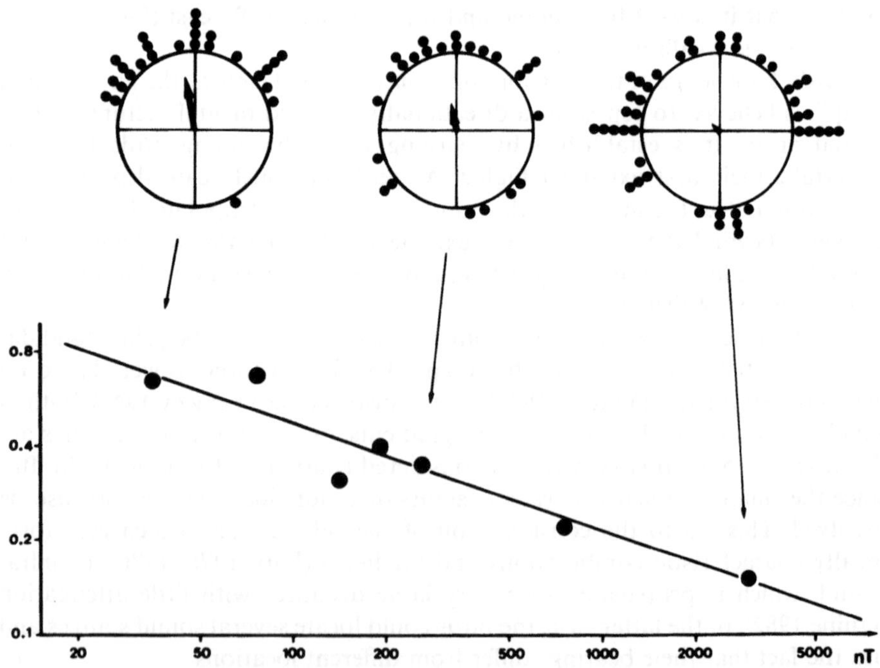

Fig. 5.8. The effect of magnetic anomalies on initial orientation. The vector lengths are negatively correlated with the steepness of the magnetic gradient towards home (given in nT/km). Note that both scales are logarithmic. In the three examples given, the *symbols at the periphery of the circle* mark the vanishing bearings of individual pigeons

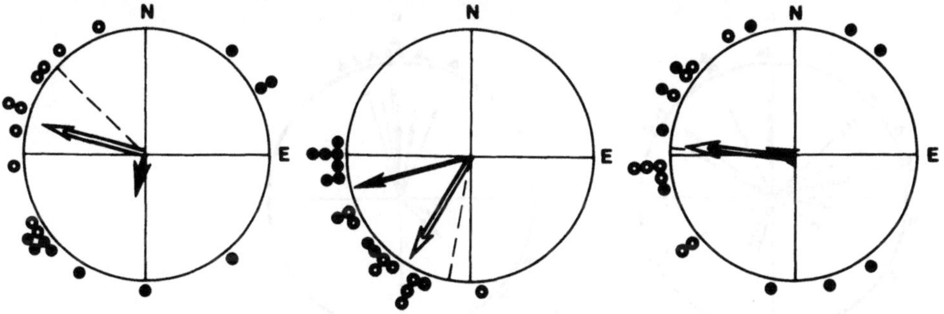

Fig. 5.9. The effect of olfactory deprivation on initial orientation of homing pigeons may consist in a deviation from the mean of controls or in disorientation. *Open symbols* controls; *solid symbols* birds under olfactory deprivation (Data from Wiltschko et al. 1986)

from the behaviour of untreated controls. The effect was regionally variable; in some areas, an increase in scatter was observed, while in other areas the scatter decreased (Schöps and Wiltschko 1990). These findings can be best interpreted in terms of a multi-cue system, where the combination of available cues and the pattern of agreement between them determines the role of any single cue in the system.

Certain lunar rhythms in the initial orientation of pigeons (Larkin and Keeton 1978) seemed to suggest the possible involvement of gravity. This would mean that birds would have to detect extremely small gravitational differences. Gravity anomalies, however, did not have a negative effect on orientation (Lednor and Walcott 1984). Landmarks have also been discussed as orientation cues for birds, yet they seem to play a role only in the immediate vicinity of home, where they ensure orientation during the final approach (Schlichte 1973).

Altogether, a diversity of cues is incorporated into the "map", with the above list probably not complete. These factors all seem to contribute to the navigational process, yet the removal of any of the cues mentioned above does not lead to regular disorientation, which suggests that any one cue can be replaced to a large extent by a combination of others. This makes it hard to assess the role of any single cue, particularly as it may vary between birds living in different regions. The variable results of attempts to repeat the Italian olfactory experiments in other countries have already been mentioned (see Keeton 1980). The fact that even identical techniques yielded different results clearly shows that the results reflect differences in the orientation system of pigeons (Wiltschko et al. 1987b). Walcott (1988) reported similar observations with respect to magnetic anomalies. Pigeons from a loft in Massachusetts were disoriented, whereas the behaviour of pigeons from a loft in upstate New York seemed largely unaffected by the same anomaly. This suggests that birds do not process the navigational information in a constant way. Various groups of birds seem to prefer certain cues over others and rate and rank them differently.

An analysis of these variations in orientation behaviour and their possible causes has just begun. First results indicate that they may be based on genetic

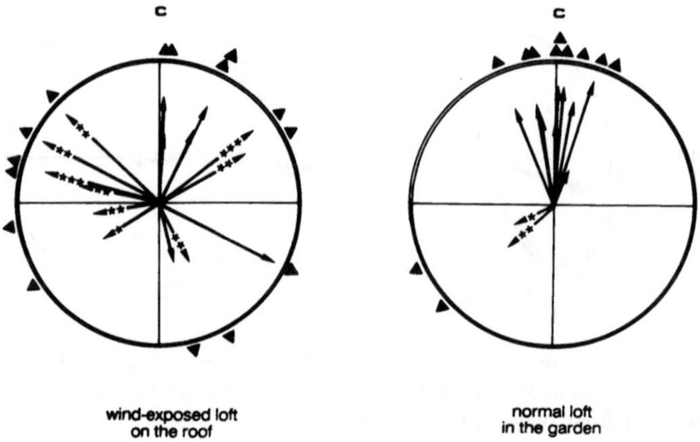

Fig. 5.10. Pigeons raised in a wind-exposed loft react to olfactory deprivation with a strong deviation from the mean of their controls, whereas their siblings raised in a wind-protected loft seldom show a reaction. The diagrams show the mean vectors of the anosmic pigeons with respect to the mean of the controls (*C* upward). *Asterisks* indicate vectors significantly different from the control vector (Data from Wiltschko and Wiltschko 1989)

differences (Benvenuti and Ioale 1988) as well as on the birds' early experience. The factors they experience and the way these factors are presented as the "map" is established appears to affect later mechanisms of orientation (see Fig. 5.10; Walcott and Brown 1989; Wiltschko and Wiltschko 1989). This should not be surprising, since the "map" is established by learning processes and pigeons may favour factors which have proved to be the most useful and reliable during their early experience. This would result in a "map" well adapted to the local situation of the region where the birds have to orient.

5.3.3 Different Strategies Supplement Each Other

The mechanisms used to determine the homeward course include a multitude of cues. Birds seem to take advantage of all available information, and to incorporate it into their navigational system. Although the use of route-specific information and site-specific local information are, in theory, two basically different strategies, they are interrelated and supplement each other. Again, most interactions are found during the development of the navigational system. Route reversal with the help of the magnetic compass appears to be the first mechanism of homing, before the "map" is established. The same mechanism may also be involved in the processes establishing the "map". By providing information on the homeward course they supply the directional information which is matched with the local values of navigational cues in order to obtain a directionally oriented mental picture of their spatial distributions.

Later, when the birds become more experienced, the "map" becomes the preferred mechanism to determine the home direction. Route reversal based on the direction of the outward journey seems to lose importance (Wiltschko and Wiltschko 1985), but it may continue to play a role in extending and updating the "map" (Grüter and Wiltschko 1990). Another type of route-specific information, namely the mechanisms of following the initial route while the birds are still in the familiar area (Papi et al. 1984), continues to be used in the homing process. However, its specific role seems to depend on local factors (Tögel and Wiltschko 1992). Both strategies are used together in a flexible way.

5.4 Determining the Migratory Direction

For migration, birds have to move to a specific region of the world, the winter quarters of their species. For young birds on their first migration, this involves moving to an area where they have never been before and which is still unknown to them.

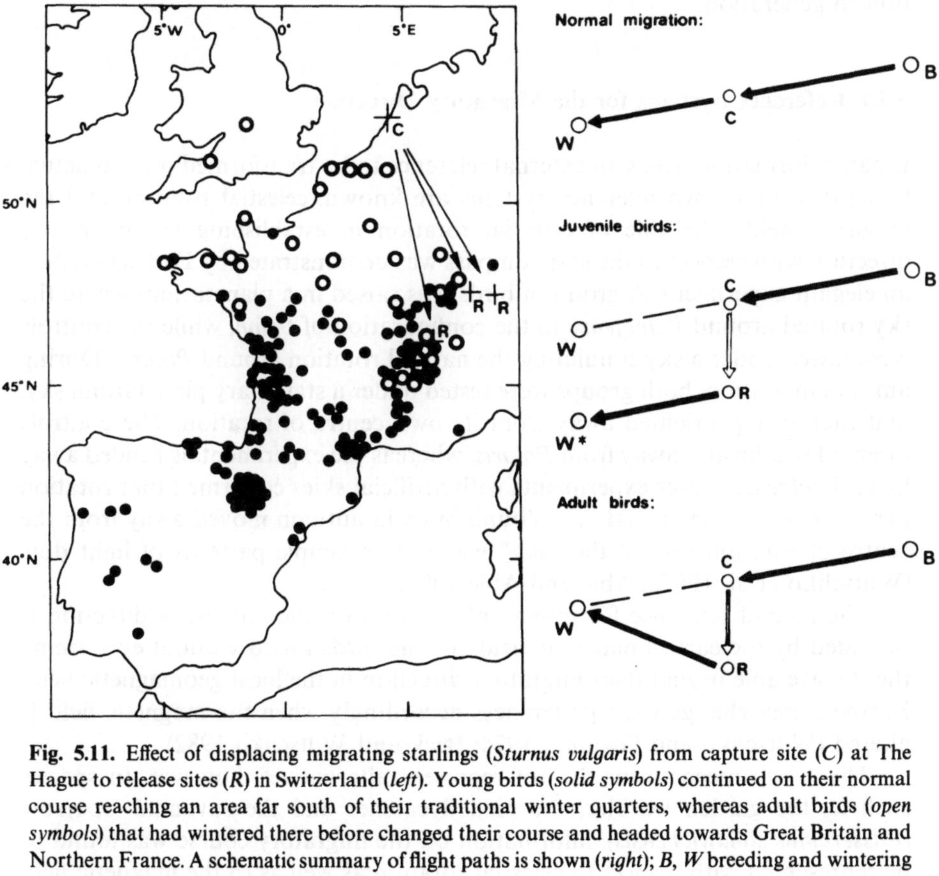

Fig. 5.11. Effect of displacing migrating starlings (*Sturnus vulgaris*) from capture site (*C*) at The Hague to release sites (*R*) in Switzerland (*left*). Young birds (*solid symbols*) continued on their normal course reaching an area far south of their traditional winter quarters, whereas adult birds (*open symbols*) that had wintered there before changed their course and headed towards Great Britain and Northern France. A schematic summary of flight paths is shown (*right*); *B*, *W* breeding and wintering areas respectively (After Perdeck 1958)

A large-scale displacement experiment by the Dutch Vogeltrek-station (Perdeck 1958) revealed the type of information on which they rely. Thousands of starlings of Baltic origin, bound for Northern France and England, were caught near The Hague and, after having been displaced about 500 km perpendicular to their normal migration route, were released in Switzerland. The sites where ringed birds were recovered are shown in Fig. 5.11. Adult birds that had already spent a winter in their wintering quarters headed toward their normal winter range. Apparently, they were able to change their migratory direction to compensate for the displacement. The young birds, however, continued on their original course, which would have brought them to their normal winter quarters if they had not been displaced. This suggested that young birds rely on innate information about their migratory course. Also, young hand-raised migrants that never had contact with experienced conspecifics, clearly preferred their species-specific migratory direction when tested during migration (e.g. Emlen 1972; Bingman 1983a; Able and Able 1990). This, too, indicates that genetically encoded information on the migratory course, or on a sequence of courses (see Gwinner and Wiltschko 1978; Helbig et al. 1989), is passed down from generation to generation.

5.4.1 Reference Systems for the Migratory Direction

Innate information needs an external reference to be transformed into an actual flying direction. Two reference systems are known: celestial rotation and the magnetic field. The role of celestial rotation in establishing the migratory direction with respect to the star compass was demonstrated by Emlen (1972) in an elegant experiment. A group of birds was raised in a planetarium where the sky rotated around *Betelgeuze* in the configuration of *Orion*, while the controls were raised under a sky simulating the natural rotation around *Polaris*. During autumn migration, both groups were tested under a stationary planetarium sky, and each group oriented away from its own centre of rotation. The controls oriented southward, away from *Polaris*, whereas the experimentals headed away from *Betelgeuze*. Later experiments with artificial skies confirmed that rotation per se provided the crucial cue. Young birds in autumn moved away from the centre of rotation even if the "sky" was only a simple patterns of light dots (Wiltschko et al. 1987a; Able and Able 1990).

The second reference for innate information on the migratory direction is provided by the earth's magnetic field. Young birds raised without ever seeing the sky are able to find their migratory direction in the local geomagnetic field. Further, they change their preferences accordingly when the magnetic field is altered (Wiltschko and Gwinner 1974; Beck and Wiltschko 1982).

These two mechanisms do not represent alternatives, however. In species such as the garden warbler, the pied flycatcher and the savannah sparrow (*Passerculus sandwichensis*), information on the migratory course was found to be represented with respect to celestial rotation as well as to the magnetic field

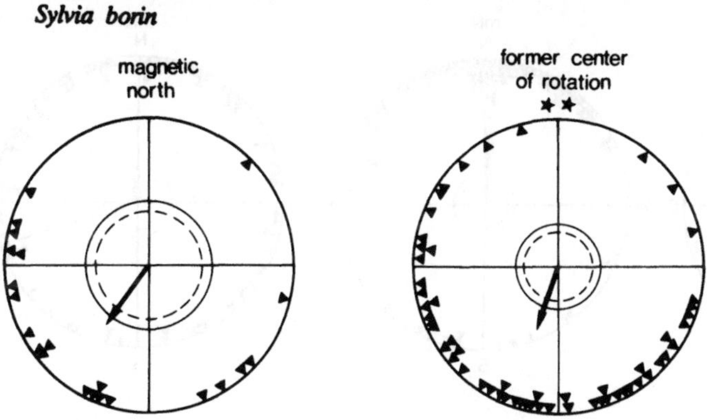

Fig. 5.12. The magnetic field and celestial rotation may both serve as references for the migratory direction in garden warblers (*Sylvia borin*). *Left* birds hand-raised and tested without ever seeing celestial cues; *Right* birds hand-raised under a rotating artificial sky and tested under the same, now stationary, sky in the absence of meaningful magnetic information (Data from Wiltschko and Gwinner 1974; Wiltschko et al. 1987a)

(Fig. 5.12). Young birds were able to locate their migratory direction as "away from the centre of rotation" with their star compass, and in relation to the magnetic field lines with their magnetic compass (for summary, see Able and Bingman 1987; Wiltschko et al. 1989). This suggests a certain redundancy, which seems surprising in the case of innate information.

The factors associated with sunset have also been discussed as a reference for the migratory direction (e.g. Katz 1985; Moore 1987). The only available data (Bingman 1983a), however, argue against such a function.

5.4.2 The Interrelation Between Celestial Rotation and the Magnetic Field During Ontogeny

Although each system by itself can ensure that the migratory direction is located, use of celestial rotation and the magnetic field are not really independent. Under natural conditions, they interact to produce one common course. Bingman (1983b) was the first to report an effect of celestial rotation on magnetic orientation. During their first summer, young birds were exposed to the natural sky in an artificial field with magnetic north shifted 90°. When these birds were later tested without visual cues, they changed their magnetic course and preferred the magnetic course which had been their true (geographic) migratory direction during ontogeny (see Fig. 5.13). An effect of the magnetic field during ontogeny on the later orientation with respect to the stars, on the other hand, could not be demonstrated (Wiltschko et al. 1987a). This suggests

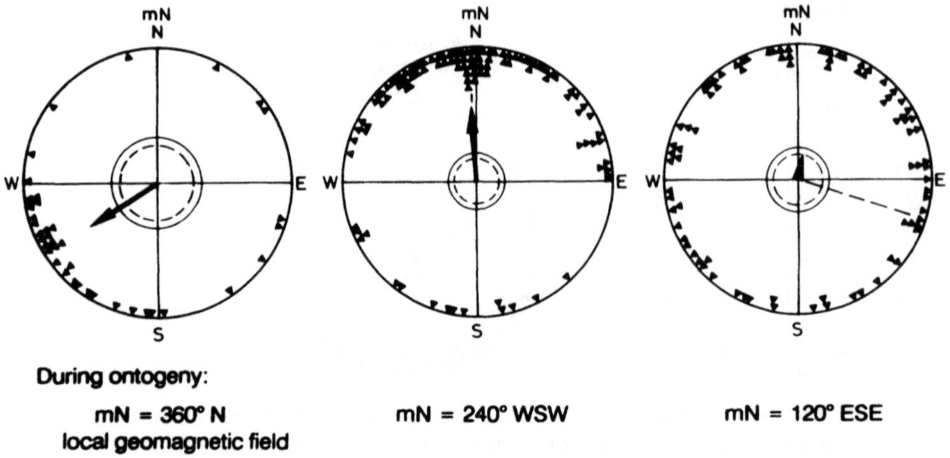

Fig. 5.13. The effect of celestial rotation during ontogeny on later migratory orientation using the magnetic compass. Hand-raised pigeons that had been exposed to different relationships between magnetic and stellar north during ontogeny were tested in the local magnetic field in the absence of visual cues during migration (Data from Prinz and Wiltschko 1992)

that when the migratory direction is first established, celestial rotation dominates over the information provided by the magnetic field.

Recent experiments indicate that the relationship between various factors during the pre-migratory period may not be as simple as it would appear from the above-mentioned results. Repeating the experiments, Prinz and Wiltschko (1992) reported that the magnetic course could be easily altered only when the change was clockwise; the mirror image situation which should have produced a counter-clockwise shift resulted in random behaviour (Fig. 5.13). The reasons for this asymmetry remain unclear. The realization of the migratory direction as an actual direction may prove to be a more complex process than anticipated, particularly as the relative importance of the stars and the magnetic field changes during actual migration, with the magnetic compass gaining dominance. At high latitudes, when the birds start their migration, magnetic declination (i.e. the difference between magnetic north and true north) may be considerable and may change relatively quickly. Here, celestial rotation is a highly reliable indicator of the true direction. As the birds move south, however, the familiar stars of their home region lose altitude and finally disappear below the horizon, while new stars appear in the southern sky. These stars can be calibrated using magnetic information. Since the magnetic field is fairly regular at lower latitudes, it provides a reliable directional reference. Likewise, the factors associated with sunset undergo considerable change as a function of geographic latitude, which can be compensated by recalibrating them with the help of the magnetic field.

5.5 Conclusion

The examples presented here show that each process in bird orientation involves a multitude of cues. Findings such as dominance of the sun compass over the magnetic compass led many authors to use terms like a "hierarchy" of cues or mechanisms (e.g. Emlen 1975; Able 1991) to characterize their relationship. These terms, however, imply a fixed hierarchy and disregard the interactions between the various cues (cf. Chap. 16). While the sum compass normally appears dominant over the magnetic compass, the directional significance of the sun is established (and possibly controlled) with the help of the magnetic field. A similar relationship might also be assumed between some of the other cues. The various factors do not represent independent sources of information which are simply added up. Instead, the information is integrated, with the directional significance of some factors being adjusted to that of others to bring the entire system into harmony. Therefore, the different factors must be seen as integrated components of a complex system, which is partially redundant, and which tries to use all suitable factors the environment provides.

The advantage to birds is obvious. Any system containing a certain redundancy is less susceptible to outside interference by adverse conditions. In addition, it can overcome situations where an important factor cannot be used. For example, at the magnetic equator, the magnetic compass becomes bimodal and cannot be used to discriminate the migratory direction from the direction opposite to it. In the laboratory, migratory birds are disoriented when tested in a horizontal magnetic field (see Wiltschko and Wiltschko 1988). However, under natural conditions, the star compass and the factors associated with sunset might easily provide information enabling migrants to continue in the right direction. In homing, too, it is important that there is some redundancy in the factors used to orient. The environment is not completely constant, and navigational factors may show certain fluctuations. One example is provided by the variability observed in the initial orientation of pigeons released at the same sites. Magnetic storms, for instance, may distort the local distribution of magnetic parameters. Strong weather fronts may add infrasound and make the infrasound picture difficult to interpret. However, it is highly unlikely that all factors involved would be severely disturbed at the same time.

A multiple cues system generally offers more safety to its users, and minimizes the effects of interference, especially when one or even more factors are not available or produce contradicting information. This could become crucial particularly in extreme situations, and it should increase the chance of reaching the desired goal. The advantages such a multiple cue system offers should lead to its widespread use in the animal kingdom (cf. Chap. 16). Even modern technology follows analogous principles. For example, ships are nowadays equipped with gyro-compasses as well as magnetic compasses, and satellite-based systems are used to determine the current position everywhere on earth within a range of a few hundred metres. Nevertheless, navigation officers are still trained in the art of astronavigation, and the ship carries a sextant and nautical tables which can be used to determine positions in the traditional way. As in bird navigation, numerous navigational factors are available for their user, and it becomes obvious that in seafaring the same principles are followed for exactly the same reasons.

References

Able KP (1991) Common themes and variations in animal orientation systems. Am Zool 31: 157–167
Able KP, Able MA (1990) Calibration of the magnetic compass of a migratory bird by celestial rotation. Nature (Lond) 347:378–380
Able KP, Bingman VP (1987) The development of orientation and navigation behaviour in birds. Q Rev Biol 62:1–29
Able KP, Cherry JD (1986) Mechanisms of dusk orientation in White-Throated Sparrows (*Zonotrichia albicollis*): clock-shift experiments. J Comp Physiol A 159:107–113
Alerstam T, Högstedt G (1983) The role of the geomagnetic field in the development of birds' compass sense. Nature (Lond) 306:463–465
Beason RC (1987) Interaction of visual and non-visual cues during migratory orientation by the bobolink (*Dolichonyx oryzivorus*). J Ornithol 128:317–324
Beck W, Wiltschko W (1982) The magnetic field as reference system for the genetically encoded migratory direction in pied flycatchers (*Ficedula hypoleuca* PALLAS). Z Tierpsychol 60:41–46
Bellrose F (1967) Radar in orientation research. In: Snow DW (ed) Proc XIV Int Ornithol Congr Oxford 1966. Blackwell, Oxford, pp 281–309
Benvenuti S, Ioale P (1988) Initial orientation of homing pigeons: different sensitivity to altered magnetic fields in birds of different countries. Experientia 44:358–359
Benvenuti S, Fiaschi V, Fiore L, Papi F (1973) Disturbances of homing behaviour in pigeons experimentally induced by olfactory stimuli. Monit Zool Ital 7:117–128
Berthold P (1988) The control of migration in European Warblers. In: Ouellet H (ed) Acta XIX Int Ornithol Congr Ottawa 1986. Univ of Ottawa Press, Ottawa, pp 215–294
Bingman VP (1983a) Importance of earth's magnetism for the sunset orientation of migratory naive Savannah Sparrows. Monit Zool Ital 17:395–400
Bingman VP (1983b) Magnetic field orientation of migratory Savannah Sparrows with different first summer experience. Behaviour 87:43–53
Bingman VP (1987) Earth's magnetism and the nocturnal orientation of migratory European Robins. Auk 104:523–525
Bingman VP, Wiltschko W (1988) Orientation of dunnocks (*Prunella modularis*) at sunset. Ethology 77:1–9
Emlen ST (1967) Migratory orientation in the indigo bunting, *Passerina cyanea*. Part II. Mechanism of celestial orientation. Auk 84:463–489
Emlen ST (1972) Migration: orientation and navigation. In: Galler SR, Schmidt-Koenig K, Jacobs GJ, Belleville RE (eds) Animal orientation and navigation. NASA SP-262, US Govt Print Office, Washington DC, pp 191–210
Emlen ST (1975) Migration: orientation and navigation. In: Farner DS, King JR (eds) Avian biology, vol 5. Academic Press, New York, pp 129–219
Ganzhorn JU, Burkhardt (1991) Pigeon homing: new airbag experiments to assess the role of olfactory information for pigeon navigation. Behav Ecol Sociobiol 29:69–75
Grüter M, Wiltschko R (1990) Pigeon homing: the effect of local experience on initial orientation and homing success. Ethology 84:239–255
Gwinner E, Wiltschko W (1978) Endogenously controlled changes in the migratory direction of the garden warbler, *Sylvia borin*. J Comp Physiol A 125:267–273

Helbig AJ (1991) Dusk orientation of migratory European robins, *Erithacus rubecula*: the role of sun-related directional information. Anim Behav 41:313–322
Helbig AJ, Berthold P, Wiltschko W (1989) Migratory orientation of blackcaps (*Sylvia atricapilla*): population specific shifts of direction during autumn. Ethology 82:307–315
Ioale P (1982) Pigeon homing: effects of differential shielding of home cages. In: Papi F, Wallraff HG (eds) Avian navigation. Springer, Berlin Heidelberg New York, pp 170–178
Katz YB (1985) Orientation behaviour of the European robin (*Erithacus rubecula*). Anim Behav 33:825–828
Keeton WT (1969) Orientation by pigeons: is the sun necessary? Science 165:922–928
Keeton WT (1971) Magnets interfere with pigeon homing. Proc Natl Acad Sci USA 68:102–106
Keeton WT (1980) Avian orientation and navigation: new developments in an old mystery. In: Nöhring R (ed) Acta XVII Congr Int Ornithol, vol 1. Dtsch Ornithol-Ges, Berlin, pp 137–158
Keeton WT, Larkin TS, Windson DM (1974) Normal fluctuations in the earth's magnetic field influence pigeon orientation. J Comp Physiol A 95:95–103
Kiepenheuer J (1978) The effect of magnetic fields inverted during displacement on the homing behaviour of pigeons. In: Schmidt-Koenig K, Keeton WT (eds) Animal migration, navigation, and homing. Springer, Berlin Heidelberg New York, pp 135–142
Kramer G (1959) Recent experiments on bird orientation. Ibis 101:399–416
Larkin T, Keeton WT (1978) An apparent lunar rhythm in the day-to-day variations in initial bearings of homing pigeons. In: Schmidt-Koenig K, Keeton WT (eds) Animal migration, navigation, and homing. Springer, Berlin Heidelberg New York, pp 92–106
Leask MJM (1977) A physiochemical mechanism for magnetic field detection by migratory birds and homing pigeons. Nature (Lond) 267:144–145
Lednor AJ, Walcott C (1984) The orientation of pigeons at gravity anomalies. J Exp Biol 111:259–263
Matthews GVT (1953) Sun navigation in homing pigeons. J Exp Biol 30:243–267
Moore FR (1985) Integration of environmental stimuli in the migratory orientation of the savannah sparrow (*Passerculus sandwichensis*). Anim Behav 33:657–663
Moore FR (1987) Sunset and the orientation behaviour of migrating birds. Biol Rev 62:65–86
Orth G, Wiltschko W (1981) Die Orientierung von Wiesenpiepern (*Anthus pratensis* L.). Verh Dtsch Zool Ges Bremen 1981:252
Papi F (1976) The olfactory navigation system of the homing pigeon. Verh Dtsch Zool Ges Ham 1976:184–205
Papi F (1986) Pigeon navigation: solved problems and open questions. Monit Zool Ital 20:471–517
Papi F, Ioale P, Fiaschi V, Benvenuti S (1984) Pigeon homing: the effect of outward-journey detours on orientation. Monit Zool Ital 18:53–87
Pardi L, Ugolini A, Faqi AS, Scapini F, Ercolini A (1988) Zonal recovering in equatorial sandhoppers: interaction between magnetic and solar orientation. In: Chelazzi G, Vannini M (eds) Behavioural adaption to intertidal life. Plenum Press, New York, pp 79–92
Perdeck AC (1958) Two types of orientation in migrating *Sturnus vulgaris* and *Fringilla coelebs* as revealed by displacement experiments. Ardea 46:1–37
Prinz K, Wiltschko W (1992) Migratory orientation of pied flycatchers: interaction of stellar and magnetic information during ontogeny. Anim Behav 44:539–546
Quine DB (1982) Infrasound: a potential cue for homing pigeons. In: Papi F, Wallraff HG (eds) Avian navigation. Springer, Berlin Heidelberg New York, pp 373–376
Schlichte H-J (1973) Untersuchungen über die Bedeutung optischer Parameter für das Heimkehrverhalten der Brieftauben. Z Tierpsychol 32:257–280
Schmidt-Koenig K (1960) Internal clocks and homing. Cold Spring Harbor Symp Quant Biol 25:389–393
Schöps M, Wiltschko R (1990) Benutzen Brieftauben Infraschall zur Orientierung? Verh Dtsch Zool Ges 83:430
Semm P, Beason RC (1990) Responses to small magnetic variations by the trigeminal system of the bobolink. Brain Res Bull 25:735–740

Semm P, Demaine C (1986) Neurophysiological properties of magnetic cells in the pigeon's visual system. J Comp Physiol A 159:619–625

Semm P, Nohr D, Demaine C, Wiltschko W (1984) Neural basis of the magnetic compass: interaction of visual magnetic and vestibular inputs in the pigeon's brain. J Comp Physiol A 155:283–288

Tögel A, Wiltschko R (1992) Detour experiments with homing pigeons: information obtained during the outward journey is included in the navigational process. Behav Ecol Sociobiol 31:73–79

Viguier C (1882) Le sens de l'orientation et ses organes chez les animaux et chez l'homme. Rev Philos Fr Etranger 14:1–36

Vleugel DA (1955) Über die Unzulänglichkeit der Visierorientierung für das Geradeausfliegen, insbesondere beim Zug des Buchfinken (*Fringilla coelebs* L.). Ornis Fenn 32:34–40

Walcott C (1978) Anomalies in the earth's magnetic field increase the scatter of pigeons vanishing bearings. In: Schmidt-Koenig K, Keeton WT (eds) Animal migration, navigation, and homing. Springer, Berlin Heidelberg New York, pp 143–151

Walcott C (1988) Homing in pigeons: are differences in experimental results due to different home-loft environments? In: Ouellet H (ed) Acta XIX Congr Int Ornithol I. Univ of Ottawa Press, Ottawa, pp 305–308

Walcott C, Brown AI (1989) The disorientation of pigeons at Jersey Hill. In: Royal Institute of Navigation (ed) Orientation and navigation – birds, humans and other animals. Cardiff 1989, Pap 8

Walcott C, Gould JL, Kirschvink JL (1979) Pigeons have magnets. Science 205:1027–1029

Wallraff HG (1974) Das Navigationssystem der Vögel. Ein theoretischer Beitrag zur Analyse ungeklärter Orientierungsleistungen. Oldenbourg, München

Wallraff HG, Foa A, Ioale P (1980) Does pigeon homing depend on stimuli perceived during displacement? II. Experiments in Italy. J Comp Physiol A 139:203–208

Wiltschko R (1981) Die Sonnenorientierung der Vögel. 2. Entwicklung des Sonnenkompaß und sein Stellenwert im Orientierungssystem. J Ornithol 122:1–22

Wiltschko R (1983) The ontogeny of orientation in young pigeons. Comp Biochem Physiol A 76:701–708

Wiltschko R, Wiltschko W (1978) Evidence for the use of magnetic outward-journey information in homing pigeons. Naturwissenschaften 65:112

Wiltschko R, Wiltschko W (1981) The development of sun compass orientation in young homing pigeons. Behav Ecol Sociobiol 2:135–141

Wiltschko R, Wiltschko W (1985) Pigeon homing: change in navigational strategy during ontogeny. Anim Behav 33:583–590

Wiltschko R, Wiltschko W (1989) Pigeon homing: olfactory orientation – a paradox. Behav Ecol Sociobiol 24:163–173

Wiltschko R, Wiltschko W (1990) Zur Entwicklung des Sonnenkompaß bei jungen Brieftauben. J Ornithol 131:1–20

Wiltschko W (1968) Über den Einfluß statischer Magnetfelder auf die Zugorientierung der Rotkehlchen (*Erithacus rubecula*). Z Tierpsychol 25:537–558

Wiltschko W, Gwinner E (1974) Evidence for an innate magnetic compass in Garden Warblers. Naturwissenschaften 61:406

Wiltschko W, Wiltschko R (1975a) The interaction of stars and magnetic field in the orientation system of night migrating birds. I. Autumn experiments with European warblers (Gen. *Sylvia*). Z Tierpsychol 37:337–355

Wiltschko W, Wiltschko R (1975b) The interaction of stars and magentic field in the orientation system of night migrating birds II. Spring experiments with European robins (*Erithacus rubecula*). Z Tierpsychol 39:265–282

Wiltschko W, Wiltschko R (1987) Cognitive maps and navigation in homing pigeons. In: Ellen P, Thinus-Blanc C (eds) Cognitive processes and spatial orientation in animal and man. Nijhoff, Dordrecht, pp 201–216

Wiltschko W, Wiltschko R (1988) Magnetic orientation in birds. Curr Ornithol 5:67–121

Wiltschko W, Wiltschko R (1991a) Der Magnetkompaß als Komponente eines komplexeren Richtungsorientierungssystems. Zool Jahrb Physiol 95:437–446

Wiltschko W, Wiltschko R (1991b) Magnetic orientation and celestial cues in migratory orientation. In: Berthold P (ed) Orientation in birds. Birkhäuser, Basel, pp 16–37

Wiltschko W, Wiltschko R, Keeton WT (1984) Effect of a "permanent" clock-shift on the orintation of experienced homing pigeons. Behav Ecol Sociobiol 15:263–272

Wiltschko W, Wiltschko R, Foa A, Benvenuti S (1986) Orientation behavior of pigeons deprived of olfactory information during the outward journey and at the release site. Monit Zool Ital 20:183–193

Wiltschko W, Daum P, Fergenbauer-Kimmel A, Wiltschko R (1987a) The development of the star compass in garden warblers, *Sylvia borin.* Ethology 74:285–292

Wiltschko W, Wiltschko R, Walcott C (1987b) Pigeon homing: different effects of olfactory deprivation in different countries. Behav Ecol Sociobiol 21:333–342

Wiltschko W, Daum-Benz P, Munro U, Wiltschko R (1989) Interaction of magnetic and stellar cues in migratory orientation. J Navig 42:355–366

Yeagley HL (1947) A preliminary study of a physical basis of bird navigation. J Appl Phys 18:1035–1063

Introduction to Section II

Like all vertebrates, birds possess complexly articulated skeletons to which large numbers of muscles are attached. How does the central nervous system control concurrently the strength and timing of contraction of all these muscles so as to produce co-ordinated behaviour that is adjusted to the environment? An important influence on research into this problem has been the concept of the "central pattern generator". Although strikingly successful in the analysis of invertebrate neural organization, it has long been recognized that this concept is not generally sufficient. It cannot cope with the phenomenon of "motor equivalence" (different patterns of muscle activity produce the same behavioural outcome), or with the fact that the effect of a muscle contraction depends upon its context. These problems were identified decades ago by Lashley (1951) and Bernstein (1967), and have been explored more recently by theorists such as Turvey et al. (1982).

One way of approaching these problems is to look for clues in the processes by which coordinated motor patterns are established during the early development of the nervous system. Section II begins with a chapter by Bekoff which reviews studies of the motor development of bird embryos up until hatching. Initial patterns of body movement are irregular, and during the first 2 weeks of development the timings of muscle contractions within and between limbs become more regular. There is clear evidence that this basic embryonic motor patterning develops in the absence of sensory input, suggesting that central pattern generators are established early in development. The later detailed differentiation of these basic patterns do require sensory input, however, and the mechanisms involved remain to be discovered.

The developmental theme is taken further in Chapter 7 by Provine, who focuses on the development of wing-flapping from early embryonic stages to adulthood. Evidence shows that sensory feedback is not needed for the development of the basic wing-flapping pattern, but is required for co-ordination between the two wings. Provine describes comparative behavioural analyses of wing-flapping in different birds, which demonstrate that loss of flight can occur in either of two ways. In the ratites, the central pattern generator controlling wing-flapping has been lost, whereas in penguins and some domesticated fowl there has been no change in the CNS, but instead flight has been lost as a result of peripheral changes in muscle mass and wing feather area.

M.N.O. Davies and P.R. Green (Eds.)
Perception and Motor Control in Birds

Provine draws these comparative findings together with neurophysiological evidence to argue for a "centripetal" hypothesis of behavioural change in the course of evolution, illustrated by the evolution of flightlessness. The hypothesis is that initial changes through natural selection in muscle mass can lead, through effects on selective motor neuron death, to eventual evolutionary changes in the CNS. The same problem of interaction between periphery and CNS in behavioural evolution is discussed again, in a different context, in Chapter 14.

A third developmental approach to motor organization is described by Deich and Balsam in Chapter 8, who are again concerned with the role of sensory feedback and learning in the development of motor patterning. They describe research on the development of pecking in ring doves and pigeons which challenges strongly the view that the motor pattern involved in the peck is controlled by a central pattern generator which requires no sensory feedback for its development. The fine detail of peck topography does not appear in its adult form in squabs, where the thrust and gape components of pecking are poorly co-ordinated. Further, the development of peck topography can be influenced by Pavlovian and operant contingencies. Deich and Balsam propose a theoretical framework for understanding the development of species-specific motor patterns, drawing on the concept of task analysis. Abstract similarities between diverse tasks such as pecking and reaching by a primate or by a robot arm reveal constraints which shape the course of development of learned behaviour patterns.

The general theme emerging from Chapters 6, 7 and 8 is that the central pattern generator concept is relevant to understanding how some basic component motor patterns develop, but that early in development complex interactions with peripheral processes begin to shape adult motor patterns. The remaining chapters in Section II focus on the patterns of muscle activity involved in adult motor control. In Chapter 9, Zeigler, Bermejo and Bout describe in detail the kinematics of jaw movements during eating and drinking in pigeons. They take up one characteristic of birds' pecks at food discussed by Deich and Balsam; the scaling of beak opening, or gape, to the size of the target. Kinematic analysis shows that gape is determined by both the velocity and the duration of beak opening. The adjustment of gape provides a good model of motor equivalence, as the relative contributions of the control of velocity and of duration vary from one peck to another while jointly achieving the same end result.

Using EMG methods, Zeigler et al. have also discovered the pattern of muscle activity underlying the control of gape; the crucial variable turns out to be the temporal relation between activity in an "opener" muscle and in a "brake" muscle. Finally, they show that gape patterns can be controlled by conditioning procedures, and argue that these provide a promising technique for analyzing the adaptive modification of patterns of muscle control.

Successful pecking involves not only control of the jaw, discussed in Chapters 8 and 9, but also of the posture of the neck, which is the subject of

Chapter 10. The bird neck contains 20 or more vertebrae, and up to 200 muscles, and the work reviewed by Zweers, Bout and Heidweiller represents a major advance in understanding the principles by which this complex system is controlled. A thorough analysis of the anatomy of the cervical column is used to deduce mechanical principles relevant to the control problem, and these are tested against observations and refined further. These principles provide means by which the problem of controlling large numbers of muscles can be made more manageable, by greatly reducing the number of degrees of freedom involved.

Zweers et al. argue that the problem of head-neck control has features in common with other cases where complexly articulated limbs must be moved accurately towards a target, and they describe neural network models of head-neck control similar to those used in simulating reaching. Taken together, the research described in Chapters 8, 9 and 10 demonstrates that the motor organization of head movement and pecking in birds provides a powerful model for tackling general problems in motor control. A major challenge in robotics is to devise systems capable of dynamic adjustment to changing environmental demands, and the principles of motor organization in birds may well be both sufficiently complex in their adaptive properties, and sufficiently tractable experimentally, to be an important future source of design principles in robotics.

References

Bernstein N (1967) The coordination and regulation of movements. Pergamon Press, Oxford

Lashley KS (1951) The problem of serial order in behaviour. In: Jeffress LA (ed) Cerebral mechanisms in behaviour. Wiley, New York, pp 112–136

Turvey MT, Fitch HL, Tuller B (1982) The Bernstein perspective I. The problems of degrees of freedom and context-conditioned variability. In: Kelso JAS (ed) Human motor behaviour: an introduction. Lawrence Erlbaum, Hillsdale, NJ, pp 239–252

6 Neuroembryology of Motor Behaviour in Birds

A. Bekoff

6.1 Introduction

Development in the bird embryo occurs within a small and well-defined environment: the egg. Nevertheless, few recent studies of the ontogeny of motor behaviour in chicks have focused on the role of environmental factors and sensory feedback. This is for good reasons, as a variety of evidence suggests that the embryo lacks sensory input at early stages and is relatively insensitive to such input when it becomes available. For example, Hamburger's (1963; Hamburger et al. 1965) pioneering work confirmed the earlier suggestion (Preyer 1885) that movement in chick embryos begins prior to the time at which the embryo responds to sensory stimulation. Furthermore, several of the studies discussed below have suggested that removal of sensory input has relatively minor effects on the development of motor behaviour. As recent studies have added to our understanding of both the development of sensory input and the ontogeny of motor patterns in chick embryos, it seems worthwhile to re-examine the role of sensory factors in the embryonic development of motor behaviours.

Gottlieb (1968) reviewed the development of sensory systems in birds and provided evidence for prenatal onset of function. Nevertheless, there has been relatively little effort made to understand how sensory input is actually used in the production of embryonic behaviours. This is partly due to the fact that until recently it has not been possible to acquire the kind of detailed kinematic and electromyographic data that would allow the evaluation and comparison of specific motor patterns in embryos (e.g. Bekoff 1976; Landmesser and O'Donovan 1984; Bradley and Bekoff 1990; Watson and Bekoff 1990). Prior to the development of these techniques it was possible to describe embryonic behaviour in general terms such as local versus total movements, and to quantify such features as periodicity, frequency and duration of movements (e.g. Hamburger 1963; Oppenheim 1975). However, it was not feasible to describe quantitatively the actual movement patterns. In addition, there is considerable evidence, which will be discussed below, to suggest that sensory input is little used in embryos.

Department of Environmental, Population and Organismic Biology and Center for Neuroscience, University of Colorado, Boulder, CO 80309-0334, USA

M.N.O. Davies and P.R. Green (Eds.)
Perception and Motor Control in Birds

The issue of what role is played by sensory input, particularly tactile and proprioceptive, in generating motor behaviours in chick embryos will be the focus of the following discussion. Questions which will be addressed are: When do sensory inputs become available? How do embryonic inputs differ from those in adults? How are they used during on-going behaviours? How might they be involved in the development of later behaviours?

6.2 The Environment Within the Egg

The bird's eggshell and shell membranes initially evolved as mechanisms to protect the embryo from desiccation in the dry terrestrial environment. They also serve to buffer the embryo from external sources of sensory stimulation and, together with the fluid within the membranes, modify incoming stimuli. For example, both visual and mechanical stimuli are likely to be severely attenuated as they pass through these layers before reaching the sensory receptors of the embryo.

Furthermore, the fact that the embryo is, at early stages at least, floating in the amniotic fluid undoubtedly influences internal stimuli as well. For example, the rocking movements due to contractions of the amnion are likely to stimulate vestibular receptors. Immersion in fluid presumably results in tactile stimulation that differs from that experienced in air, and stimuli due to weight support are certainly absent under these conditions. As the embryo grows larger, it becomes more closely confined within the eggshell. This will impede the full range of movements to the extent that the movements may even become isometric at later stages. In addition, the embryo is confined in a particular postural configuration, which will also have effects on the sensory input. With the exception of Oppenheim (1972b) few, if any, studies have related the unique features of the embryonic environment to the development of motor behaviour. In fact, even the study by Oppenheim (1972b) showed that the frequency of embryonic movements was not related to amniotic contractions.

6.3 Embryonic Motor Behaviours

Embryonic movements have been extensively described by Hamburger and his colleagues (Hamburger 1963; Hamburger and Balaban 1963; Hamburger et al. 1965; Hamburger and Oppenheim 1967; see also Sect. 7.2). The first movements begin at 3.5 days of incubation and consist of bending of the anterior body to the side. Next the movements spread to more posterior levels and look like S-waves travelling up or down the length of the body. Shortly thereafter the movements break up and appear to become random and disorganized. By 5.5 to 6 days, the wings and legs also participate in the activity. Later, eye and beak movements begin. Jerky and apparently uncoordinated movements of all parts of the body, termed Type I embryonic motility, continue throughout the remainder of the incubation period, until the onset of hatching on day 20 or 21.

6.3.1 Type I Embryonic Motility

Type I embryonic movements are organized into bouts that occur periodically, with inactivity periods alternating with activity periods. Over the incubation period the frequency of movements changes (Oppenheim 1972b). Low levels (3–4 movements per minute) are seen at early stages, then they increase to a high level (20–21 movements per minute) at around 13 days, when the embryos appear to be almost constantly in motion. The frequency drops again to around 4 movements per minute toward the time of hatching (21 days). In addition, the durations of the activity and inactivity periods change.

Within an activity period, every part of the body that can move appears to participate. However, at the behavioural level, there is no apparent relationship between the movement of one part and that of any other (Hamburger 1963). Nevertheless, the three studies in which coordination between body parts has been examined in more detail have identified the emergence of some interlimb coordination. For example, Provine (1980; see also Sect. 7.2), used behavioural observations to show that synchronous movements of the two wings increased while synchronous movements of wing and leg decreased over the incubation period. He correlated these changes with the fact that in the adult, wing movements are bilaterally synchronous while ipsilateral leg movements and wing movements do not occur together. Cooper (1983) used electromyogram (EMG) recordings from the sartorius muscles in the two legs to show that whereas irregular patterns of coordination were seen at 7 days, by 8–10 days clear and consistent alternation was common. Finally, Watson and Bekoff (1990) used kinematic analysis of leg movements to show that at 9 days of incubation alternation was the most common (47%) pattern of interlimb coordination, although interlimb synchrony (18%) and isolated limb movements (35%) were also seen.

Within a leg, even tighter coordination is seen during Type I embryonic motility, despite the fact that behavioural observations do not indicate its presence. For example, a variety of studies using EMG recordings have shown that antagonist muscles alternate while agonists are active synchronously (Bekoff 1976; Landmesser and O'Donovan 1984; Bradley and Bekoff 1990). These results initially appeared difficult to reconcile with the behavioural observations. However, recent kinematic data have resolved this apparent conflict (Watson and Bekoff 1990). Detailed analysis of hip, knee and ankle movements in 9 and 10 day embryos showed that when more than one joint moved, activity typically began and ended synchronously. Furthermore, all active joints extended or flexed together and the movements were symmetrical. These data are consistent with the EMG results and confirm the presence of a simple and symmetrical pattern of embryonic leg movements, which has been called the "basic pattern" (Bekoff 1988).

Several variable features of the embryonic movements were also identified, which appear to account for the disorganized and jerky appearance of the behaviour. For example, there was variation in the durations of the movements and in the number of joints that participated in each movement. Movements could begin with either flexion or extension. In addition, the limb could return to its rest position between movements, or move continuously to produce sequences of variable lengths. This variability has recently been shown to be reflected in the EMG recordings from embryonic leg muscles as well (Bradley and Bekoff 1990).

6.3.2 Type II and Type III Embryonic Motility

Type II motility was initially characterized by Hamburger and Oppenheim (1967) as consisting of rapid, vigorous, jerky movements described as startles and wriggles that involve the entire body. These movements are interspersed with the Type I motility beginning at about day 11. Very little further work has been done to characterize this type of activity.

Type III motility includes movements that are characterized by very vigorous, but slow and smooth wriggles of the whole body (Hamburger and Oppenheim 1967). They include head, trunk and wing movements showing a rotatory component, and may include leg movements as well. Usually, during the performance of Type III motility, Types I and II motility are suspended (Hamburger and Oppenheim 1967; Bekoff 1976). The Type III behaviours have been described by Hamburger and Oppenheim (1967) and are usually grouped as the prehatching behaviours: tucking, membrane penetration and pipping and hatching behaviour, also called climax or Type IV behaviour.

Tucking is the behaviour in which the head is turned to the right and tucked underneath the right wing in order to attain the hatching position. It can be further divided into pretucking, tucking and post-tucking phases. It usually begins on day 17 and is completed toward the end of day 18.

Membrane penetration, the behaviour in which the beak pierces the egg membranes so that it protrudes into the air space at the blunt end of the egg, can occur anywhere between the middle of day 18 and the middle of day 20. There does not appear to be a specific behaviour designed to accomplish this. Rather it occurs as a result of the protracted performance of a variety of movements that ultimately wear a hole in the membranes.

Pipping is the cracking of the eggshell by the beak. The behaviour consists of rapid, vigorous thrusts of the head and neck. Eventually one of these cracks a hole in the shell. This may occur anywhere from day 19 to the onset of hatching.

Hatching usually begins rather abruptly and continues for about 45 to 90 minutes, until the chick has escaped from the shell. During hatching the chick thrusts its beak back against the shell while it pushes against the opposite end with its legs and rotates to the right. A typical hatching episode consists of one or two synchronous extension-flexion sequences of the legs along with head thrusts (Bekoff and Kauer 1982), lasting 1 to 3 s. These episodes are separated by periods of inactivity lasting about 20 s. The behaviour results in cracking the shell progressively further and further around until the cap is loosened enough to be pushed off.

6.4 Role of Sensory Information During Ongoing Embryonic Behaviours

The issue of what role(s) sensory information plays in embryonic motor development is an important one for understanding the mechanisms involved in the production of embryonic as well as later behaviours. This is because sensory information could potentially play one or more of several different roles in developing embryos. For example, it could be involved in the actual production of the movements and muscle activation patterns underlying specific embryonic behaviours. It could be used to trigger the onset of new behaviours, whether or not it is involved in producing those behaviours (see Sect. 6.5). In addition, it could be important in the development of behaviours that do not appear until after hatching independent of its role at earlier stages (see Sect. 6.6). Alternatively, sensory input could play no role at all in the development of motor behaviours. The first step in examining this issue is to discuss the availability of sensory information during embryonic stages.

6.4.1 What Sensory Information Is Available?

The earliest embryonic movements occur prior to the establishment of functional reflex arcs (for a review, see Hamburger 1963). Furthermore, recent work has shown that motor activity with many normal features can be recorded from embryos that have been either acutely (Landmesser and O'Donovan 1984) or chronically (Hamburger et al. 1966) deafferented to remove all tactile and proprioceptive sensory input to the spinal cord, or chronically immobilized to prevent movement-related sensory feedback (Landmesser and Szente 1986). Nevertheless, as will be discussed below, it has been shown that sensory input is available from day 7.5 of the 21-day incubation period (Lee et al. 1988; Davis et al. 1989). This is about the same time that reflexes can first be elicited (Windle and Orr 1934; Hamburger and Levi-Montalcini 1949; Hamburger and Balaban 1963; Oppenheim 1972a). These results raise several questions. For example, are there any deficits when sensory input is removed? Are any changes in ongoing motor behaviour seen when specific sources of sensory input first become available? In other words, how is the sensory input that is available actually used by the embryo? As will be seen, we are still far from being able to answer these questions.

Oppenheim (1972a) found that responses to proprioceptive stimuli appear at 7.5 days in all regions of the body that were tested (e.g. distal and proximal wings and legs, beak and dorsal trunk). Light tapping or flipping was used as a proprioceptive stimulus. More recent anatomical and electrophysiological studies have concentrated on the lumbosacral, or leg-innervating, region of the spinal cord. These studies have found that sensory fibres first reach lumbosacral motor neuron dendrites at 7.5 days (Davis et al. 1989), and that monosynaptic EPSPs can first be detected in the lumbosacral ventral roots after stimulation of the sensory afferents at this stage (Lee et al. 1988). These findings correlate well with the behavioural results.

Based on observations of the behavioural response to stroking the skin, cutaneous sensitivity to tactile input first appears in the beak and oral region in chick embryos at 7.5 days (Oppenheim 1972a). Cutaneous sensitivity was detected at 8.5 days in the other regions of the body tested, including the legs. Dense projections to the dorsal laminae of the lumbosacral spinal cord, which are thought to represent cutaneous afferents, have not been seen until a somewhat later stage (13 days; Davis et al. 1989). The time at which motor neuron responses to cutaneous afferents are first seen has not yet been determined in chick embryos.

6.4.2 How Is Sensory Information Used?

While it is true that sensory input is available from at least 7.5 days of incubation, it is not clear in what way, or to what extent, it is used by the embryo. Oppenheim (1972a, b) has shown that stimulation that would be expected to activate both tactile and proprioceptive receptors does not increase the frequency, periodicity or other qualitative aspects of embryonic movements between embryonic days 5 and 19. There is, in fact, a strong tendency for the embryo to stop responding to a stimulus after only one or two consecutive presentations (Oppenheim 1972a). These results suggest that tactile and proprioceptive stimuli do not play a role in eliciting movements in the normal embryo. On the other hand, the effects on other characteristics of embryonic movements, such as amplitude, duration, sequential organization, and their role in shaping the precise movements or muscle activation patterns, have not been examined (Oppenheim 1972b). Furthermore, this issue has not yet been addressed during the prehatching and hatching behaviours at later stages of incubation.

Limb deafferentation (Hamburger et al. 1966; Landmesser and O'Donovan 1984) and chronic immobilization (Landmesser and Szente 1986) studies have shown that many aspects of embryonic motor behaviour are maintained in the absence of sensory input. For example, the behavioural study showed that periodicity and gross appearance of the movements were unchanged until at least 15 days of incubation (Hamburger et al. 1966). The studies in which EMGs were recorded in an isolated spinal cord-hind limb preparation showed that extensor and flexor muscles exhibited normal patterns of alternation (Landmesser and O'Donovan 1984; Landmesser and Szente 1986). Nevertheless, these EMG studies examined only a small range of ages, from 9 to 12 days.

It is also clear that motor output recorded from isolated spinal cord preparations (Landmesser and O'Donovan 1984; O'Donovan and Landmesser 1987), which lack normal patterns of sensory and descending input, is not identical to that recorded from intact embryos in ovo (Bekoff 1976; Landmesser and O'Donovan 1984; Bradley and Bekoff 1990). For example, synchronous bursts in all muscles begin nearly every cycle of muscle activity in the isolated spinal cord preparations, whereas they occur less often in intact embryos *in ovo*. Many other aspects of the motor patterns, including duration and phasing of muscle bursts are more complex in intact embryos. In addition, spontaneous episodes of activity occur less frequently in vitro (O'Donovan 1987). Since descending input has been shown to have a much smaller role in modulating motor output in post-hatching chicks during walking than does sensory input (Bekoff et al. 1987, 1989) we might suspect that it is the lack of sensory input that is primarily responsible for the differences seen in embryos as well. However, further experimental analysis will be necessary to determine whether this is the case. Other factors, such as differences in temperature between the perfusion medium used in the isolated spinal cord preparation and the blood of the intact embryo, may also play a role.

Another finding which is consistent with the view that sensory input may play a role in shaping embryonic motor patterns is the observation that the motor patterns of the embryos become more complex toward the middle of the incubation period (Bekoff 1976). As discussed earlier, this is after sensory input has been shown to be available.

6.5 Role of Sensory Input at Transitions in Behaviour

Nothing is known about what might trigger the onset of the Type III prehatching behaviours at around 17 days of incubation. While there appears to be relatively regular alternation of periods of Type I motility with bouts of tucking on day 17 (Bekoff 1976), nothing is known about control of this alternation. These are fertile areas for further investigation.

There is some information available on the control of the initiation of hatching, which normally occurs on days 20 or 21 of incubation. For example, bending the neck to the right as it is in the normal hatching position results in asymmetric stimulation of neck proprioceptors and this has been shown to turn on hatching in post-hatching chicks (Bekoff and Kauer 1982; Bekoff and Sabichi 1987). This signal is effective even in older chickens, up to 61 days post-hatching (Bekoff and Kauer 1984). Furthermore, releasing the neck from the hatching position, which normally occurs when the chick pushes the cap of the eggshell off and emerges from the egg, terminates hatching.

Nevertheless, because the chick achieves the hatching position several days before the onset of hatching behaviour (Hamburger and Oppenheim 1967; Kovach 1968; Oppenheim 1973), bending of the neck cannot be the only signal required for the initiation of hatching. It seems likely that a neurochemical signal is used to facilitate or disinhibit the input from the neck on day 20 or 21 (Bekoff and Kauer 1982).

6.6 Role of Prior Sensory Input in Development of Later Behaviours

A few studies have altered sensory input at early stages of development in chick embryos and have assessed the impact on motor behaviours seen at later stages. For example, Landmesser and Szente (1986) eliminated movement-related sensory feedback by immobilizing chick embryos with d-tubocurarine from embryonic days 6 to 12. They assessed motor output in an isolated spinal cord-hind limb preparation. Flexor and extensor muscles were found to alternate as they did in controls. Thus the basic pattern of muscle activation appears to develop normally in the absence of movement-related sensory input. Whether other aspects of the pattern, which are seen in the more complex activity recorded from intact embryos in ovo, are also retained has not yet been examined. This issue is complicated by the fact that immobilization results in ankylosis of the joints and atrophy of the muscles of the limbs (Drachman and Sokoloff 1966; see also Sect. 7.4). Therefore, sensory feedback might be abnormal in these chicks even if there is no impact of the immobilization on the development of other aspects of the neural pattern generating circuitry. Furthermore, drug studies are subject to the criticism that the drug may have interfered with central synapses (Landmesser and Szente 1986).

Another approach is to alter the sensory experience of a chick in other ways and observe the effect on later behaviours. For example, if a chick is removed from the shell just prior to the onset of hatching, does walking develop normally? Preliminary results from our lab suggest that there may be a slight delay in the appearance of adult walking, but that there are no permanent deficits (Bekoff and Nicholl unpubl. results). On the other hand, if weight support is eliminated for 24 to 48 h post-hatching, there appear to be long-lasting alterations in the walking motor pattern (Bekoff and Herrera unpubl. results). It is not yet clear whether these alterations are permanent or not. Further research in this area will undoubtedly be useful.

6.7 Conclusions

It is clear that both proprioceptive and tactile sensory input as well as motor patterns develop very early in ontogeny in bird embryos. It is also evident that the early motor patterns develop prior to the onset of sensory function. Furthermore, at least some aspects of the motor patterns are maintained in the absence of sensory input. Nevertheless, it is still not known how, or if, sensory input is used when it is available. Work on adults increasingly suggests that motor behaviour is produced by multifunctional neuronal circuits, in which sensory input is often included as an integral part (Harris-Warrick and Johnson 1989). It therefore seems unlikely that sensory input and motor behaviour develop independently of one another. Understanding the earliest integration of sensory input and motor output in bird embryos remains a challenging area for future research.

Acknowledgements. My recent work has been generously supported by NIH grants NS20310 and HD28247.

References

Bekoff A (1976) Ontogeny of leg motor output in the chick embryo: a neural analysis. Brain Res 106:271–291

Bekoff A (1988) Embryonic motor output and movement patterns: Relationship to postnatal behavior. In: Smotherman WP, Robinson SR (eds) Behaviour of the fetus. Telford, Caldwell, NJ, pp 191–206

Bekoff A, Kauer JA (1982) Neural control of hatching: Role of neck position in turning on hatching leg movements in post-hatching chicks. J Comp Physiol 145:497–504

Bekoff A, Kauer JA (1984) Neural control of hatching: Fate of the pattern generator for the leg movements of hatching in posthatching chicks. J Neurosci 11:2659–2666

Bekoff A, Sabichi AL (1987) Sensory of the initiation of hatching in chicks: Effects of a local anesthetic injected into the neck. Dev Psychobiol 20:489–495

Bekoff A, Nusbaum MP, Sabichi AL, Clifford M (1987) Neural control of limb coordination. I. Comparison of hatching and walking leg motor output patterns in normal and deafferented chicks. J Neurosci 7:2320–2330

Bekoff A, Kauer JA, Fulstone A, Summers TR (1989) Neural control of limb coordination. II. Hatching and walking motor output patterns in the absence of input from the brain. Exp Brain Res 74:609–617

Bradley NS, Bekoff A (1990) Development of coordinated movement in chicks: I. Temporal analysis of hindlimb synergies at embryonic days 9 and 10. Dev Psychobiol 23:763–782

Cooper MW (1983) Development of interlimb coordination in the embryonic chick. Thesis, Yale University, New Haven

Davis BM, Frank E, Johnson FA, Scott SA (1989) Development of central projections of lumbosacral sensory neurons in the chick. J Comp Neurol 279:556–566

Drachman DB, Sokoloff L (1966) The role of movement in embryonic joint development. Dev Biol 14:401–420

Gottlieb G (1968) Prenatal behavior of birds. Q Rev Biol 43:148–174

Hamburger V (1963) Some aspects of the embryology of behavior. Q Rev Biol 38:342–365

Hamburger V, Balaban M (1963) Observations and experiments on spontaneous rhythmical behavior in the chick embryo. Dev Biol 7:533–545

Hamburger V, Levi-Montalicini R (1949) Proliferation, differentiation and degeneration in the spinal ganglia of the chick embryo under normal and experimental conditions. J Exp Zool 111:457–502

Hamburger V, Oppenheim RW (1967) Prehatching motility and hatching behavior in the chick. J Exp Zool 166:171–204

Hamburger V, Balaban M, Oppenheim R, Wenger E (1965) Periodic motility of normal and spinal chick embryos between 8 and 17 days of incubation. J Exp Zool 159:1–14

Hamburger V, Wenger E, Oppenheim R (1966) Motility in the absence of sensory input. J Exp Zool 162:133–160

Harris-Warrick RM, Johnson BR (1989) Motor pattern networks: Flexible foundations for rhythmic

pattern production. In: Carew TJ, Kelley DB (eds) Perspectives in neural systems and behavior. MBL Lectures in Biology, vol 10. Liss, New York, pp 51–71
Kovach JK (1968) Spatial orientation of the chick embryo during the last five days of incubation. J Comp Physiol Psychol 73:392–406
Landmesser LT, O'Donovan MJ (1984) Activation patterns of embryonic chiek hindlimb muscles recorded in ovo and in an isolated spinal cord preparation. J Physiol (Lond) 347:189–204
Landmesser LT, Szente M (1986) Activation patterns of embryonic chick hindlimb muscles following blockade of activity and motoneurone cell death. J Physiol (Lond) 380:157–174
Lee Mt, Koebbe MJ, O'Donovan MJ (1988) The development of sensorimotor synaptic connections in the lumbosacral cord of the chick embryo. J Neurosci 8:2530–2543
O'Donovan MJ (1987) In vitro methods for the analysis of motor function in the developing spinal cord of the chick embryo. Med Sci Sports Exercise 19:S130–S133
O'Donovan MJ, Landmesser LT (1987) The development of hindlimb motor activity studied in the isolated spinal cord of the chick embryo. J Neurosci 7:3256–3264
Oppenheim RW (1972a) An experimental investigation of the possible role of tactile and proprioceptive stimulation in certain aspects of embryonic behavior in the chick. Dev Psychobiol 5:71–91
Oppenheim RW (1972b) Embryology of behaviour in birds: a critical review of the role of sensory stimulation in embryonic movement. Proc Int Ornithol Congr 15:283–302
Oppenheim RW (1973) Prehatching and hatching behavior: a comparative and physiological consideration. In: Gottlieb G (ed) Behavioral embryology. Academic Press, New York, pp 163–244
Oppenheim RW (1975) The role of supraspinal input in embryonic motility: A re-examination in the chick. J Comp Neurol 160:37–50
Preyer W (1885) Spezielle Physiologie des Embryo. Grieben, Leipzig
Provine RR (1980) Development of between-limb movement synchronization in the chick embryo. Dev Psychobiol 13:151–163
Watson SJ, Bekoff A (1990) A kinematic analysis of hindlimb motility in 9- and 10-day old chick embryos. J Neurobiol 21:651–660
Windle WF, Orr DW (1934) The development of behaviour in chick embryos: Spinal cord structure correlated with early somatic motility. J Comp Neurol 60:287–308

7 Pre- and Postnatal Development of Wing-Flapping and Flight in Birds: Embryological, Comparative and Evolutionary Perspectives

R.R. Provine

7.1 Introduction

Bird flight has attracted scientific and popular interest for generations. Substantial progress has been made toward understanding the aerodynamics, mechanics, anatomy and physiology of avian flight (e.g. Greenewalt 1962, 1975; Tucker 1969; Ruppell 1977; Rayner 1979; Viscor and Fuster 1987; Norberg 1989; see Chap. 11). The phylogenetic problem of how flight evolved has been more controversial and less susceptible to analysis (Ostrom 1979; Feduccia 1980; Caple et al. 1983; Lewin 1983; Norberg 1989). The primary concern of the present chapter is the question of how bird flight develops within the life span of the individual. This ontogenetic problem involves the consideration of evidence from embryology, neurophysiology, neuroanatomy, genetics, natural history and comparative/evolutionary studies, and the answer has implications beyond the immediate context of bird flight (Provine 1986, 1988). The developmental analysis of wing-flapping is a good starting point from which to investigate general issues of how behaviour emerges during embryogenesis and how it is shaped by artificial and natural selection.

7.2 Prenatal Development of Spontaneous Wing-Flapping

The embryo of the white leghorn chicken (*Gallus domesticus*) was selected for prenatal analysis of wing-flapping because it is widely available, its development is well understood, its behaviour is easy to observe, and its movements reflect internal neural events (Provine 1972a; Ripley and Provine 1972). The first embryonic movements are lateral nods of the head performed at 3.5 to 4 days after laying, 17 days prior to hatching (Hamburger 1963; Oppenheim 1974; Provine 1980). At this time, the limbs are only small buds on the lateral trunk, and are incapable of independent movement. Other body parts become active in a roughly rostrocaudal order, as in many other vertebrate embryos (see Chap. 6). As more body parts become motile, the entire trunk performs slow lateral

Department of Psychology, University of Maryland Baltimore County, Baltimore, MD 21228-5398, USA

M.N.O. Davies and P.R. Green (Eds.)
Perception and Motor Control in Birds

flexions..These flexions may be initiated at any level along the long axis of the trunk and then may be propagated either toward the head, the tail, or in both directions simultaneously.

At 6 to 7 days after laying, the stubby, paddle-like wings and legs become motile, trunk movements become more complex, and the earlier flexions are replaced by faster and jerkier generalized movement that seems to involve all parts of the body in irregular sequences (Provine 1980). The jerky movements predominate until a few days before hatching, when they alternate with the smooth and co-ordinated movements involved in breaking out of the shell (Hamburger and Oppenheim 1967; Provine 1972b). The frequency of the spontaneous, jerky movements increases to a peak at the middle of the incubation period and then declines until hatching (Fig. 7.1).

At the earliest stages of limb activity, movements of pairs of wings or legs are no more synchronized than those of ipsilateral wing and leg (Fig. 7.2). Thus, a wing is no more likely to move synchronously with the other wing than with the leg on the same side. During embryonic development, synchronization between pairs of legs and wings increases, while that between ipsilateral wing and leg decreases, especially after 15 to 17 days (Fig. 7.2). The relatively late onset of coordination between the limbs may help to reconcile Bekoff's (1976) discovery, by electromyography, of early synchronization within and among the joints with the visual impression of jerkiness in most embryonic movements (Hambur-

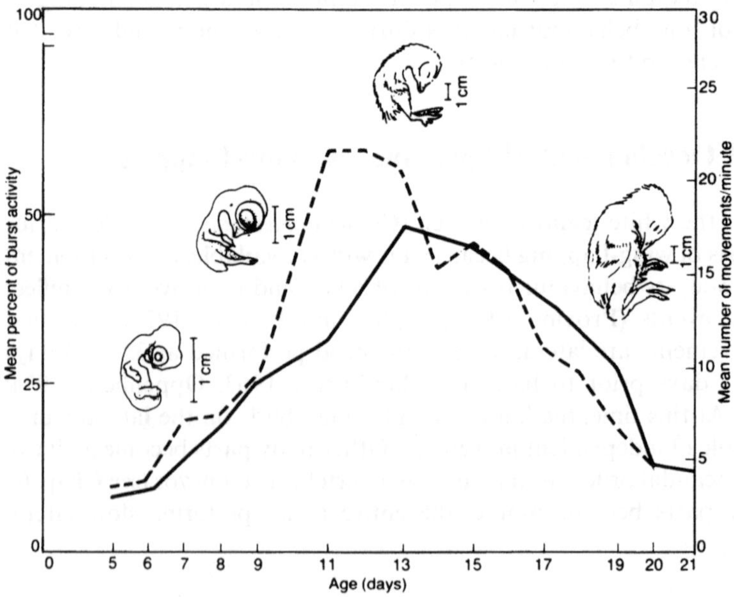

Fig. 7.1. *Solid line* the percentage of time during which spinal cord burst discharges were present at different ages of prenatal chicks; *dotted line* the mean number of body movements reported by Oppenheim (1974) (Provine 1972a)

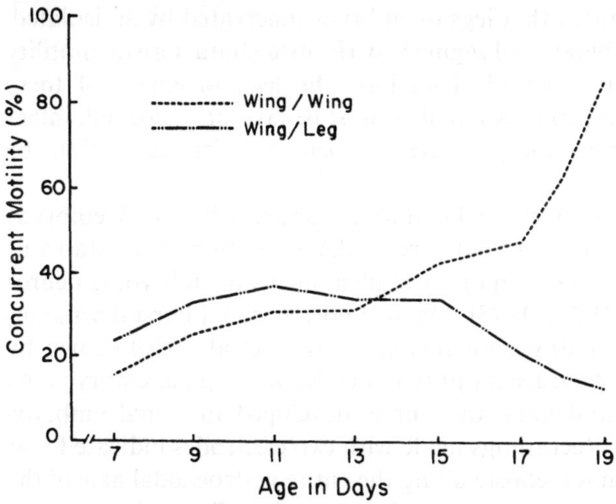

Fig. 7.2. Synchronization between limb pairs as percentages of the total duration of limb movement, as a function of embryonic age. *Dashed line* ipsilateral wing and leg; *dotted line* right and left wings (Provine 1980)

ger and Oppenheim 1967). Coordination of muscles within limbs may appear before coordination of muscles between limbs. By the time the chick hatches, synchronized leg movements that resemble walking and wing movements that resemble flapping are present. Some such movements are performed as early as a week or more before hatching (Provine 1980).

Further consideration of limb motility should take place within the broader context of embryonic neuromuscular development. Embryos are not miniature adults. Because the mechanisms and contexts of behaviour are different before and after hatching[1], caution should be exercised when extending research (e.g. studies of coordination) from the postnatal into the prenatal period. Many functional and structural properties of embryos are adaptations to the prenatal environment that do not have postnatal counterparts (Oppenheim 1981; Provine 1988, 1993).

7.3 Neural Basis of Embryonic Behaviour

William Preyer (1885) discovered that a chick embryo moves several days *before* a tactile reflex can be evoked. Preyer correctly speculated that embryonic movements originate in spontaneous activity, an idea confirmed by Hamburger

[1] Unlike much prenatal behaviour, the function of which may be the mere act of moving, hatching requires a degree of coordination and precision that approximates that of post-natal behaviour. Shell-related feedback, for example, controls the duration of the rotatory component of hatching, maximizing the probability of escape from the shell (Provine 1972b). This theme was developed in subsequent work by Bekoff (see Chap. 6).

et al. (1966), who demonstrated that legs of embryos innervated by an isolated, deafferented lumbosacral spinal cord segment performed characteristic motility during most of the prenatal period. Therefore, the leg movement of these deafferented, spinal embryos, and presumably most behaviour of normal, intact embryos, is both spontaneous and generated by neuronal circuitry within the spinal cord.

Electrical recordings from paralyzed (curarized) and moving chick embryos confirmed the contribution of the spinal cord to the generation of spontaneous movement and allowed the detection of novel phenomena of embryonic neurophysiology (Provine 1971, 1972a, 1973). Massive polyneuronal burst discharges were identified within the ventral cord region (Provine et al. 1970) shown by Hamburger et al. (1966) to produce motility in deafferented spinal embryos. As suggested by the behavioural data, the bursts developed in spinal embryos (Provine and Rogers 1977). Recordings made with two electrodes indicated that bursts occurred almost simultaneously along the entire rostrocaudal axis of the cord (Provine 1971). The bursts occurred at a frequency (Fig. 7.1) and periodicity typical of embryonic movement. The bursts were synchronized with embryonic movements in motile embryos and with motor nerve discharges in curarized preparations (Ripley and Provine 1972). When a burst occurred, the embryo moved; embryonic movements never occurred in the absence of a simultaneous burst.

The spinal cord burst discharges and associated phenomena are probably unique to the embryo, as noted by Landmesser and O'Donovan (1984) and O'Donovan and Landmesser (1987) in their rediscovery of some of these phenomena. Indeed, after hatching, spontaneous cord bursting and correlated muscular twitches would produce chaotic behaviour upon which reflexes and voluntary movements would be superimposed (Provine 1988, 1993). Such seizure-like movements would be lethal because of their interference with pulmonary respiration and other adaptive movements.

Electromyographic (EMG) investigations are useful in understanding how the spinal cord bursts translate into the contraction of specific muscles (Bekoff et al. 1975; Bekoff 1976; Landmesser and O'Donovan 1984; see Chap. 6). However, such studies have tended to neglect the unique neurophysiological process generating the activity that they observe. EMG data indicate that embryonic spinal cord circuitry produces non-random patterns of motor outflow to the leg at pre-reflex stages that is driven by the massive and widely distributed spinal cord burst discharges described here. The precision of the pattern of motor outflow continues to increase and its "noisiness" decreases during development.

Although embryonic behaviour is driven by massive spontaneous cord discharges, the EMG work by Bekoff and others indicates that motor outflow is shaped in varying degrees by central pattern generating circuits. The most convincing support for a spinal locus for the pattern generator for wing-flapping comes from elegant and technically difficult transplantation studies by Straznicky (1963) and Narayanan and Hamburger (1971). At 2.5 days of embryonic age, they removed the lumbosacral segment of spinal cord that normally innervates the legs and substituted a segment of brachial cord, taken from a donor embryo, that normally innervates the wings. Both before and after hatching, recipient chicks moved their wings and legs synchronously as in flapping, demonstrating that flapping is produced by circuits within the spinal cord adjacent to the wings, which are not modified by feedback from the limb. Furthermore, these neural circuits must be irreversibly determined soon after the closure of the neural tube, the time when the cord segments were transplanted.

7.4 Effect of Spontaneous Embryonic Behaviour on Muscle and Joint Development

Embryonic movement has novel morphological consequences that are often overlooked by behavioural investigators. For example, as little as 24 to 48 h of curare-induced paralysis produces permanent malformation of joints and atrophy of muscles in the chick embryo (Drachman and Sokoloff 1966; see Sect. 6.6). Therefore, early movement is necessary for the normal development of muscles and joints. This result also suggests that the precision of fit between the components of ball and socket joints is the result of a sculpting process produced by the constant movement of the joint while it is being formed. Behaviourally driven processes of this sort force a re-evaluation of what it means for a behavicur, structure, or process to be "innate" or "genetically determined." Joint development is an epigenetic process driven by the nervous system and its product, behaviour.

At a more molecular level, motor neuron activity influences the differentiation of muscle types. Adult vertebrates have slow-twitch and fast-twitch muscles that differ in contraction properties, morphology, biochemistry, and pattern of innervation. Slow-twitch muscles are important for the maintenance of posture; fast-twitch muscles are used for bursts of rapid movement. In both birds and mammals, the nerves innervating the embryonic muscles influence the expression of fast and slow muscle properties (Vrbova et al. 1978). For example, the cross-innervation of fast muscles with nerves that typically innervate slow muscles redirects differentiation such that the prospective fast muscles assume the properties of slow muscles and vice versa. Synaptic activity and/or muscle contraction at the myoneural junction are implicated because the modification of such activity produces similar changes in muscle properties in the absence of cross-innervation.

7.5 Naturally Occurring Motor Neuron Death

During normal vertebrate embryonic development, thousands of apparently normal spinal cord motor neurons start to develop and then die (Hamburger and Oppenheim 1982). In the chick embryo, about 40% of limb-innervating lateral motor neurons die during the first third of incubation. This phenomenon is not unique to avians, because it is also present in the phylogenetically remote snapping turtle (McKay et al. 1987).

The proportion of motor neurons that die spontaneously during normal development can be manipulated by increasing or decreasing the size of skeletal muscle mass being innervated. The removal of a limb bud at early developmental stages results in the death of almost all limb-innervating motor neurons. The complementary procedure of adding of supernumerary (extra) limb-bud saves many of the motor neurons that normally would have died. A given motor neuron is not destined to die; dying motor neurons presumably lose out in a competition between motor neurons for limited muscle innervation sites or trophic agents (Hamburger and Oppenheim 1982). In contrast, post-mitotic spinal interneurons providing input to motor neurons and other neurons do not die spontaneously or in response to the removal of either the target of innervation or their afferents (McKay and Oppenheim 1991).

Synaptic activity and perhaps behaviour play a role in the determination of motor neuron numbers (Pittman and Oppenheim 1978, 1979). The pre-synaptic and post-synaptic blockade of neuromuscular transmission with curare and curare-like drugs during the period of normal motor neuron death prevents most motor neuron loss. A complementary study showed that motor neuron loss can be enhanced by experimentally increasing neuromuscular activity by the direct electrical stimulation of peripheral nerves and limb muscles (Oppenheim and Nunez 1982). Further, more specific evidence for the involvement of behaviour in the regulation of motor neuron numbers is suggested by the coincidence of the onset of motor neuron death and limb movement (cf. Sect. 7.11).

7.6 Comparative Development of Wing-Flapping and Flight: Effects of Domestication

By the end of the embryonic development of wing movements described in Sect. 7.2, the wings produce primarily smooth, low amplitude, bilaterally synchronized movements. However, the relation between these prenatal movements and mature wing-flapping is uncertain. The developmental analysis of wing movement has been extended to postnatal stages by a comparative investigation of the emergence of lateral flight and wing-flapping, conducted in parallel with a developmental study of wing morphology (Provine 1981a; Provine et al. 1984).

In these studies, the development of four fowl was compared to reveal common development phenomena and to evaluate the effects of domestication (Lorenz 1935/1970, 1940, 1954/1971; Boice 1972, 1973; Desforges and Wood-Gush 1975a, b, 1976; Heaton 1976; Miller 1977). These were the red jungle fowl (*Gallus gallus*), white leghorn and Cornish X rock strains of domestic chickens (*Gallus domesticus*), and the Japanese quail (*Coturnix japonica*). The red jungle fowl (JF) is a light-bodied chicken presumed to be ancestral to contemporary domestic chickens (Darwin 1892; Beebe 1926; Zeuner 1963). The white leghorn (WL) and Cornish X rock (CR) are, respectively, light- and heavy-bodied domestic chickens. The Japanese quail (JQ) is also a domestic strain, but is much smaller and a much better flyer than any of the chickens.

The development of flight in these four fowl was first assessed by measuring the lateral distance flown by birds of different ages when dropped from a height of 1.7 m (Provine 1981a; Provine et al. 1984). Lateral flight first developed in WL, JF, and JQ between 7 and 9 days and was well established by 13 days (Fig. 7.3). The lateral flight of CR was feeble and late-appearing. These lateral flight scores for cage-reared birds probably underestimate the flight performance of birds allowed to roam freely and obtain more wing-flapping and flight experience and exercise.

Lateral flight of the JQ, WL and JF developed most rapidly during the second week of postnatal life. The development of drop-evoked wing-flapping was observed during the period of emerging flight. Wing-flapping was evoked by dropping birds 1.9 m using a tethering device that provided a controlled and equal time of descent for all birds (approximately 540 ms), and flapping frequency was determined strobophotographically. Vigorous, bilaterally symmetrical wing-flapping was evoked in all four bird types when tested on the day of hatching, at least 7 days before it is functional in producing lateral flight. All chickens (WL, JF, CR) nearly doubled their wing-flapping rates during the 2 weeks after hatching (Fig. 7.4), whereas the JQ had a high initial flapping rate that was maintained during the first 3 weeks. Despite this higher initial flapping

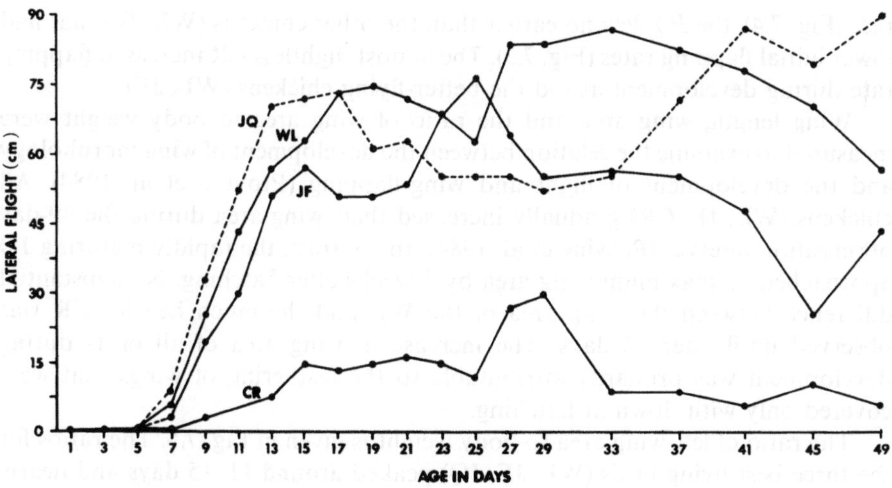

Fig. 7.3. Mean lateral flight distances as a function of age in young birds dropped from a height of 1.7 m. *JQ* Japanese quail; *WL* white leghorn chicken; *JF* red jungle fowl; *CR* Cornish X rock chicken. The asymptote in lateral flight scores of the three best fliers (*WL, JF, JQ*) after 13–21 days is in part due to a ceiling in the scoring system and the tendency of some birds to flutter straight down instead of demonstrating their maximum lateral flight capacity (Provine et al. 1984)

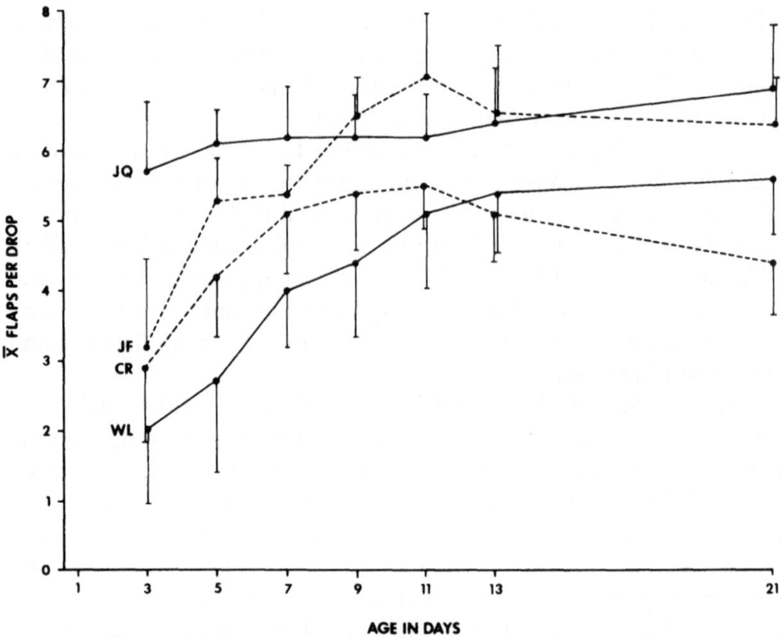

Fig. 7.4. Mean drop-evoked flapping rate measured as flaps per drop (ca. 540 ms) of the left wing, as a function of age. *Bars* represent standard deviations; *JQ, WL, CR, JF* see caption to Fig. 7.3 (Provine et al. 1984)

rate (Fig. 7.4), the JQ flew no earlier than the other chickens (WL, JF) that had lower initial flapping rates (Fig. 7.3). The almost flightless CR increased flapping rate during development as did the better-flying chickens (WL, JF).

Wing length, wing area and the ratio of wing area to body weight were measured to examine the relation between the development of wing morphology and the development of flight and wing-flapping (Provine et al. 1984). All chickens (WL, JF, CR) gradually increased their wing area during the 49-day observation interval (Provine et al. 1984). In contrast, the rapidly maturing JQ approached its maximum wing area by 3 weeks after hatching. No substantial difference between the wing area of the WL and the much heavier CR was observed until after 33 days. The increase in wing area of all birds during development was primarily attributable to the feathering of wings that were covered only with down at hatching.

The ratio of left-wing area to body weight is given in Fig. 7.5. The ratios for the three best flying birds (WL, JF, JQ) peaked around 11–15 days and nearly overlapped at most ages. The low ratio of wing area to body weight in the very heavy CR precluded effective lateral flight despite vigorous wing-flapping (Fig. 7.4).

These studies revealed several general features of the development of flight in fowl. All birds performed drop-evoked, bilaterally synchronized wing-flapping

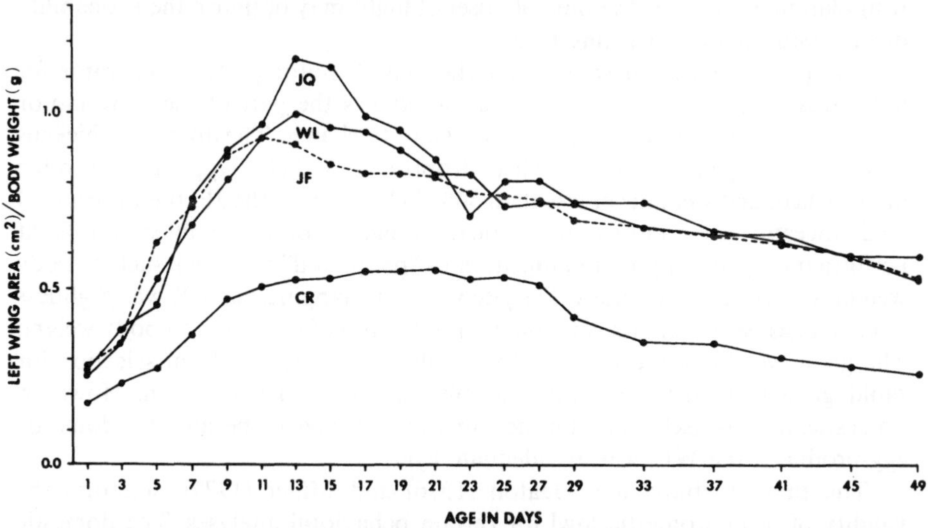

Fig. 7.5. Mean ratio of left-wing area (cm^2) to body weight (g) as a function of age. *JQ*, *WL*, *CR*, *JF* see caption to Fig. 7.3 (Provine et al. 1984)

on the day of hatching, in accordance with other evidence that practice and functional significance are not necessary for the development of wing-flapping (Narayanan and Malloy 1974a, b; Provine 1979, 1980, 1981a, b, 1982a) and other motor behaviour (Hamburger et al. 1966; Bentley and Hoy 1970; Fentress 1973; Bekoff et al. 1975; Grillner 1975; Bekoff 1976). Another general finding was that the wing-flapping frequency of the three chickens (JF, WL, CR) increased approximately twofold during the first 2–3 weeks after hatching (see also Provine 1981a). This is consistent with reports that limb-stroke frequency increases during the development of swimming in rats (Bekoff and Trainer 1979) and wing-beating in locusts (Altman 1975; Kutsch 1976; Gewecke and Kutsch 1979; Kutsch and Gewecke 1979), moths (Kammer and Rheuben 1976; Kammer and Kinnamon 1979), crickets (Bentley and Hoy 1970), and grasshoppers (Kutsch 1976). The present finding that the wing-flapping rate of the JQ remained stable during the 2–3 weeks after hatching may not be an exception; the increase in the output frequency of the pattern generator for wing-flapping may have increased, unobserved, at prenatal stages.

Another general result was that the onset of lateral flight was correlated with a burst in the development of wing area relative to body weight. The increase in area was primarily due to the appearance of flight feathers on the previously down-covered wings. Peaks in the ratio of wing area to body weight in JQ, JF, and WL occurred at 11–15 days when lateral flight was becoming prominent and when vigorous and rapid wing-flapping rates were present. The CR was an exceptional case because its ratio of wing area to body weight was so low that flight was impossible at most ages. The peaking of the area to weight ratios and wing-flapping rate near the time of onset of flight may optimize the probability of successful flight in a young bird.

The present research shows that the neural motor pattern generator for flight was affected little by domestication, and is the part of the wing motor system that is most conservative and resistant to change. The domestic chickens (WL, CR) flapped vigorously when dropped and their flapping rates often approached, and were never less than 30% below that of the JF, their presumed wild ancestor. In contrast, domestication had a dramatic affect on flight performance, primarily by modification of "peripheral" variables such as body weight and wing morphology. Despite vigorous flapping, the CR was flightless at most ages because of its very unfavourable ratio of wing area to body weight. The great weight of the CR was the result of intensive artificial selection for rapid growth, high meat yield, and efficiency of feed conversion. However, domestication by itself had little effect upon flight capacity because the domestic egg-producer, the WL, was an adequate flyer.

The present study, like Heaton (1976) and Miller (1977), supports the validity of using domestic fowl for certain behavioral analyses. The domestic chicken is not necessarily behaviourally degenerate, as Lorenz (1935/1970, 1940, 1954/1971) supposed, nor is its use in the study of wing-flapping a "bad habit" (Lockhard 1968). This conclusion supports Boice's (1973, p. 26) contention, based upon rat studies, that "domestication is not necessarily equivalent to degeneracy".

These data also have implications for studies of the evolution and aerodynamics of avian flight. For example, by describing the combinations of wing-flapping frequency, wing area and body weight present at the development of flight onset, insights can be gained about the values critical for flight. In addition, performance data about birds at different stages after flight onset provide a novel opportunity to evaluate the flight-related effects of a wide variety of interacting behavioural and morphological variables within the life of a *single individual.* Previous approaches to such problems have resorted to comparative analyses and modelling (e.g. Norberg 1989).

7.7 Experimental Studies of the Postnatal Development of Wing-Flapping and Flight

Sensory input and feedback are not required for the development and performance of movement in many vertebrate and invertebrate species (Provine 1986). In the case of the chick embryo, the evidence outlined in Sect. 7.3 shows that motility originates within the central nervous system, and that endogenous spinal cord processes play an important part in its generation. However, few comparable studies evaluate the extent to which experiential factors influence the development of adult motility patterns in the chick or other avians (Spalding 1875; Grohmann 1939; Krischke 1983). The studies described in this section evaluate the consequence of various experimental interventions on post-natal wing behaviour.

First, the influence of post-hatching wing movement on the development of wing-flapping and flight of the chick was evaluated by immobilizing both wings with an elastic bandage from the day after hatching until day 13. At this age, the wing-flapping rates and lateral flight of restrained chicks were not significantly different from control values shown in Figs. 7.3 and 7.4. Thus, post-hatching wing movement is not necessary for the development of typical amounts of lateral flight and wing-flapping. This finding agrees with the results of similar studies conducted with other avian species by Spalding (1875), Grohmann (1939), and Krischke (1983). The possibility that isometric contractions occurred in the flight muscles of the immobilized birds does not alter this conclusion, although the contractions may reduce atrophy of the immobilized musculature.

Second, the role of peripheral wing structures, sensory input, and practice in the development of wing-flapping was evaluated by comparing the "wing-flapping" rate of 13 day-old chicks that had undergone bilateral wing amputation on the day after hatching with that of intact control chicks (Provine 1979). "Wing-flapping" of wingless chicks was made visible by attaching prostheses (lengths of soda straws) to the stumps of the amputated limbs.

Stroboscopic photographs summarize the flapping behaviour of an intact 13 day-old control chick (Fig. 7.6, left) and a chick whose wings were amputated on day 1 and replaced by wing prostheses immediately before testing (Fig. 7.6, right). No significant difference was detected between the flapping rates of winged and wingless chicks, showing that post-natal flapping experience and

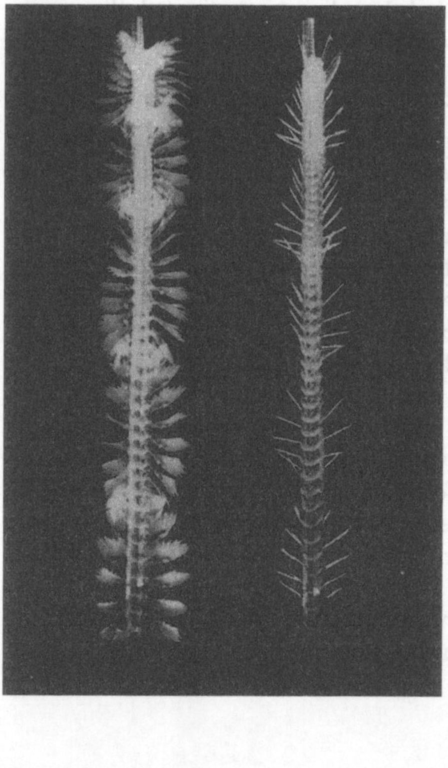

Fig. 7.6. Stroboscopic photographs (100 Hz) of drop-evoked wing-flapping at 13 days post-hatching of an intact chick (*left*) and a wingless chick with wing prosthesis (*right*). (From Provine 1979)

trophic and sensory influences from the peripheral wing are not necessary for the development of basic, bilaterally coordinated, drop-evoked wing-flapping movements. Practice and learning do not play significant roles in the development of flapping because the movement emerges at the normal time in wingless chicks where it has no obvious behavioural consequence. This does not deny a possible impact of these variables upon other more complex and subtle motor patterns involved in flight.

In a third experiment, two lines of chickens, flightless because they lacked flight features, were used to evaluate the influence of flight-related experience and adaptive significance on the development of wing-flapping. Using such flightless chickens provides a non-traumatic alternative to amputation, deafferentation, and immobilization (Provine 1979, 1981b). The two lines investigated were "scaleless" (Abbott and Asmundson 1957) and "delayed feathering" (Somes 1969). The scaleless condition results from the action of an autosomal recessive gene (sc) with full penetrance. Homozygous (sc/sc) scaleless birds were born almost completely naked and remained so throughout the study. "Delayed feathering" refers to chickens possessing a dominant, sex-linked gene at the K locus (K^n) which controls feather growth rate. All K^n chickens hatched with normal down but failed to develop normal feathers during the study.

Chicks from both featherless, flightless lines did not differ from controls in their rates of wing-flapping at 13 days. At this age, normal chicks flap at adult rates, and the development of wing-flapping therefore cannot depend upon flight-related behavioural consequences.

Two additional experiments with flightless lines investigated the mechanisms which initiate drop-evoked wing-flapping. Flightless chicks were dropped either in darkness or with their eyes taped shut, and were found to flap vigorously, at the same rate as controls, in both cases. Thus, visual cues associated with falling were unnecessary stimuli for wing-flapping. Although central motor programmes probably orchestrate the basic patterning of motor outflow to the muscles, sensory factors must play an important role in fine-tuning wing movement to the ever-changing patterns of body attitude, wing position, wind velocity and wind direction that occur during flight. Gewecke and Woike (1978) reported that receptors associated with the breast features are involved in the initiation, maintenance, and modulation of wing-flapping in tethered siskins flying in a wind tunnel (see Sect. 11.4). However, the present finding of drop-evoked wing-flapping in naked, scaleless mutants argues against a flap-evoking role for breast feathers in chickens. The naked chickens also performed drop-evoked wing-flapping when deprived of the visual cues produced by dropping. The chick's loss of foot contact with the ground or a perch was not a necessary stimulus because the legs were not supported before a drop. Thus, by exclusion, wing-flapping was probably initiated by vestibular processes activated by dropping.

Both lines of flight-featherless mutant chickens are interesting "experiments of nature." Since many other feather mutations produce flightlessness, such "experiments" are not especially rare (Abbott and Asmundson 1957). Given a consistently warm climate, made necessary by the loss of most of the insulating feathers, and minimal predation, both of the present feather mutants may survive and proliferate. The prospects for survival would be better for birds with feather mutations which yield adequate body insulation even though the feathers will not sustain flight. The case of dominant feather mutants is particularly interesting. In such cases, a major population of flightless birds may suddenly emerge in only a few generations.

Flightlessness may even confer a selective advantage upon birds in environments where bipedal locomotion, which is more energy efficient than flight, is sufficient. Additional modifiers may appear which would increase viability and extend the range of the birds. The loss of flight capacity may eventually lead to selection against the size of the wings and against the massive, cumbersome, and energetically expensive pectoral apparatus. Flightlessness in birds such as the ratites (e.g. ostrich, emu, rhea, cassowary, kiwi) may have had its origin in flighted birds which underwent sudden mutations of feathering. All of these birds currently possess feathering that makes flight impossible (Feduccia 1980). The appearance of anatomical changes such as reductions in wing and pectoral structure during the evolutionary history of the ratites and some other flightless birds (Feduccia 1980) may have been only gradual secondary adaptations to flightlessness that was the result of a feather mutation. The hypothesis that the flightlessness of some birds originated in feather mutations has not been proposed previously. However, another possible explanation can be offered for the high incidence of abnormal feathering in flightless birds. The mutant feathering may have appeared after the evolution of flightlessness. The loss of flight feathers would be less critical to the survival and reproductive success of flightless than of flying species. Therefore, a relatively higher incidence of feather anomalies would be expected in flightless species.

7.8 Bilateral Wing Coordination: Studies of Induced Bilateral Asymmetry

In stable flapping flight by birds, each wing must produce lift, and thrust and balance must be maintained between the flapping movements and posture of the two wings (Greenewalt 1962, 1975; Brown 1963; Pennycuick 1975; Rayner 1979). To meet these criteria, wing-flapping is probably controlled by a neuronal mechanism that provides a cyclic, bilaterally symmetrical pattern of motor output to the wings, and a feedback system that compensates for differences in the performance of the two wings (Weis-Fogh 1964; Wilson 1968).

The presence of spontaneous, bilaterally synchronized wing movements late in embryogenesis (see Sect. 7.2) and immediately after hatching (see Sect. 7.6) suggests that a mechanism which bilaterally synchronizes wing movement develops prenatally and is well established at hatching. However, the presence of bilateral wing synchronization in normal birds does not indicate whether or not a feedback mechanism actively synchronizes the flapping of the wings. In order to test for such feedback mechanisms, three different techniques were used to induce bilateral asymmetry in feedback from the wings (Provine 1982a). Wing-flapping was recorded using strobophotography (see Sect. 7.6).

First, the effect of immobilizing the right wing on the drop-evoked flapping rate of the contralateral left wing was observed in 3 to 21 day-old chicks (Provine 1982a). The wing was immobilized in the normal folded position by gluing it to the side of the bird's body with fast drying adhesive. Chicks with immobilized right wings flapped their left wings at a lower rate than controls, at all ages except 5 days. A few chicks decreased the flapping rate of the free wing relative to the control wing, but otherwise flapped normally. Others flapped at normal rates but at amplitudes below normal. However, most had marked anomalies in their wing-flapping pattern, particularly at the 13- and 21-day stages. The anomalous patterns were horizontal wing extension and slow, low-amplitude flapping or a fluttering of the wing while held in an overhead position. These results of unilateral wing immobilization indicate the presence of a bilateral synchronization process acting across the midline that is already in place at pre-flight stages.

In a second experiment, the effect of weighting the right wing on the drop-evoked flapping rates of both wings of 5, 7, and 13 day-old chicks was observed (Provine 1982a). Testing involved a control trial in which a chick with normal wings was dropped and a subsequent experimental trial in which a small lead weight was attached to the chick's right wing with quick drying adhesive immediately before testing. Weighting the right wing slowed the rate of drop-evoked flapping of both wings but did not disrupt the bilateral symmetry of the wing-stroke. The presence of slowed but bilaterally symmetrical flapping is further evidence for the presence of a flap-coordinating mechanism that acts across the body midline.

In the third experiment, the right wing of day-old chicks was amputated, in order to create a bilateral asymmetry by eliminating the sensory and motor processes of one wing. The flapping frequency on the side of the amputation was visualized by attaching a prosthesis to the stump of the amputated limb. Amputation had little effect on the development of normal flapping rates of the intact left wing and/or the right wing prosthesis at 7 and 13 days (Fig. 7.7). At both stages, the intact left wings of unilateral amputees without prostheses flapped at the same rate as the left wings of bilaterally intact controls (Fig. 7.7a, c). Furthermore, a bilaterally synchronized pattern of motor outflow to the wings was maintained in the unilateral amputees, as indicated by the bilaterally symmetrical envelopes formed by the tips of the flapping left wing and the right-wing prosthesis in strobophotographs at 7 and 13 days (Fig. 7.7b, d). Although the flapping pattern of the intact left wing of the unilateral amputees often appeared normal, abnormal flapping was present in many cases, especially at 13 days (Fig. 7.7e, f). These flapping patterns often resembled those of chicks with one immobilized wing (see first experiment of this series). This atypical behaviour is a reminder that even "stereotyped" acts such as wing-flapping can

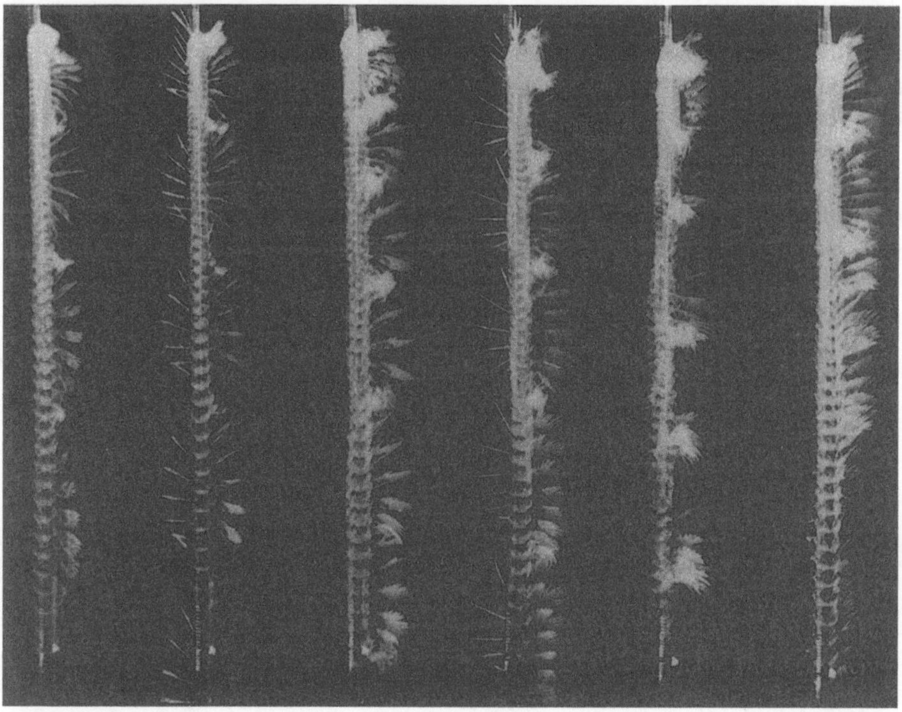

a b c d e f

Fig. 7.7. Stroboscopic photographs (100 Hz) of drop-evoked wing-flapping of 7-day-old (**a, b**) and 13-day-old (**c, d, e, f**) chicks whose right wings were amputated on the day of hatching. Shown are: **a** normal flapping of the left wing of a chick without a right wing; **b** bilaterally symmetrical flapping of the left wing and of a limb prosthesis attached to the stump of the amputated right wing; **c** flapping of the left wing of a chick without a right wing; **d** bilaterally symmetrical flapping of the left wing and a limb prosthesis attached to the stump of the amputated right wing; **e, f** abnormal flapping of chicks with amputated right wings (Provine 1982a)

be modified in response to environmental contingencies (see Sect. 8.3 for comparable evidence in the case of pecking).

Overall, these results indicate that bilaterally symmetrical wing-flapping is maintained by an *active* bilateral coordinating mechanism that utilizes feedback produced by wing movement to match the flapping frequencies of the two wings. The bilaterally symmetrical flapping by chicks with unilateral wing weighting or amputation demonstrates further that synchronization is maintained under conditions of increased or decreased wing loading. This result is consistent with the finding of slowed flapping of the free wing when the contralateral wing is immobilized.

Similar active coordinating processes play an important role in the production and control of movement in vertebrates (Grillner 1975) and invertebrates (Wilson 1968; Wendler 1974). In the present study, the flapping rate of the faster wing always decreased to match that of the slower wing. Although it is unknown if matching can occur in the other direction, it is logical to match the flapping frequency of the slower wing, because it is most often the rate-limiting member of a wing pair.

The bilateral wing coordinating mechanism is present by days 3 to 5, but lateral flight does not appear until days 7 to 9 (Provine 1981a). Thus, the coordination system develops in anticipation of function and does not require exercise during flight for its maturation. The finding of "wing-flapping" on the wingless side of the unilateral amputees demonstrates further that the sensory and trophic periphery associated with a given wing is not necessary for the post-natal development and performance of normal rates of bilaterally synchronized flap-related motor outflow. Fentress (1973) reached a similar conclusion concerning the ontogeny of grooming in mice with unilaterally amputated forelimbs. In the present case, the relatively normal rates of wing-flapping of the unilateral amputees may depend on bilaterally intact shoulder joints and breast muscles that produce the strokes of the wings, and their associated sensory receptors. These structures escaped amputation and thus remained capable of providing relatively normal patterns of bilaterally symmetrical wing-related proprioceptive feedback during the drop-evoked flapping episodes. The receptors of the intact shoulder joint may provide important flap-produced sensory input. However, such sensory receptors may not be necessary. Trendelenburg (1906) reported bilaterally symmetrical wing-flapping (and flight) in pigeons with unilateral wing deafferentation.

7.9 Development of Wing-Flapping and Flight in Dystrophic Chickens

Chicken dystrophy (Wilson et al. 1979) is an interesting disorder from the perspective of the genetics of the avian wing motor system (Wood-Gush 1971; Siegel 1979; Ehrman and Parsons 1981) because this nonlethal pathology is almost exclusive to the muscles that power the wings in flight. Dystrophic chickens are therefore flight mutants (see Sects. 7.10, 7.11). Chicken muscular dystrophy is a nonlethal, single-gene defect whose expression is modifiable by "background" genes (Asmundson and Julian 1956; Asmundson et al. 1966). The primary effect of the disorder is upon fast-twitch alpha-white muscle fibres that are eventually destroyed (Wilson et al. 1979).

The present study examined how drop-evoked wing-flapping and lateral flight developed in homozygous dystrophic (University of California, Davis, line 413) and closely related normal control (Davis, line 412) chickens (Provine 1983). In contrast to control chicks, dystrophic birds never developed substantial lateral flight. Most dystrophic chicks fell straight down when dropped and received lateral flight scores of zero. As with controls, drop-evoked wing-flapping developed in dystrophic chicks several days before it could play a role in flight. Both dystrophic and control chicks performed vigorous, bilaterally symmetrical, drop-evoked wing-flapping within a few hours after hatching. At 13 days, the age at which wing-flapping rate peaks in white leghorn chickens, dystrophic chicks flapped at an even higher rate than normal controls, suggesting that some aspect of flight function other than flapping rate must be responsible for at least the initial deficiency in lateral flight performance. After the first 2–3 weeks of development, the angular extent through which the wings of dystrophic chicks moved during a flap cycle decreased gradually but dramatically. By 49 days, when a dystrophic chick was dropped, it extended its wings laterally and fluttered them through a range of only a few centimetres. The present research also found that dystrophic chickens of all ages had difficulty in righting themselves, a confirmation of previous work by Wilson et al. (1979) and others.

7.10 Wing-Flapping in Flightless Birds: Evolutionary Insights

Having evaluated the effects of generations of domestication on wing-flapping and flight (Provine et al. 1984), we will now consider the stability of the neurobehavioural mechanism of wing-flapping during the millennia of natural flightlessness in birds presumed to have evolved from flighted ancestors (Provine 1984; but see McGowan 1984). Tests were carried out for both spontaneous and drop-evoked wing-flapping in a variety of flightless species. Young birds were tested by dropping them while they remained cradled in the observer's hands. The spontaneous behaviour of free-ranging young and adult birds was also observed.

Table 7.1 shows first the results from five giant ostrich-like birds of the ratite group; emus (*Dromaius novaehollandiae*), common (*Rhea americana*) and Darwin's (*Pterocnemia pennata*) rheas, cassowarys (*Casuarius*) and ostriches (*Struthio camelus*). Little has been reported previously about how and when these birds move their wings (Storer 1960b; Feduccia 1980). The present results show that none flapped their wings at any of the ages tested. However, other wing movements were observed. Free-ranging rheas and ostriches occasionally performed isolated unilateral wing movements alternating wing movements, and bilaterally symmetrical wing extensions. These birds used their wings individually or in concert while standing at rest or for balancing manoeuvres while running or dodging. The graceful, bilateral wing-extensions of the male ostrich during courtship are well known. Movements of the much smaller wings of the emu and cassowary were difficult to observe. However, one aviculturist noted that some emus may extend both wings ventrolaterally and maintain this posture when "tense," "curious," or "aggressive."

Table 7.1 shows results from Humboldt (*Spheniscus humboldti*)), black-footed (*Spheniscus demersus*), Adeli (*Pygoscelis adeliae*), and king (*Aptenodytes patagonicus*) penguins. These birds are flightless in air, but they perform powerful, bilaterally synchronized wing-flapping movements in wing-propelled swimming and diving (Simpson 1946; Storer 1960a).

Table 7.1. Wing-flapping in flightless birds

Species	Age	Number of birds	Spontaneous flapping[a]	Drop-evoked flapping[a]
Emu	2 weeks	4	0	0
	5 weeks	6	0	0
	8 weeks	2	0	0
	Adult	5	0	NT
Common rhea	4–5 days	3	0	0
	Adult	7	0	NT
Darwin's rhea	Adult	10	0	NT
Ostrich	Adult	2	0	NT
Cassowary	2 weeks	1	0	0
	Adult	2	0	NT
Black-footed penguin	4 weeks	3	+	0
	8 weeks	4	+	0
	> 12 weeks	4	+	0
Humboldt penguin	> 12 weeks	8	+	0
King penguin	4–6 weeks	2	+	0
	Adult	6	+	NT
Adelie penguin	> 12 weeks	7	+	0
Steamer duck	Adult	1	+	+ +

[a] + indicates that flapping was observed at least once during free-ranging activity. + + indicates that flapping was evoked on at least one of the three dropping trials in all birds in each category. 0 indicates that wing-flapping was never observed. NT means "not tested"

Young penguins performed vigorous wing-flapping when stretching, struggling, or attempting to right themselves. When they are older and their down is replaced by water repellant feathers, they enter the water and use wing-flapping for submarine "flight" and diving. Unlike all birds capable of aerial flight that have been tested, none of the penguins of any age performed drop-evoked wing-flapping. However, penguins will occasionally extend their wings or flap them in apparent balancing manoeuvres when falling or jumping into the water.

One rare, flightless steamer duck (*Tachyeres brachytterus*) was tested (see Livezey and Humphrey 1986, for a thoughtful treatment of *Tachyeres* morphology and evolution). Although this heavy-bodied duck is flightless, it still uses its wings in a bilaterally symmetrical fashion to paddle through the water in a manner reminiscent of its namesake, the old-fashioned side-wheel steam boat. The single adult tested performed vigorous spontaneous and drop-evoked wing-flapping.

If the penguin's ancestor once performed drop-evoked wing-flapping, the response was lost some time after the evolution of penguins during the Eocene (Storer 1960a, b; Simpson 1976). The absence of the drop response may be due to the loss or inactivation of the neural (probably vestibular) circuitry that triggers wing-flapping. The circuitry responsible for wing-flapping by penguins is clearly intact because it is used in submarine flight. The failure of ratites to perform either spontaneous or drop-evoked wing-flapping indicates that the neurobehavioural mechanisms that produced the behaviours were lost or became inactivated sometime after the appearance of these birds during the Pleistocene and Eocene (Storer 1960a, b; Feduccia 1980). Both spontaneous and drop-evoked wing-flapping remain in the steamer duck, whose flightlessness is probably attributable to an unfavourable ratio of wing-area to body weight as in the Cornish X Rock chicken (see Sect. 7.6).

Paleo-ornithologists frequently cite the effects of selection pressure against the pectoral apparatus and other morphological features of birds during the evolution of flightlessness (Feduccia 1980). The present findings from ratites and penguins indicate that the evolution of flightlessness also influenced the function and structure of the neural circuitry for aerial flight.

The major neural, feather, and other adaptations of the penguins and ratites of this study suggest a specialization for flightlessness that is increasingly irreversible. The ratites, for example, all have great body weight, a massive skeleton specialized for running, lack the pectoral musculature sufficient for powering the now vestigial wings in flight, lack the neural pattern generator (or the capacity to activate it) for wing-flapping, and have fluffy, down-like feathers, unsuited for flight. Many transformations would be required for them to re-evolve flight. Such losses of the motor pattern for flight during evolution are examples of "behavioural extinction" (Provine 1984).

In contrast, the re-acquisition of flight by the domestic Cornish X rock chicken, and perhaps the steamer duck, would involve relatively minor adjustments in such variables as the ratio of wing area to body weight, because the neurobehavioural apparatus and feathering necessary for flight remain in place. These more timid natural experiments with flightlessness, where there is no "extinction" of the flight motor pattern, may be relatively common among avians. Among marine birds, both the northern and southern hemispheres have wing-propelled divers (i.e. murres, razor bills, puffins, diving petrels) that retain marginal aerial flight without crossing the threshold into flightlessness as did the penguins and the now extinct great auk (*Pinguinus impennis*).

Much more could be learned about the evolution and neurobehavioural consequences of flightlessness if we increased our sample of flightless birds, many of which are endangered species. It is already too late for the recently extinguished dodo (*Raphus cucullatus*) and the great auk. Other candidates for study are the kiwi (*Apteryx spp.*), the flightless owl parrot (*Strigops habroptilus*), the Galapagos flightless cormorant (*Phalacrocarax harrisi*), the Auckland Islands teal (*Anas* [*a.*] *aucklandica*; Livezey 1990), and a variety of flightless rails and other birds that if not completely flightless are nearly so.

In Olson's (1973) analysis of flightless rails of the South Atlantic islands, he noted the ease and rapidity with which species can lose the ability to fly. Birds often become flightless in relatively hospitable environments because an environment that does not select for flight will select against the heavy and energy-consuming flight apparatus. For example, the ratites have lost much of the massive pectoral and other musculature that presumably powered their smaller ancestors in the air (Feduccia 1980). Less obvious, but perhaps more intriguing, are the secondary consequences that selection against these muscles may have on the nervous system: selection against muscles also selects against the related motor neurons in the spinal cord. Such processes have important implications for the evolution of the nervous system and behaviour.

7.11 Centripetal Hypothesis of Neurobehavioural Evolution

The centripetal hypothesis describes how selection acting upon the organ of locomotion, the muscle, can be a pivotal event in the process of neurobehavioural evolution that is precise, rapid, and efficient (Provine 1982b, 1984, 1993). The elegance of this process comes from the relatively direct manner in which the environment selects and sculpts behaviour by acting on muscles; it does not depend upon micro-increments of random, central, synaptic changes. The hypothesis is called centripetal (outside-in) because selection for or against muscles modulates the pattern of naturally occurring motor neuron death, an event that can shape the qualitative and quantitative nature of motor outflow (muscle → motorneuron → interneuron). The process is an informative example of the interplay between ontogenetic and phylogenetic events.

The power and rapidity of muscle selection are demonstrated by the meat industry's use of artificial selection to increase dramatically the muscle mass of cattle and chickens over relatively few generations, not millennia. Natural selection against muscle can be equally rapid because of the high energetic cost of excess muscle mass (Provine 1984). The selection for or against muscles has the secondary consequence of selecting for or against related motor neurons in the spinal cord, through the cell death processes considered in Sect. 7.5. The number of motor neurons is adjusted during development by competition for limited innervation sites or trophic agents (Hamburger 1977; Hamburger and Oppenheim 1982). This process may serve as a buffer mechanism to provide adequate innervation of a muscle (Katz and Lasek 1978). Also, it has other, more dramatic consequences for neurobehavioural evolution (Provine 1982b, 1984).

Selection for or against motor neurons is a quantitative process that has qualitative consequences. For example, if the usual motor neuronal target is lost through cell death, its interneuronal pattern generating circuits probably remain intact (McKay and Oppenheim 1991) and may synapse with novel motor neurons that innervate different muscles, and thus produce novel movements. Also, novel patterns of muscle contraction could arise through increased numbers of motor neurons made available through reduced cell death produced by an increase in muscle mass. Either scenario demonstrates how a centripetal process can be the engine that generates novel behaviour.

Observations of wing movements in ratites (Sect. 7.10) provide a preliminary test of one aspect of the centripetal hypothesis. The vestigial wings of these birds are powered by very small pectoral muscles, innervated by a small number of brachial motor neurons (Kappers et al. 1936). The absence of both spontaneous and drop-evoked wing-flapping (Provine 1984) may be the consequence of a centripetal chain reaction that began with selection against pectoral muscles and their motor neurons. The motor outflow from the central pattern generator may have been lost through inhibition, loss of activating input, or degeneration of the pattern generator. In contrast, penguins have massive pectorals needed to "fly" through the dense medium of water, and their spontaneous wing-flapping is consistent with the centripetal hypothesis. Penguins have, however, lost the drop-evoked wing-flapping reflex presumed present in their flighted ancestors, gull-like birds of the Southern Hemisphere (Provine 1984).

Another evolutionary trend is seen in steamer ducks and domestic meat chickens (Cornish X rock), which have massive pectoral muscles and perform both spontaneous and drop-evoked wing-flapping (Provine 1984). Neither of these heavy birds is able to fly because of their unfavorable ratio of wing area to body mass. Penguins, steamer ducks, and meat chickens are all "flightless" birds that have maintained their pectoral mass and, consistent with the centripetal process, conserved their wing-flapping behaviour. Only the penguin seems to have experienced a transformation of the wing motor system, losing the non-adaptive antigravity response, but not the pattern generator for flapping.

The evolution of motor processes is conservative. The motor pattern generators evolve slowly and linger long after they cease to be adaptive, perhaps becoming part of a library of subroutines from which future motor acts can be constructed (Provine 1984). Obsolete motor acts may remain in a quiescent state, or stay active as behavioural "relics." The grasp and Babinski reflexes of human neonates are leading candidates for human relic behaviour. Both reflexes disappear during early infancy, reappearing in adulthood only as symptoms of central nervous system pathology. Their potential for reappearance, although in an abnormally stereotyped form (Touwen 1984), suggests that the Babinski and grasp reflexes are maintained in an intact but inhibited state after infancy. Evidence for other higher and lower level "relic" behaviours and conservatism in behavioural evolution is widespread among animals (Kavanau 1990). The conservation of pattern generators is consistent with the lack of spontaneous cell death among spinal interneurons, the principal neuronal elements in such circuits (McKay and Oppenheim 1991).

Acknowledgements. The author's research was supported by grant HD 11973 from the National Institute of Child Health and Human Development and by grants MH 28476 and MH 36474 from the National Institute of Heath.

References

Abbott VK, Asmundson VS (1957) Scaleless, an inherited ectodermal defect in the domestic fowl. J Hered 48:63–70

Altman JS (1975) Changes in the flight motor pattern during the development of the Australian plague locust, *Chortoicetes terminifera.* J Comp Physiol 97:127–142

Asmundson VS, Julian LM (1956) Inherited muscle abnormality in the domestic fowl. J Hered 47:258–262
Asmundson VS, Kratzer FH, Julian LM (1966) Inherited myopathy in the chicken. Ann NY Acad Sci 138:49–58
Beebe W (1926) Pheasants, their lives and homes. Doubleday, New York
Bekoff A (1976) Ontogeny of leg motor output in the chick embryo: A neural analysis. Brain Res 106:271–291
Bekoff A, Trainer W (1979) The development of interlimb coordination during swimming in postnatal rats. J Exp Biol 83:1–11
Bekoff A. Stein PSG, Hamburger V (1975) Coordinated motor output in the hindlimb of a 7-day chick embryo. Proc Natl Acad Sci USA 72:1245–1248
Bentley DR, Hoy RR (1970) Postembryonic development of adult motor patterns in crickets: a neural analysis. Science 170:1409–1411
Boice R (1972) Some behavioural effects of domestication in Norway rats. Behaviour 42:198–231
Boice R (1973) Domestication. Psychol Bull 80:215–230
Brown RHJ (1963) The flight of birds. Biol Rev 38:460–489
Caple G, Balda RP, Willis WR (1983) The physics of leaping animals and the evolution of preflight. Am Nat 121:455–467
Darwin C (1892) The variation of animals and plants under domestication. Appleton, New York
Desforges MF, Wood-Gush DGM (1975a) A behavioural comparison of domestic and mallard ducks: habituation and flight reactions. Anim Behav 23:692–697
Desforges MF, Wood-Gush DGM (1975b) A behavioural comparison of domestic and mallard ducks: spatial relationships in small flocks. Anim Behav 23:698–705
Desforges MF, Wood-Gush DGM (1976) Behavioural comparison of Aylesbury and mallard ducks: sexual behaviour. Anim Behav 24:391–397
Drachman DB, Sokoloff L (1966) The role of movement in embryonic joint development. Dev Biol 14:401–420
Ehrman L, Parsons PA (1981) Behavior genetics and evolution. McGraw-Hill, New York, pp 254–265
Feduccia A (1980) The age of birds. Harvard Univ Press, Cambridge, MA
Fentress JC (1973) Development of grooming in mice with amputated forelimbs. Science 169:95–97
Gewecke M, Kutsch W (1979) Development of flight behaviour in maturing adults of *Locusta migratoria*. I. Flight performance of wing-stroke parameters. J Insect Physiol 25:249–253
Gewecke M, Woike M (1978) Breast feathers as an air-current sense organ for the control of flight behaviour in a songbird (*Carduelis spinus*). Z Tierpsychol 47:293–298
Greenewalt CH (1962) Dimensional relationships for flying animals. Smithson Misc Collect 144(2):1–46
Greenewalt CH (1975) The flight of birds. Trans Am Philos Soc 65:1–65
Grillner S (1975) Locomotion in vertebrates: central mechanisms and reflex interaction. Physiol Rev 55:247–304
Grohmann S (1939) Modifikation oder Funktionsreifung? Ein Beitrag zur Klärung der Wechselseitigkeit Zwischen Instinkthandlung und Erfahrung. Z Tierpsychol 2:132–144
Hamburger V (1963) Some aspects of the embryology of behaviour. Q Rev Biol 38:342–365
Hamburger V (1977) The developmental history of the motoneuron. Neurosci Res Program Bull (Suppl) 15:1–37
Hamburger V, Oppenheim RW (1967) Prehatching motility and hatching behaviour in the chick. J Exp Zool 166:171–204
Hamburger V, Oppenheim RW (1982) Naturally occurring neuronal death in vertebrates. Neurosci Commentaries 1:39–55
Hamburger V, Wenger E, Oppenheim RW (1966) Motility in the chick embryo in the absence of sensory input. J Exp Zool 162:133–160
Heaton MB (1976) Developing visual function in the red jungle fowl embryo. J Comp Physiol Psychol 90:53–56

Kammer AE, Kinnamon SC (1979) Maturation of the flight motor pattern without movement in *Manduca sexta*. J Comp Physiol 130:29–37

Kammer AE, Rheuben MB (1976) Adult motor patterns produced by moth pupae during development. J Exp Biol 65:65–84

Kappers CVA, Huber CG, Crosby EC (1936) Comparative anatomy of the nervous system of vertebrates, including man. Macmillan, New York (reprinted by Hafner, New York, 1960)

Katz MJ, Lasek RJ (1978) Evolution of the nervous system: role of ontogenetic buffer mechanisms in the evolution of matching populations. Proc Natl Acad Sci USA 75:263–288

Kavanau JL (1990) Conservative behavioural evolution, the neural substrate. Anim Behav 39:758–767

Krischke N (1983) Beiträge zur Ontogenese der Flug- und Manövrierfähigkeit der Haustaube (*Calumba livia var. Domestica*). Behaviour 84:265–286

Kutsch W (1976) Post-larval development of two rhythmical behavioural patterns: flight and song in the grasshopper, *Omocestus veridulus*. Physiol Entymol 1:255–304

Kutsch W, Gewecke M (1979) Development of flight behaviour in maturing adults of *Locusta migratoria*. II. Aerodynamic parameter. J Insect Physiol 25:299–304

Landmesser LT, O'Donovan MJ (1984) Activation patterns of embryonic chick hind limb muscles recorded in ovo and in an isolated spinal cord preparation. J Physiol 347:189–204

Lewin R (1983) How did vertebrates take to the air? Science 221:38–39

Livezey BC (1990) Evolutionary morphology of flightlessness in the Auckland Islands teal. Condor 92:639–673

Livezey BC, Humphrey PS (1986) Flightlessness in steamer ducks (Anatidae: Tachyeres): its morphological bases and probable evolution. Evolution 40:540–558

Lockhard RB (1968) The albino rat: a defensible choice or a bad habit? Am Psychol 23:734–742

Lorenz K (1935) Companions as factors in the bird's environment. In: Martin R (Translator) Studies in animal and human behaviour, vol 1. Harvard University Press, Cambridge, Mass (1970), pp 101–258

Lorenz K (1940) Durch Domestikation verursachte Störungen arteigenen Verhaltens. Z Angew Psychol Char 59:2–82

Lorenz K (1954) Psychology and phylogeny. In: Martin R (Translator) Studies in animal and human behaviour, vol 2. Harvard University Press, Cambridge, Mass (1971), pp 196–245

McGowan C (1984) Evolutionary relationships of ratites and carinates: evidence from ontogeny of the tarsus. Nature 307:733–735

McKay SE, Oppenheim RW (1991) Lack of evidence for cell death among avian spinal cord interneurons during normal development and following removal of targets and afferents. J Neurobiol 22:721–733

McKay SE. Provine RR, Oppenheim RW (1987) Naturally occurring motoneuron death in the spinal cord of the turtle embryo (*Chelydra serpentina*). Soc Neurosci Abstr 13 (2):922

Miller DB (1977) Social displays of mallard ducks (*Anas platyrhynchos*): effects of domestication. J Comp Physiol Psychol 91:221–232

Narayanan CH, Hamburger V (1971) Motility in chick embryos with substitution of lumbosacral by brachial and branchial by lumbosacral spinal cord segments. J Exp Zool 178:415–432

Narayanan CH, Malloy RB (1974a) Deafferentation studies on motor activity in the chick. I. Activity pattern of hindlimbs. J. Exp Zool 189:163–176

Narayanan CH, Malloy RB (1974b) Deafferentation studies on motor activity in the chick. II. Activity pattern of wings. J Exp Zool 189:177–188

Norberg UM (1989) Vertebrate flight: mechanics, physiology, morphology and evolution. Springer, Berlin Heidelberg New York

O'Donovan MJ, Landmesser L (1987) The development of hindlimb motor activity studied in the isolated spinal cord of the chick embryo. J Neurosci 7:3256–3264

Olson SL (1973) Evolution of the rails of the South Atlantic islands (Aves: Rallidae). Smithson Contrib Zool 52:1–53

Oppenheim RW (1974) The ontogeny of behaviour of the chick embryo. In: Lehrman DS, Hinde RA,

Shaw E, Rosenblatt J (eds) Advances in the study of behaviour, vol 5. Academic Press, New York, pp 133–172
Oppenheim RW (1981) Ontogenetic adaptations and regressive processes in the development of the nervous system and behavior: a neuroembryological perspective. In: Connolly K, Prechtl HFR (eds) Maturation and development. Lippincott, Philadelphia, pp 73–109
Oppenheim RW, Nunez R (1982) Electrical stimulation of hindlimb increases neuronal cell death in chick embryos. Nature 295:57–59
Ostrom JH (1979) Bird flight: how did it begin? Am Sci 67:46–56
Pennycuick CJ (1975) Mechanics of flight. In: Farnell DS, King JR, Parkes KD (eds) Avian biology, vol 5. Academic Press, London, pp 1–75
Pittman R, Oppenheim RW (1978) Neuromuscular blockade increases motoneuron survival during normal cell death in chick embryo. Nature 271:364–366
Pittman R, Oppenheim RW (1979) Cell death of motoneurons in the chick embryo spinal cord. IV. Evidence that a functional neuromuscular interaction is involved in the regulation of naturally occurring cell death and the stabilization of synapses. J Comp Neurol 187:425–446
Preyer W (1885) Specielle Physiologie des Embryo. Griebens, Leipzig
Provine RR (1971) Embryonic spinal cord: synchrony and spatial distribution of polyneuronal burst discharges. Brain Res 29:155–158
Provine RR (1972a) Ontogeny of bioelectric activity in the spinal cord of the chick embryo and its behavioral implications. Brain Res 41:365–378
Provine RR (1972b) Hatching behavior of the chick (*Gallus domesticus*): plasticity of the rotatory component. Psychon Sci 29:27–28
Provine RR (1973) Neurophysiological aspects of behavior development in the chick embryo. In: Gottlieb G (ed) Behavioral embryology. Academic Press, New York, pp 77–102
Provine RR (1979) Wing-flapping develops in wingless chicks. Behav Neural Biol 27:233–237
Provine RR (1980) Development of between-limb movement synchronization in the chick embryo. Dev Psychobiol 13:151–163
Provine RR (1981a) Development of wing-flapping and flight in normal and flap-deprived chicks (*Gallus domesticus*). Dev Psychobiol 14:279–291
Provine RR (1981b) Wing-flapping develops in chicks made flightless by feather mutations. Dev Psychobiol 14:481–486
Provine RR (1982a) Pre-flight development of bilateral coordination in the chick (*Gallus domesticus*): Effects of induced bilateral wing asymmetry. Dev Psychobiol 15:245–255
Provine RR (1982b) Evolution of flightlessness: neurobehavioral effects. Soc Neurosci Abstr 8:612
Provine RR (1983) Chicken muscular dystrophy: an inherited disorder of flight. Dev Psychobiol 16:23–27
Privine RR (1984) Wing-flapping during development and evolution. Am Sci 72:448–455
Provine RR (1986) Behavioral neuroembryology: motor perspectives. In: Greenough WT, Juraska J (eds) Developmental neuropsychobiology. Academic Press, New York, pp 213–239
Provine RR (1988) On the uniqueness of embryos and the difference it makes. In: Smotherman WP, Robinson SC (eds) Behavior of the fetus. Telford Press, Caldwell, NJ, pp 35–46
Provine RR (1993) Prenatal behavior development: Ontogenetic adaptations and non-linear processes. In: Savelsbergh GJP (ed) The development of coordination in infancy. Elsevier, Amsterdam, pp 203–236
Provine RR, Rogers L (1977) Development of spinal cord bioelectric activity in spinal chick embryos and its behavioral implications. J. Neurobiol 8:217–228
Provine RR, Sharma SC, Sandel TT, Hamburger V (1970) Electrical activity in the spinal cord of the chick embryo in situ. Proc Natl Acad Sci USA 65:508–515
Provine RR, Strawbridge C, Harrison BJ (1984) Comparative analysis of the development of wing-flapping and flight in the fowl. Dev Psychobiol 17:1–10
Rayner JMV (1979) A new approach to animal flight mechanics. J Exp Biol 80:17–54
Ripley KL, Provine RR (1972) Neural correlates of embryonic motility in the chick. Brain Res 45:127–134
Ruppell G (1977) Bird flight. Van Nostrand-Reinhold, New York

Siegel PB (1979) Behaviour genetics in chickens: a review. World's Poult Sci J 35:9–19
Simpson GG (1946) Fossil penguins. Bull Am Mus Nat Hist 87:1–100
Simpson GG (1976) Penguins: past and present, here and there. Yale Univ Press, New Haven
Somes RG (1969) Delayed feathering, a third allele at the K locus of the domestic fowl. J Hered 60:281–286
Spalding D (1875) Instinct and acquisition. Nature 12:507–508
Storer RW (1960a) Evolution of the diving birds. In: Bergman G, Donner KO, Haartman L (eds) Proc XII Int Ornithological Congr, Helsinki, 1958. Tilgmannin Kirjapaino, Helsinki, pp 694–707
Storer RW (1960b) Adaptive radiation in birds. In: Marshall AJ (ed) Biology and comparative physiology of birds, vol 1. Academic Press, New York, pp 15–55
Straznicky K (1963) Function of heterotopic spinal cord segments investigated in the chick. Acta Biol (Budapest) 14:145–155
Touwen BCL (1984) Primitive reflexes: Conceptual or semantic probelm? In: Prechtl HFR (ed) Continuity of neural functions from prenatal to postnatal life. Lippincott, Philadelphia, pp 115–125
Trendelenburg W (1906) Über die Bewegung der Vögel nach Durchschneidung hinterer Rückenmarkswurzeln. Arch Anat Physiol (Physiol Abt, Leipzig), pp 1–126
Tucker VA (1969) The energetics of bird flight. Sci Am 220:70–78
Viscor G, Fuster JF (1987) Relationship between morphological parameters in birds with different flying habits. Comp Biochem Physiol 87A:231–249
Vrbova G, Gordon T, Jones R (1978) Nerve-muscle interaction. Chapman and Hall, London
Weis-Fogh T (1964) Control of basic movements in flying insects. Symp Soc Exp Biol 18:343–361
Wendler G (1974) The influence of proprioceptive feedback on locust flight coordination. J Comp Physiol 88:173–200
Wilson BW, Randall WR, Patterson GT, Entrikin RK (1979) Major physiologic and histochemical characteristics of inherited dystrophy of the chicken. Ann NY Acad Sci 317:224–246
Wilson DM (1968) The nervous control of insect flight and related behaviour. In: Beament JWL, Treherne JE, Wigglesworth VB (eds) Adv Insect Physiol, vol 5. Academic Press, London, pp 289–338
Wood-Gush DGM (1971) The behaviour of the domestic fowl. Heinemann, London
Zeuner FE (1963) A history of domesticated animals. Harper & Row, New York

8 Development of Prehensile Feeding in Ring Doves (*Streptopelia risoria*): Learning Under Organismic and Task Constraints

J.D. Deich and P.D. Balsam

8.1 Introduction

Birds use their beaks as visually guided prehensile (reaching/grasping) effectors, employing the beak for functions critical to survival. The beak is used to pick up and ingest food, preen the feathers, gather nesting material and assemble the nest. In many altricial species, beak-mediated prehension is involved in the delivery of food to the young, either by grasping the beaks of the young and delivering food regurgitatively, or by using the beak to seize food and carry it back to the nest. In aggressive interactions the beak is used to grasp forcefully and to twist the skin and feathers of the opponent. The beak is also involved in critical non-prehensile tasks; notably, song production, drinking, and aggressive jabbing. As well as showing variation in beak use across functions, some species must show flexibility in their beak movements within a task, such as feeding.

Any species confronted with variation in food sources may need to adapt its behaviour to changes in food types or feeding locales. Such adaptation may require changes in feeding responses. Columbidae species are very adept at adjusting to these changes, as evidenced by their success at occupying a variety of niches ranging from Sahelian plains to urban canyons. The present work reports on investigations of feeding behaviour in two Columbidae species, the pigeon (*Columba livia*) and the ring dove (*Streptopelia risoria*), which demonstrate that prehensile feeding is a skilled response that shows considerable plasticity. Feeding develops gradually, and experience plays an essential role in producing functional adult prehensile feeding.

We first describe an analysis of the ontogeny of prehensile feeding in Columbidae and of the experiences necessary for development of this response in terms of Pavlovian and operant conditioning. We then use these data to illustrate a general approach to predicting and understanding response form regardless of its origin, which we call task analysis. Finally, we consider the implications of task analysis for the development of prehensile feeding and other responses whose development is strongly dependent on experience.

Psychology Department, Barnard College, 3009 Broadway, New York, NY 10027, USA

M.N.O. Davies and P.R. Green (Eds.)
Perception and Motor Control in Birds

8.2 Thrusting and Grasping During Feeding in the Adult

Prehension in avians, like that in primates (Jeannerod 1981, 1984), requires the coordination of two response components (Zeigler et al. 1980; Klein et al. 1985; for discussion see Deich et al. 1988); the effector must be transported to the target object and then closed around it. In primates the transport component is typically called reaching or simply transport; for avians we prefer the term thrust. Closing around the target and securing it is called grasping. A successful peck requires that thrusting and grasping are well coordinated (see Chaps. 9 and 10).

The duration of thrusting and the accompanying opening phase of grasping is so short in the Columbidae (about 50–120 ms; see Deich et al. 1985b; Bermejo et al. 1989) that either high speed motion picture photography (Klein et al. 1985) or direct transduction by attached sensors is necessary to measure these movements precisely. A system for the measurement of the gape component was developed in the pigeon (Deich et al. 1985a). Here, a small rare-earth magnet is glued to the bird's lower beak and a small magnetic sensor to the upper beak. The intensity of the magnetic field detected by the sensor decreases as the beak opens and can be recorded in real time (see Sect. 9.4).

We employed this system to record the gape of adult ring doves. Figure 8.1 shows a sample record of a successful peck at a piece of grain (from Deich and

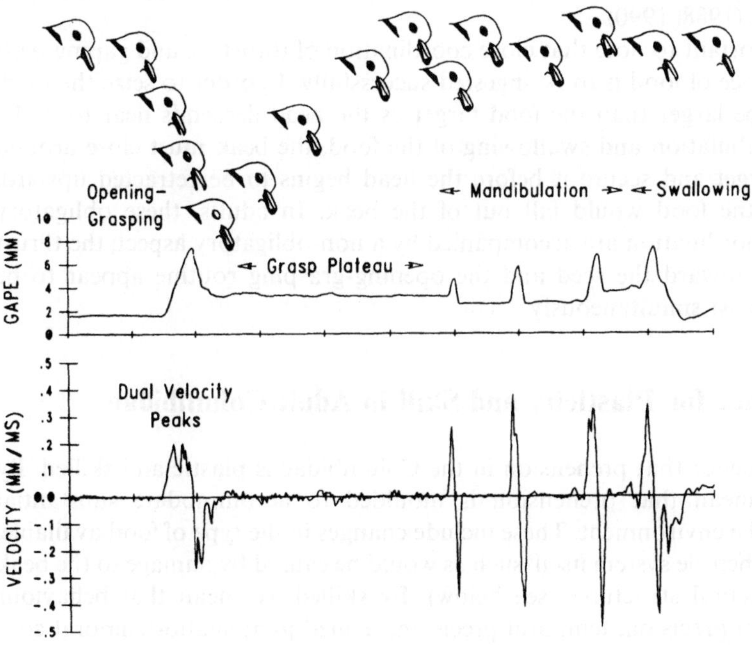

Fig. 8.1. Gape record of a successful peck at seed in a ring dove. The *upper panel* shows the changes in the distance between the upper and lower beak tips (gape). The *lower panel* shows the changes in velocity of gape. Negative velocities indicate reclosure of the beak

Balsam 1993). This figure shows the movement of the head synchronized with that of the beak. The gape sensor record below is also synchronized with the drawings. As an ingestive sequence begins, the head is briefly fixated above the seed (Goodale 1983) and then drives downward toward it. During this thrusting movement the beak opens, and gape reaches its maximum while the head is being thrust toward the seed. As the beak reaches the seed it is reclosed around it creating a plateau in the record. After reclosure with the seed secured at the beak tip, there is a set of higher velocity beak opening movements that typically have only a single velocity peak and result from moving the seed to the back of the beak for swallowing. The movement of the seed backwards in the beak is called mandibulation, and is often assisted by the tongue which sticks to the seed and pulls it backwards (Zweers 1982; Van Gennip 1988; see also Chaps. 9 and 10). At the end of the ingestive sequence the beak opens one final time. This opening appears in the gape record as a peak with rounded shoulders. According to Van Gennip (1988), this epoch of the signal is caused by movement of the tongue to allow swallowing.

The velocity curve of the initial opening phase of prehensile feeding in the adult ring dove typically has two (or more) clear peaks. These dual velocity peaks (see Fig. 8.1), and the moderate velocity that characterizes the opening phase, are taken as prima facie evidence of a feedback-sensitive motor system by Brooks (1974). In a similar vein, the dual velocity peaks also resemble the type of profile that would occur with corrective submovements, as hypothesized by Meyer et al. (1988, 1990).

It is important to note that close coordination of thrusting and gaping *must* occur if a piece of food is to be ingested successfully. In order to seize the food, gape must be larger than the food target as the head descends near to it. To allow mandibulation and swallowing of the food, the beak must close around the food target and secure it before the head begins to be retracted upward. Otherwise, the food would fall out of the beak. In adults, these obligatory aspects of coordination are accompanied by a non-obligatory aspect; the thrust of the head toward the seed and the opening-grasping routine appear to be initiated almost simultaneously.

8.3 Evidence for Plasticity and Skill in Adult Columbidae

There is evidence that prehension in the Columbidae is plastic and skilled. By plastic we mean that prehension is moulded to accommodate substantial changes in the environment. These include changes in the type of food available, or in the prehensile system itself, such as would be caused by damage to the beak or related neural structures (see below). By skilled we mean that behaviour shows spatial precision, temporal precision, refined joint spatio-temporal control, and coordination of its movement components, as illustrated here by thrust and gape (Sect. 8.2). To be sure, some degree of plasticity may be necessary for the development of skill. We include both labels to emphasize that prehension in

Columbidae can adjust both to minor task variations that do not require major movement reorganization and to substantial task alterations that do require substantial movement reorganization. We now examine several lines of evidence that lead us to assert that the peck of Columbidae is skilled and plastic.

First, in the pigeon (Zeigler et al. 1980; LaMon and Zeigler 1984; Deich et al. 1985b; see Sect. 9.4.2) and the ring dove (Deich and Balsam 1993) the beak is adjusted to the diameter of various size food targets, reaching a maximum distance between beak tips (gape) before contact with the food (see Fig. 8.2). Maximal gape before grasping the pellet is adjusted to about 2–4 mm greater than the size of the pellet. Individual subjects show consistency in gapes directed at food pellets of a given size (see Fig. 8.2). This adjustment demonstrates the sensitivity of the prehensile system to changes in the environment, and reflects the skilled nature of prehensile feeding. This precise adjustment of behaviour to seed size begins at the start of the movement, as different opening velocities to different seed sizes are observed in the earliest parts of the movement (Bermejo et al. 1989; see Sect. 9.4.2). Thus, it seems likely that the decision(s) supporting this adjustment are made before the movement begins; perhaps while the bird's head is fixated above the food target (Goodale 1983; see also Sects. 2.4 and 3.5).

Second, Deich et al. (1985b) found that adult pigeons could compensate for efferent nerve sections that rendered the lower beak unable to move by increasing the movement time (opening duration) of the upper beak. After the nerve section, these pigeons attempted to eat mainly the smaller food pellets that could be accommodated by this strategy. Further, Premock and Klipec (1981) removed the front half of their birds' upper beaks rendering the species-typical adult movement pattern ineffective in feeding. The result was that most of these pigeons were able to learn alternative feeding response topographies, such as "shoveling" movements of their beaks. The adjustment to such injuries as

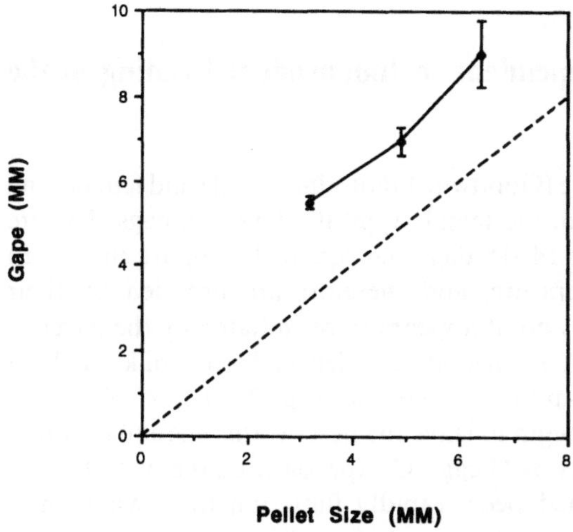

Fig. 8.2. Maximal gape before contact as a function of the size of a target food pellet. *Circles* represent the means from three birds, and means of the standard deviations for each bird are shown by *vertical bars*. The *dotted line* shows the relationship predicted if gape equalled pellet size

administered in Deich et al. (1985b) and Premock and Klipec (1981) illustrates the plasticity of beak-mediated feeding.

The third and final study (Deich et al. 1988) suggests substantial plasticity of the adult's prehensile behaviour by showing that the gape component is influenced by differential reinforcement. Pigeons direct the same general sort of prehensile responses to a response key when reinforced with grain for pecking the key as they do to grain itself (Wolin 1968; Jenkins and Moore 1973). Deich et al. (1988) set daily gape criteria based on the distribution of gapes made to the response key on the previous day. Only pecks at the response key with gapes that exceeded this criterion were reinforced by the presentation of a food-pellet reward. Thus, in separate phases, gapes smaller (downwards differentiation of gape) or larger (upwards differentiation of gape) than most of the previous day's gapes were reinforced. Successive changes in the criterion were successful in moving gape size upwards and downwards. Thus, reinforcement was used to move the gape component of key-directed pecking toward both larger and smaller values, demonstrating considerable plasticity in this response.

The thrust component of the peck also shows plasticity. The pigeon's keypeck has been a focus of study in the area of animal learning and conditioning. Work on conditioning has demonstrated that several aspects of key-directed pecking (thrusting) are influenced by differential reinforcement. These include the spatial locations of pecks (Barrera 1974; Eckerman et al. 1978), the time of occurrence of pecks (Deich and Wasserman 1977), the force of pecks (Chung 1965; Cole 1965), and the duration of pecks (Ziriax and Silberberg 1978). The nature of the targets at which thrusts are directed can also be altered by experience. Localized visual cues associated with food come to evoke thrusts (Brown and Jenkins 1968) and observation of other birds pecking at novel objects can result in thrust being directed at new targets (Palameta and Lafebvre 1985).

It was against this background of adult plasticity and skill that we began our study of the development of prehensile feeding in the ring dove.

8.4 The Transition from Dependent to Independent Feeding in the Ring Dove

As is the case for most Columbidae (Goodwin 1960), after a male and female ring dove (Lehrman 1955) have mated, the female typically lays two eggs that are incubated by both parents. After 14–15 days the squabs hatch, are unable to maintain their own body temperature, and therefore are brooded by their parents. Dependent feeding at this point appears to be initiated by the parents, who grasp the squab's beak and regurgitatively deliver "crop milk", a high protein substance sloughed off from the lining of the crop (Patel 1936). At about 8 days post-hatching, the squabs begin to leave the nest. At this age squabs often initiate regurgitative feeding bouts by "begging"; approaching the parent while making a stereotyped high-pitched peep, rapidly fluttering their wings, and

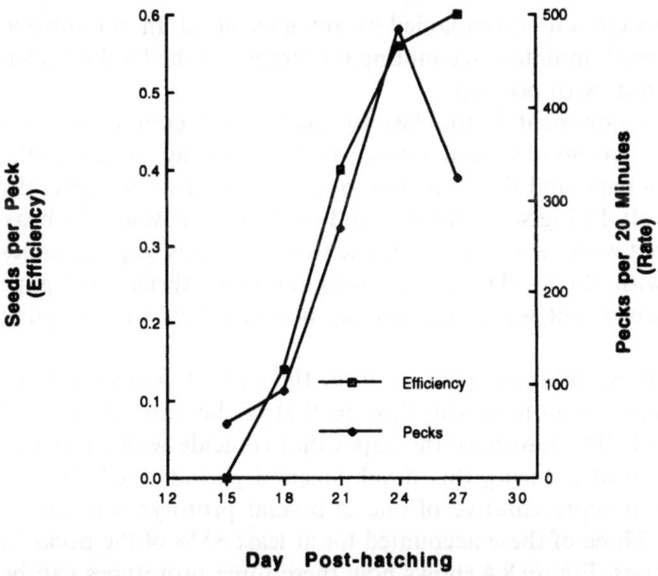

Fig. 8.3. Improvement of pecking skill in ring dove squabs as a function of age. *Open squares* show median pecks in a daily 20-min test session (*right axis*). *Closed circles* show efficiency of pecking as median seeds per peck (*left axis*). The data presented are aggregated from multiple experiments

thrusting their beaks in the direction of the parent's beak. At about day 14 to 16 post-hatching, squabs make their first food-directed thrusts. As shown in Fig. 8.3, the number of pecks made increases steadily until about day 30 post-hatching, when it is near mature levels. The success of ingesting grain also increases steadily during this period (also see Fig. 8.3). Regurgitative feeding concomitantly decreases (Wortis 1969; Hirose and Balsam submitted).

8.5 Development of Pecking

We adapted the system developed by Deich et al. (1985a) for the measurement of adult avian gape for use in squabs as young as day 10 (Deich and Balsam 1993). The presence of the measurement system did not appear to affect substantially the squab's interaction with its parents or the development of feeding. In order to ascertain the behavioural context in which gape occurred, we simultaneously videotaped all experimental sessions. The videotaping also allowed assessment of the thrust component of prehension. Unfortunately, the capture rate of the video camera was too slow to allow detailed analysis of the coordination of the thrust and gape components.

Of central interest was the form of the earliest grain-directed pecks. We found that these early pecks differed from the adult response in important ways (Deich and Balsam 1993). All subjects showed many instances of head thrusts directed at grain that were not accompanied by any gape at all. In one subject, this no-gape pattern predominated, accounting for almost all the food-directed thrusts in its first session with pecking.

While the thrust component is too fast to characterize completely with videotape alone, close analysis of several individual sequences suggested that the initial coordination of gape and thrust is poor and variable. For example, the beak fails to open at all during some thrusts, and on other occasions the beak opens late, after contact with the substrate. As the squab matures, gape becomes better coordinated with thrust. Thus, the coordination of thrust and gape emerges during ontogeny, but we cannot yet say precisely how this coupling occurs.

In a related study, we followed pecking from the earliest pecks until the topography of the gape component was close to that of the adult on day 32 (Deich et al. submitted). The nature of the gapes that coincide with thrusts at seed changed systematically during this developmental period. Seed-directed pecks were classified as representative of one of several prototypes to aid in tracing development. Three of these accounted for at least 85% of the pecks in all individual daily cases. Figure 8.4 shows how these three prototypes can be easily distinguished by their velocity profiles. The three types are: (1) a gape signal with a *single* velocity peak whose velocity was typically high (greater than $0.7\ \mathrm{mm\,ms^{-1}}$); (2) a gape signal with a nearly *square* velocity profile that reached a much lower maximum velocity; and (3) a gape signal with *dual* or multiple velocity peaks. The frequency of these different patterns is shown for two squabs in Fig. 8.5. The single velocity peak form predominated early in the acquisition of feeding. At an intermediate stage, the square velocity profile predominated. In the last stages of development and into adulthood, the multiple velocity peak was the predominant form.

While it will take further study to define the advantage that the bird gains by employing the adult movement form, with dual or multiple velocity peaks, three possibilities can be proposed. First, it may allow finer control of opening and grasping, so that seeds of various sizes are handled appropriately. Second, the velocity profiles early in development may represent movements that are not under sensory control ("continuous movements"; Brooks 1974), whereas the adult movement permits some sensory monitoring of opening. Third, the adult velocity profile may represent a compromise between the two earlier developmental forms, the single-peak profile that often has a high velocity but may have low accuracy, and the square wave form which results in greater accuracy but at the cost of slower feeding.

8.5.1 Behavioural Analysis of the Development of Pecking

The slow increase in the rate of pecking at food, the gradual increase in the efficiency of this pecking, and the changes in the types of pecks seen during development, together suggest the operation of an acquisition process. These

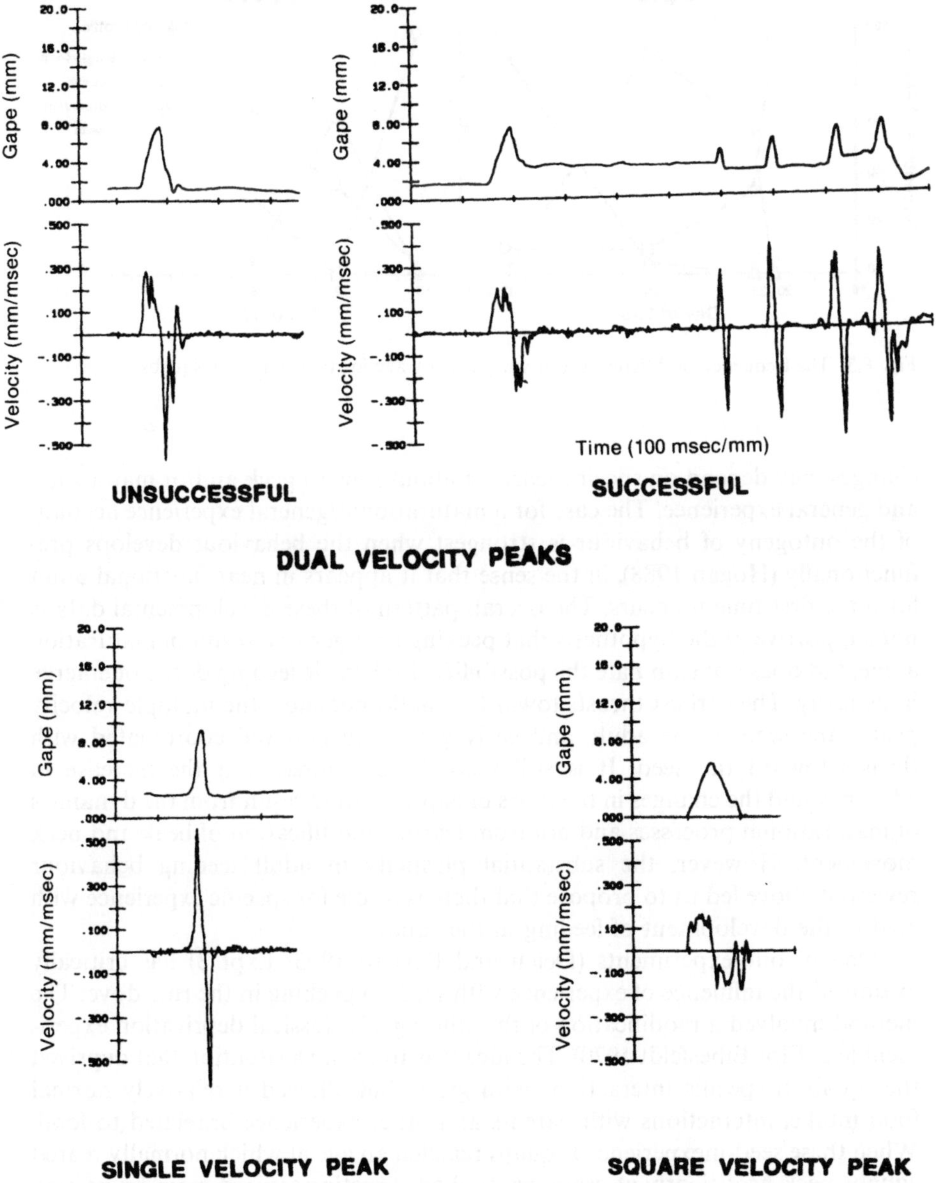

Fig. 8.4. The *top row* shows ring dove gape and velocity records from typical unsuccessful (*left side*) and successful (*right side*) juvenile (near adult) pecks. In each case the velocity profile has dual peaks. The left-hand peck appeared to be unsuccessful because of a spatial targeting error. The *lower row* shows two of the most common velocity profile types that precede this dual velocity peak form in development; a single velocity peak (*lower left*) and a square velocity profile (*lower right*)

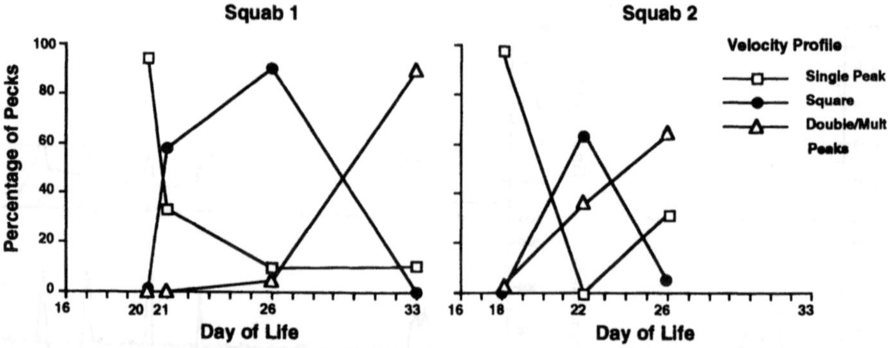

Fig. 8.5. The frequency of different prototype gapes by age for two ring dove squabs

changes may depend on specific learning about how to peck and/or maturation and general experience. The case for a maturational/general experience account of the ontogeny of behaviour is strongest when the behaviour develops pre-functionally (Hogan 1988), in the sense that it appears in near-functional adult form the first time it occurs. The overall pattern of these developmental data is not supportive of the hypothesis that pecking emerges as a result of maturation alone, but does not eliminate the possibility. Prehensile feeding does not emerge holistically. The earliest thrusts toward grain do not show the multiple velocity peak gape seen in the adult, and early gapes are not well coordinated with thrusts toward the seed. It is still possible, of course, that the increase in efficiency and the changes in the types of gapes emitted result from the dynamics of maturational processes and not from learned modification of beak and neck movement. However, the substantial plasticity in adult feeding behaviour reviewed above led us to propose that there is a role for specific experience with seed in the development of feeding in the squab.

One of our experiments (Deich and Balsam 1993, Expt. 3) has critically examined the influence of experience with seed on pecking in the ring dove. The method involved a modification of the ethologist's classical deprivation experiment (e.g. Eibl-Eibesfeldt 1970). The idea was to create a situation that deprived the squab of specific interactions with grain, but allowed it relatively normal food intake, interactions with parents and other experience unrelated to food. When these seed-inexperienced squabs reached an age at which normally reared squabs peck proficiently at grain, we evoked thrusting toward grain by pairing the sight of grain with hand feeding (Balsam et al. 1992b). A maturational account of the changes described during development would have been supported if Pavlovian pairings of the sight of grain (the CS) with hand feeding (the US) had quickly evoked adult responses similar to those of normally reared age mates. The advantage of using this type of control to assay the effects of maturation is that it deprives the squab of experience with whole grain and during the test directs the thrust component toward grain. It is reasonable to think that if the prehensile feeding response of the adult is preformed, then a similar response will be evoked when thrusts occur towards the seed, the species typical target for prehension. In any case, the Pavlovian pairings at least guarantee that the seed will be attended to. These aspects of the Pavlovian procedure make it an attractive method for searching for maturational effects.

These pairings were first administered with the squabs gently restrained a few centimetres above a grain filled floor. The CS period was 20 min and the US was the administration of a mush-like food via a needleless syringe. This procedure did not generate a substantial amount of pecking in our squabs, so we used a revised procedure, in which the squabs were allowed to move freely around the test chamber. After an intertrial period that averaged 60 s, fresh seed was presented to each subject and was followed by hand feeding. This revised procedure did evoke thrusting directed toward the grain. Contrary to the prediction based on a maturational account, the gapes produced by these birds (see Fig. 8.6) did not appear to be similar to those of either normally reared squabs of the same age or of adults. Many of the gapes that did occur had single peak or square velocity profiles. Some head thrusts included no gape at all. Overall, the gape records of these subjects appeared much more variable than those of adults and showed some similarity to the early gapes described above. The results argue that maturation alone is inadequate to produce the stereotyped adult gape topography described previously, but the ability of this

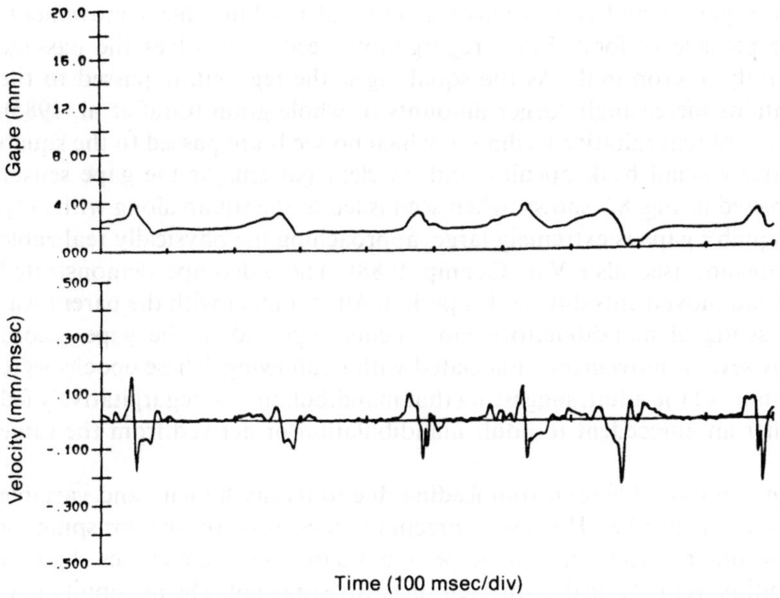

Fig. 8.6. A sample of the gapes generated in ring doves by the Pavlovian procedure (Deich and Balsam, 1993)

Pavlovian training to produce some gaping as well as some thrusting does suggest that partially formed components of the adult prehensile pattern may be prefunctional.

Given that maturation alone does not appear to provide a full account of the development of prehension, other contributions must be sought. In addition to the moulding of the adult response itself, juvenile patterns that precede prehensile feeding in development are a potential source of the mature feeding response. Begging, regurgitative feeding, and preening could serve as predecessors of the adult prehensile feeding response. These behaviours could also contribute more generally to the development of the adult response, for example, by contributing sensory-motor skills, strengthening specific muscles, or causing motor units to be more finely differentiated.

However, none of these juvenile patterns appear to be likely *direct* predecessors of prehensile feeding (Deich and Balsam 1993). When begging, the squab extends its head toward its parent's beak, flutters its wings and sometimes makes a high pitched "peeping" vocalization. Begging invariably involves thrusts either with the beak closed or with very small gapes accompanying begging vocalizations.

Regurgitative feedings, with or without the passage of seed, also do not appear similar to the critical opening/grasping phase of adult prehensile feeding. During regurgitative feeding, the squab's beak is placed inside its parent's beak. Early in development, placement is accomplished by the parent grasping the squab's beak. Later, the squab is more active in thrusting toward the parent's beak and the placement appears cooperative. Once the beaks are "locked" together, the parent and squab make a series of rhythmic head movements during the passage of food. Early regurgitative feeding involves the passage predominantly of crop milk. As the squab ages, the regurgitate passed to the squab contains increasingly larger amounts of whole grain (Graf et al. 1985). Gape records of regurgitative feeding in which no seeds are passed to the squab show relatively small beak opening and no clear pattern. As the gape sensor record depicted in Fig. 8.7 shows, when seed is fed to the squab along with crop milk, the squab's gape is extremely large, approaching the physically realizable limits of opening (see also Van Gennip 1988). The videotape demonstrated rhythmic head movements during this period. After contact with the parent was broken, a string of mandibulatory movements appeared in the gape record followed by several movements associated with swallowing. These epochs were similar to those of the adult, suggesting that mandibulation of regurgitatively fed seed is either an antecedent to adult mandibulation or derived from the same source.

Preening appears different from feeding due to its rhythmicity and variable opening velocity profile. However, preening does involve the grasping of feathers and other objects, and the gape signal based on it sometimes shows a dual or multiple velocity peak with each successive opening. The possibility that preening does facilitate the development of prehensile feeding is interesting and worthy of experimental examination.

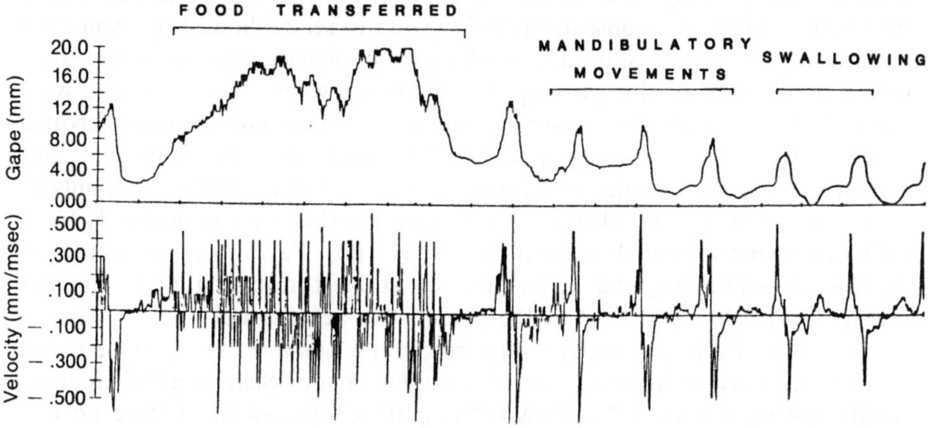

Fig. 8.7. A sample gape sensor record of regurgitative feeding in the ring dove that involved the passage of seed. The squab's beak is inside the parent's beak during the portion of the curve marked *food transferred*. The two birds have disengaged during *mandibulatory movements* and *swallowing*

While none of these developmental patterns appears to be a direct predecessor of adult prehensile feeding, they might be related to one of the early types of pecks. Evaluating this hypothesis shows that none of the forms closely resembles the developmental prototypes illustrated in Fig. 8.4, although begging is similar to the early food-directed pecks made with the beak closed. Further, it appears that the juvenile gape patterns that precede the earliest grain-directed thrusts do not later accompany grain-directed thrusts, or if they do they are quickly abandoned in favour of gape patterns that resemble the first developmental prototype shown in Fig. 8.4.

In other behavioural work (Balsam et al. 1992b), it has been found that the opportunity to make contact with, manipulate, and swallow grain augments the effects of Pavlovian pairings of the sight of grain with hand feedings, yielding higher rates of thrusting at grain. This outcome, in conjunction with the progression in the types of gapes and the increase in efficiency seen during development, supports the idea that there is a successive modification of prehensile feeding based on its successful completion. Thus, operant contingencies are likely to underlie the moulding of the gape into its adult form.

8.6 Behavioural Processes Underlying Development of Prehensile Feeding

Work in our lab (reviewed in Balsam et al. 1992b) has demonstrated that even squabs reared on powdered food have some unconditioned tendency to peck (thrust) at grain on day 14 post-hatching. However, the rate of pecking here is quite low and pecking ceases unless this experience with seed is followed closely by positive ingestional consequences. Exposure to seed followed by immediate reunion with the parents, usually resulting in immediate regurgitative feeding, increased the rate of this pecking. In conditions where these reunions were delayed, thus introducing a delay between the grain and parental feeding, pecking at grain was not acquired. In other experiments we found that substantial increases in pecking were produced when subjects were allowed only to see the seed, but not to peck at or to eat them, and then were immediately hand fed by the experimenter (Balsam et al. 1992b). Thus, Pavlovian conditioning is capable of supporting the acquisition of at least thrusting at grain, and regurgitative delivery of food by the parents can act as a US.

Therefore, Pavlovian conditioning becomes a possible causal factor in the normal ontogeny of thrusting toward grain. One major element of this analysis is still missing, however. Just how is the grain CS presented in more normal developmental situations? In caged families, whole grain is constantly available. Thus, it is unclear how seed (CS) presentation could be predictive of ingestional consequences. One possibility is that the parent directs the squab's attention to the grain by pecking at grain prior to feeding the squab. In accord with this analysis, Hirose and Balsam (submitted) found that early squab pecks tended to occur at the same time as parent's pecks. Furthermore, the squabs were usually regurgitatively fed within a few minutes after the parent pecked at seed.

However, the Pavlovian experience alone is not sufficient for generating the adult level of pecking (Balsam et al. 1992b) and, as shown above, it is not sufficient for inducing the adult gape topography. We hypothesize that the earliest gapes may be generated by the Pavlovian-induced thrusting in a manner similar to the head extension induced by the visual cues associated with movement toward the floor in pigeon squabs (Davies and Green 1989). The movement itself, or correlated visual cues, induces the early gape forms, and the success and failure of these movements shape the gape toward the adult form. The contribution of reinforcing feedback to the ontogeny of the response is suggested by the fact that, although the sight of seed followed by feeding facilitated pecking, squabs that were allowed to see seed, handle it, and ingest it prior to immediate feedings pecked at even higher rates (Balsam et al. 1992b). This result suggests that successful ingestion of seed may serve as a reinforcer for pecking. Additionally, the gradual increase in pecking efficiency, and the series of different forms of gape (see Fig. 8.5), are compatible with the hypothesis that the success and failure of different peck topographies mould the behaviour into the adult form. We have also observed some highly unusual and idiosyncratic pecking topographies when powder-reared subjects are first exposed to seed (Tankoos et al. in prep.). These topographies, described in more detail later, seem much more appropriate for scraping or scooping up powdered food than for picking up seed. This, too, suggests that prehensile feeding is moulded by the success and failure of specific actions. Lastly, such operant effects during development are certainly possible because adult pecking is highly susceptible to differential reinforcement (Deich et al. 1988).

In summary, our experiments suggest two ways in which experience is important in the ontogeny of pecking. First, Pavlovian contingencies direct the thrusting component to appropriate targets. Second, either because of previous experience with prehensile movement or because of general tendencies to gape, those gape movements that are successful will be selected over those that are less successful. These operant contingencies transform the gape into the adult form and are the basis for the tight coordination of the gape and thrust found in the adult.

8.7 The Viewpoint That Prehensile Feeding Is a Preorganized Response

The evidence that we have reviewed suggests that experience plays an important role in the development of prehensile feeding in the Columbidae and that in adults this feeding is plastic and skilled. A sharply contrasting viewpoint was prevalent among behavioural researchers during the 1970s and 1980s. According to this viewpoint, the form of prehensile feeding in many avians including the Columbidae is prefunctional. More specifically it was suggested that the peck is biologically determined, preorganized, and probably controlled by self-differentiating central neural mechanisms that require no environmental interactions to develop (Jenkins and Moore 1973; Moore 1973; Williams 1981, especially pp. 72–73). This alternative viewpoint holds that such prehensile feeding is subject to only very limited experiential modification. The idea was expressed clearly by Moore (1973, p. 169):

> ". . . while the pigeon's response topographies are sometimes modifiable, they are more often highly resistant to differentiation. The mere nature of the reinforcer typically affects at once countless dimensions of response topography. Implicit and explicit reward contingencies, by contrast, are less likely to prove effective, and when effective, they often act slowly or weakly and affect single topographic dimensions. When changes are obtained, they often arise because the animal has begun to direct the old response pattern towards some new stimulus. When real topographic changes occur, the pigeon has often merely shifted from one to another species-specific pattern."

We take the time to address the preformation hypothesis for prehensile feeding here, because of its preeminence as an analysis of feeding in Columbidae. Several findings contributed to this viewpoint about feeding. First, at an overt level of analysis, the pecking response does appear to emerge holistically in the precocial domestic chick (*Gallus gallus*) (Hunt and Smith 1967; Hogan 1973, 1984; Woodruff and Starr 1978) and in gulls (Margolis et al. 1987; Alessandro et al. 1989). Second, there is evidence that thrust (Williams and Williams 1969; see Locurto 1981 for a review) and gape topographies (Moore 1973) induced by Pavlovian contingencies are partially resistant to change by operant contingencies. Last, there is also evidence that prehensile feeding shows overt stereotypy (Jenkins and Moore 1973; Moore 1973). Stereotypy of a behaviour, within and between members of a species, is taken as evidence that a response is under strong biological control (Lorenz 1981).

We feel that interpreting these results as support for the hypothesis that gape is prefunctional is unjustified, when three points are taken into consideration. First, the studies with chicks that were taken as evidence for the holistic emergence of the adult form only measured whether or not thrusting occurred. But studies of Columbidae (Zeigler et al. 1980; Klein et al. 1985) show that the coordination of gaping and thrusting is crucial for successful adult performance. To support the idea that a response is preorganized, the response should be examined at as detailed a level as possible, and this was not done in the developmental studies of pecking in chicks. In contrast, the detailed analysis in doves described in Sect. 8.5 shows that gape and its coordination with thrust is not revealed holistically. The developmental progression involves three main squab peck topographies (see Fig. 8.4) that precede acquisition of the "stereotyped" adult peck (see Fig. 8.1). Further, the Pavlovian study above argued that the adult peck does not arise because of maturation.

Second, evidence that operant contingencies have only limited effects on key-directed pecking established and maintained by Pavlovian contingencies is best seen just as evidence about the nature of interactions between these two forms of conditioning when applied to the same response. Also, in all cases where such joint control was examined there have been substantial operant effects on pecking (see Locurto 1981). However, Deich et al. (1988) did find an interesting asymmetry in their direct study of the effects of operant contingencies on the gape component of pecking. Gape could be moved upward to near its physical limit by differential reinforcement, but could be moved downward only to values short of physically realizable closure. Further, movement upward appeared to be easier and quicker. It seems possible that this asymmetry results from adaptation to the prehensile feeding task. A gape that is smaller than the target piece of food always results in a failed peck, as the food object is struck by one or both beaks during head descent and is knocked away, whereas a gape even much larger than the target always has some chance of success. This task asymmetry may have resulted in a biased response system that adjusts itself toward larger gapes more easily than toward smaller gapes.

Last, it seems that stereotypy has often been considered as a characteristic that strongly marks a behaviour as predominantly under biological determination. Such responses have variously been called fixed action patterns (Lorenz 1937, 1981), action patterns (Timberlake and Lucas 1988), preorganized, and so forth. The extent to which each of these terms has been used to emphasize that the response is genetically determined and under limited experiential influence has varied substantially (see Timberlake and Lucas 1988, for one of the more flexible approaches). However, there is no reason to believe that stereotypy in behaviour might not reflect consistency of experiences across individuals and across time. In fact, we suggest that the stereotypy of the pecking response arises from common elements in the environment in which development occurs. If these common elements are not present, the stereotypy may disappear. For example, when subjects are reared in environments with different food sources such as grain versus powdered food they show remarkably different feeding responses.

Since we do not think it correct to attribute response topographies solely to genetic influences, it is important to suggest alternative ways of looking at this problem. Below, we propose a framework for understanding how specific response topographies are acquired.

8.8 Task Analysis

Consider a problem: ring dove squabs do not exhibit the same gape as part of all feeding responses directed at localizable cues. Although pecking at seed frequently involves beak opening, begging involves no gape until after entry into the parent's beak. Why do these different forms occur? One useful approach to this sort of problem we call *task analysis.*

Task analysis proposes that the fit between the organism's effectors and the environment is so constrained that there are few ways for a given behaviour to be efficient and successful. The elements of task analysis are the relevant effectors, any other objects involved, and a description of the desired end. For example, the effector could be a hand and the task to move a block from one location to another.

A task analysis can be performed at several levels of abstraction. Abstract task analysis makes predictions based on the nature of the task variables, not the organism and its niche. An abstract task analysis of prehension might examine the grasping of spherical objects with a movable manipulator that has opposable (pincers-like) prehensile elements. The advantage of doing a task analysis at this level is that movement aspects dictated by an efficient and successful performance of the task thus defined can be anticipated largely independently of the particular agent performing the task. Here, we would predict that the pincers would open during movement to the target in order to be efficient. Further, the pincers would need to be opened to a larger aperture than the object that is to be moved, and then reclosed securely around it. Location of and movement toward the object require visual input for efficiency. Tactile input of the position of the object in the pincers would confirm a secure grasp, thus reducing errant attempts to move the object.

These predictions of the abstract task analysis are confirmed by the operational similarity of reaching and grasping by primate hands (Jeannerod 1981, 1984), by a robotic device called a "bin picker" (Kelley et al. 1983), and by the avian beak (Zeigler et al. 1980; Zweers 1982; Deich et al. 1985b; Klein et al. 1985; see Sect. 9.4.2). In all three cases, the configuration of the manipulator is adjusted to the target as the manipulator is transported (thrust) toward the target, the maximum opening of the manipulator occurring before contact and reclosure to secure the target. Also, at least part of the response is made under visual control, and in each case the manipulator is richly supplied with tactile receptors that aid in securing the target. This operational similarity is clearly an example of convergent adaptation to a task. It is the nature of the task itself, and the functional and structural similarity of the manipulators, that lead to highly similar operation across these obviously diverse systems.

Returning to our problem, we can apply task analysis to food ingestion in the squab by comparing begging with independent prehensile feeding. We suggest that, during begging, it is optimal for the squab to keep its beak closed and thus in a wedge-like configuration, in order to facilitate regurgitative feeding. This configuration will allow the squab's beak to enter the parent's beak, even if the parent's beak is not open wide enough to admit it otherwise. In contrast, feeding on grain requires that the beak is opened to a gape wider than the target seed in order to permit grasping and mandibulation.

Task analysis suggests the nature of effective responses, but not how these are to be attained in any specific case. Either species or individual adaptation might lead a response toward more effective forms. Since it seems clear there is a very substantial contribution of experience and/or learning (hence, individual adaptation) to the ontogeny of prehensile feeding in the ring dove, it is particularly important to consider the possibility that learning, possibly including forms of learning specific to development, under task (manipulator/object goal) contingencies is a primary determinant of the form of the ring dove's feeding response.

If task variables combine with a relatively plastic motor system in the ring dove to determine the form of feeding, then changing the nature of the task should result in changes in the feeding response. One of our experiments (Tankoos et al. in prep.) has examined this issue indirectly. One group of squabs was raised from hatching with only powdered seed available to them and their parents. A second group of squabs was raised on a diet of whole grain. Starting on day 20, all squabs were placed in a chamber with whole grain spread on the floor. Two squabs from each group were also attached to the magnetic gape measurement system and videotaped. If task contingencies affect what is learned, then the powder-reared squabs should differ from the seed-reared squabs in their response form on day 20.

During the first day on which the powder-reared squabs attempted to feed on grain, their response topographies were very unusual compared to any seen in seed-reared squabs or adults. One squab brought its head comparatively slowly to the floor, left its beak open for periods as long as 500 ms and scraped its lower beak forward along the floor with little lateral movement. The beak was moved along the floor both by the neck and by propelling the entire body forward. Another powder-reared squab also moved its head downward slowly and made scraping movements at the floor, but in contrast to the first squab there were both vertical and lateral movements of the head, both the top and bottom parts of the beak were used to scrape the floor, and the squab did not move its whole body with its beak touching the floor. Further, this squab adopted an unusual posture during these feeding attempts. The squab's body was inclined unusually steeply toward the floor so that the bird's posterior was held very high. The tail, which normally angles upward during feeding, angled downward toward the floor. The posture appeared to allow the squab to scratch the floor forcefully with its beak. In both powder-reared squabs, visual targeting of the grain on the floor appeared to be very poor. Both squabs made thrusting movements into sections of the floor one or more centimeters away from the nearest piece of grain.

These squabs showed feeding topographies that are more appropriate to consuming powdered food than whole grain. Further, these unusual feeding topographies were individualized. This result is congruent with the idea that a plastic motor system interacts with different task demands, as illustrated here by ingesting powdered seed or whole seed, so producing different response forms. Note that the changes resulting from the altered task encompass not just the gape and thrust components, but the whole of the squab's body and its movements. While task analysis does correctly predict that the change in the task induced by the presence of powdered food would lead to substantial changes in response topography, it does not clearly predict that the altered movements will be as strongly individualized as they were. Task analysis might view this individualization as evidence that there are two equally effective solutions to the task, and that chance differences in experience between the individuals account for the different response topographies. It seems possible that the use of an effector for purposes for which it is clearly not well suited, such as eating powder, is likely to produce multiple and possibly unstable task solutions.

Another way to change the nature of the feeding task is to change the effector system used rather than the food eaten. Two studies reviewed in Sect. 8.3 showed this sort of change by damaging parts of the beak prehensile system. Premock and Klipec (1981) removed the front of the upper half of pigeons' beaks and found a drastically altered feeding pattern. Deich et al. (1985b) cut efferents to the pigeon's lower beak and found compensatory adjustments to the movement time of the upper beak.

One strong implication of the view that specific topographies can be acquired by individual adaptation through task-guided learning is that it is problematic to use stereotypy as a criterion for arguing that control of a behaviour is the result of phylogenetic adaptation. Stereotypy may result from task variables and the resources available to perform the task acting to guide learning rather than from "endogeneously generated inherited central patterns" (Williams 1981; Lorenz 1981).

Even using the distribution of species typical behaviours among related species (Lorenz 1981, p. 108; see also Brown 1975, pp. 4–6, 491–492) as an indicator of strong genetic control of this behaviour at a central neural level is problematic, since different species often use anatomically different effectors to perform similar tasks. Species differences in response form could be due to the presence of anatomically different manipulators functioning under the control of similar, flexible, central mechanisms that have therefore been subject to different experiential histories, rather than strong differences between species in the underlying central control mechanisms themselves (cf. a similar argument developed in Sect. 7.11). Brown (1975, p. 6) notes in discussing Van Tets' (1965) work on the distribution of social communication behaviours in Pelecaniformes:

"It is interesting that the distribution of behavioural traits conforms better to a classification based on morphology than to one based on biochemical traits as well. Is this because the morphology influences the behaviour? For example, the absence of kink-throating in the Sulidae and its presence in the Phalacrocoracidae might be due to the morphological specializations of the neck for plunge-diving in the Sulidae and for surface-diving in the Phalacrocoracidae. Similar explanations can be imagined for several of the other behavioural differences."

Brown's speculation about morphological (effector and body) differences leading to species differences in behaviours fits task analysis well. If indeed morphological differences predict behavioural differences better than biochemical differences, this suggests that task-guided learning may be present in many cases.

In cases where there appears to be no self-differentiating central neural mechanism, and task-guided learning accounts for the response form, what is the role for the organism's genetic endowment, and what is the likely nature of the interaction between genetic and experiential influences on behaviour? A simple, if profound, role of the organism's genetics may be to substantially determine the anatomy of the effectors, and of the organism's body as a whole. Task analysis sees this anatomy as a critical element in determining response form since it imposes task constraints. The nature of the sensory input to the animal is critical and the animal's genetics are likely to have strong influences here. For example, it has been suggested that the tactile fields in the pigeon's beak are strongly tuned to grain-like objects (Witkovsky et al. 1973). The adaptation of sensory systems for particular stimuli is another likely source of genetic influence, although experiential factors are known to impact in sensory development as well.

In summary, we believe that task analysis provides a useful framework for understanding why particular topographies emerge during ontogeny. Task analysis is a way of understanding the fit between environment and behaviour, but it does not dictate that the fit is accomplished mainly at a species or an individual level of adaptation. Of course both of these levels contribute and these contributions interact.

8.9 Summary

There is a substantial influence of individual experience on the ontogeny of pecking in ring doves. While other influences are possible and the developmental picture is not yet complete, our research strongly suggests that ring dove squabs learn what to peck at through a Pavlovian association between objects and positive ingestional consequences, and that the form of the gape as well as the coordination of gape and thrust, appear to be moulded by the operant contingencies established by the success and failure of various movements. Further, we suggest that these learning processes operate in a highly constrained domain revealed by a task analysis of prehensile feeding. The constraints placed on behaviour by the nature of the beak, the jaw musculature, and the body and neck of a ring dove, in conjunction with the demands of the task, limit the range of potential feeding topographies.

These processes give rise to a high level of consistency in adult pecking within and between individuals as a function of time and across individuals. Developmental analysis makes clear that such consistency arises from the similarity of the effectors and the similarity of experiences for all individuals during development. In general terms, this illustrates that stereotyped, species-typical behaviour need not reflect an inherited central motor mechanism. Finally, task analysis appears useful in understanding the form of behaviours regardless of whether they are attained on the basis of individual adaptation, species adaptation or their interaction.

Acknowledgements. We thank Alice Deich and Katharine Iskrant for their comments on and help with the manuscript. We thank Jenine Tankoos for her vital role in conducting and analyzing this research. This work was supported by NSF grant BNS-8919231.

This chapter is dedicated to the memory of Eileen Shrager, my friend and colleague, who left us too soon – JD.

References

Alessandro D, Dollinger J, Gordon JD, Mariscal SK, Gould JL (1989) The ontogeny of the pecking response of herring gull chicks. Anim Behav 37:372–382

Balsam PD, Deich JD, Hirose R (1992a) The role of experience in the transition from dependent to independent feeding in ring doves. In: Turkewitz G (ed) Developmental psychobiology. New York Academy of Sciences, New York, pp 16–36

Balsam PD, Graf JS, Silver R (1992b) Operant and Pavlovian contributions to the ontogeny of pecking in ring doves. Dev Psychobiol 25:389–410

Barrera FJ (1974) Centrifugal selection of signal-directed pecking. J Exp Anal Behav 22:341–355

Bermejo R, Houben D, Allen RW, Deich JD, Zeigler HP (1989) Prehension in the pigeon I: descriptive analysis. Exp Brain Res 75:569–576

Brooks VB (1974) Some examples of programmed limb movements. Brain Res 71:299–308

Brown JL (1975) The evolution of behavior. Norton, New York

Brown P, Jenkins HM (1968) Auto-shaping of the pigeon's keypeck. J Exp Anal Behav 11:1–8

Chung S (1965) Effects of effort on response rate. J Exp Anal Behav 8:1–7

Cole JL (1965) Force gradients in stimulus generalization. J Exp Anal Behav 8:231–241

Davies MNO, Green PR (1989) Visual head extension: transitional head co-ordination in the pigeon squab (*Columba livia*). Dev Psychobiol 22:477–488

Deich JD, Balsam PD (1993) The form of early pecking in the ring dove squab (*Streptopelia risoria*): An examination of the preformation hypothesis. J Comp Psychol 107:1–15

Deich JD, Wasserman EA (1977) Rate and temporal patterning of keypecking under autoshaping and omission schedules. J Exp Anal Behav 27:399–405

Deich JD, Houben D, Allan RW, Zeigler HP (1985a) "On-line" monitoring of jaw movements in the pigeon. Physiol Behav 35:307–311

Deich JD, Klein BG, Zeigler HP (1985b) Grasping in the pigeon: mechanisms of motor control. Brain Res 337:362–367

Deich JD, Allan RW, Zeigler HP (1988) Conjunctive differentiation of gape during food-reinforced keypecking in the pigeon. Anim Learn Behav 16:268–276

Deich JD, Tankoos J, Balsam PD, Tracing the development of the gape component of prehensile feeding in the ring dove. (submitted)
Eckerman DA, Hienz RD, Stern S, Kowlowitz V (1978) Shaping the location of the pigeon's peck: effects of rate and size of shaping steps. J Exp Anal Behav 33:299–310
Eibl-Eibesfeldt I (1970) Ethology: the biology of behavior. Holt-Rhinehart-Winston, New York, pp 19–32
Goodale J (1983) Visually guided pecking in the pigeon (*Columba livia*). Brain Behav Evol 22:22–41
Goodwin D (1960) Comparative ecology of pigeons in inner London. Br Birds 53:201–212
Graf JS, Balsam PD, Silver R (1985) Associative factors and the development of pecking in ring doves. Dev Psychobiol 18:447–460
Hirose R, Balsam PD, Parent-squab interactions during the transition from dependent to independent feeding in the ring dove, *Streptopelia risoria.* (submitted)
Hogan JA (1973) How young chicks learn to recognize food. In: Hinde RA, Stevenson-Hinde J (eds) Constraints on learning. Academic Press, London, pp 119–139
Hogan JA (1984) Pecking and feeding in chicks. Learn Motiv 15:360–376
Hogan JA (1988) Cause and function in the development of behavioral systems. In: Blass EM (ed) Handbook of behavioral neurobiology, vol 9. Developmental psychobiology and behavioral ecology. Plenum Press, New York, pp 63–106
Hunt GL, Smith WJ (1967) Pecking and initial drinking responses in young domestic fowl. J Comp Physiol Psychol 64:230–236
Jeannerod M (1981) Intersegmental coordination during reaching at natural visual objects. In: Long J, Baddely A (eds) Attention and performance IX. Erlbaum, Hillsdale, NJ, pp 153–169
Jeannerod M (1984) The timing of natural prehension movements. J Mot Behav 16:234–254
Jenkins HM, Moore BR (1973) The form of autoshaped reinforcers. J Exp Anal Behav 20:163–181
Kelley RB, Birk JR, Martins HAS, Tella R (1983) A robot system which acquires cylindrical workpieces from bins. In: Pugh A (ed) Robot vision. Springer, Berlin Heidelberg New York, pp 285–294
Klein BG, Deich JD, Zeigler HP (1985) Grasping in the pigeon: final common path mechanisms. Behav Brain Res 18:201–213
LaMon B, Zeigler HP (1984) Grasping in the pigeon (*Columba livia*): stimulus control during conditioned and consummatory responses. Anim Learn Behav 12:223–231
Lehrman DS (1955) Interaction between interval and external environments in the regulation of the reproductive cycle of the ring dove. In: Beach FA (ed) Sex and behaviour. Wiley, New York, pp 355–380
Locurto CM (1981) Contributions of autoshaping to the partitioning of conditioned behavior. In: Locurto CM, Terrace HS, Gibbon J (eds) Autoshaping and conditioning theory. Academic Press, New York, pp 101–135
Lorenz K (1937) Über die Bildung des Instinktbegriffes. Naturwissenschaften 25:289–300, 307–318, 324–331
Lorenz K (1981) Foundations of ethology. Springer, Berlin Heidelberg New York
Margolis RA, Mariscal S, Gordon J, Dollinger J, Gould JL (1987) The ontogeny of the pecking response in laughing gull chicks. Anim Behav 35:191–202
Meyer D, Abrams RA, Kornblum S, Wright CE, Smith JEK (1988) Optimality in human motor performance: ideal control of rapid aimed movements. Psychol Rev 95:340–370
Meyer DE, Smith JEK, Kornblum S, Abrams RA, Wright CE (1990) Speed-accuracy tradeoffs in aimed movements: toward a theory of rapid voluntary action. In: Jeannerod M (ed) Attention and performance, XIII. Motor representation and control. Erlbaum, Hillsdale, NJ, pp 173–226
Moore BR (1973) The role of directed Pavlovian reactions in simple instrumental learning in the pigeon. In: Hinde RA, Stevenson-Hinde J (eds) Constraints on learning. Academic Press, New York, pp 159–188
Palameta B, Lefebvre L (1985) The social transmission of a food-finding technique in pigeons: what is learned? Anim Behav 33:892–896
Patel MD (1936) The physiology of the formation of 'pigeon's milk'. Physiol Zool 9:29–152

Premock M, Klipec WD (1981) The effects of modifying consummatory behavior on the topography of the autoshaped pecking response in pigeons. J Exp Anal Behav 36:277–284

Tankoos J, Deich JD, Balsam PD Effects of experience with seed on feeding topography. (in prep.)

Timberlake W, Lucas GA (1988) The basis of superstitious behavior: chance contingency, stimulus substitution, or appetitive behavior? J Exp Anal Behav 44:279–299

Van Gennip EMSJ (1988) A functional-morphological study of the feeding system in the pigeon (*Columba livia* L.): behavioural flexibility and morphological plasticity. Thesis, University of Leiden, Netherlands.

Van Tets GF (1965) A comparative study of some social communication patterns in the Pelecaniformes. Ornithol Monogr 2:1–88

Williams DR (1981) Biconditional behavior: conditioning without constraint. In: Locurto CM, Terrace HS, Gibbon J (eds) Autoshaping and conditioning theory. Academic Press, New York, pp 55–100

Williams DR, Williams H (1969) Automaintenance in the pigeon: sustained pecking despite contingent nonreinforcement. J Exp Anal Behav 12:511–520

Witkovsky P, Zeigler HP, Silver R (1973) The nucleus basalis of the pigeon: a single unit analysis. J Comp Neurol 147:119–127

Wolin BR (1968) Difference in manner of pecking a key between pigeons reinforced with food and with water. In: Catania AC (ed) Contemporary research in operant behaviour. Scott, Foresman, Glenview, III (Reprint of paper read at Conf on Experimental analysis of behavior, 1948), p 283

Woodruff G, Starr MD (1978) Autoshaping of initial feeding and drinking reactions in newly hatched chicks. Anim Learn Behav 6:265–272

Wortis RP (1969) The transition from dependent to independent feeding in the young ring dove. Anim Behav Monogr 2:1–54

Zeigler HP, Levitt P, Levine RR (1980) Eating in the pigeon (*Columba livia*): movement patterns, stereotypy and stimulus control. J Comp Physiol Psychol 94:783–794

Ziriax JM, Silberberg A (1978) Discrimination and emission of different key-peck durations in the pigeon. J Exp Psychol Anim Behav Processes 4:1–21

Zweers GA (1982) Pecking of the pigeon (*Columbia livia* L.). Behaviour 81:173–230

9 Ingestive Behaviour and the Sensorimotor Control of the Jaw

H.P. Zeigler[1], R. Bermejo[1] and R. Bout[2]

9.1 Introduction

In its natural habitat, the pigeon's ingestive behaviour involves both transport of a prehensile effector organ (the beak) towards food or water, and the generation of jaw movement (gape) patterns with a topography appropriate to the stimulus properties of the nutrients and the motivational state of the animal. The transport component is mediated by the neck muscles (see Chap 10) and the gape component by the jaw muscles. This review focuses primarily upon the jaw motor system (Zeigler 1989).

Ingestive jaw movements reflect a variety of adaptive requirements and include both episodic and rhythmic movements. They are guided by a combination of visual and somatosensory inputs, and their sensorimotor control is likely to involve feed-forward, feedback and central pattern generating mechanisms. The pigeon's jaw movements may be studied within their normal ingestive context (Zeigler et al. 1980; Zweers 1982a, b; Klein et al. 1983; Bermejo and Zeigler 1989) or they may be brought under experimental control using classical and operant conditioning techniques. (Mallin and Delius 1983; LaMon and Zeigler 1984, 1988; Deich et al. 1988; Allan and Zeigler 1993; Remy and Zeigler 1993; see Chap. 8.) Moreover, jaw movements may be elicited in immobilized pigeons, so that neurophysiological analyses may be carried out in awake, behaving birds (Bermejo et al. 1992).

Such analyses are facilitated by several simplifying features of the pigeon's jaw motor system. As in mammals, the jaw is controlled by a small number of muscles (van Gennip 1986), but, unlike mammals, ingestive jaw movements are confined largely to a single plane. The pigeon's jaw movement system is not burdened with the added complexity of mastication, has only a minor postural function, and deals with a relatively constant load, thus eliminating the need for the complex systems of load and disturbance compensation characteristic of limb muscles. However, it shares with limb (and eye) muscles the ability to make rapid, controlled and amplitude-scaled movements in an open-loop fashion.

[1] Biopsychology Program, Department of Psychology, Hunter College, City University of New York, 695 Park Ave, New York, NY 10021, USA
[2] Department of Biology, Leiden University, The Netherlands

M.N.O. Davies and P.R. Green (Eds.)
Perception and Motor Control in Birds

Like the oculomotor system (Robinson 1986), which has often been viewed as a "model system" for the study of movement, its central neural elements are contained entirely within the cranium so that one can, in principle, record from all its neurons.

9.2 Ingestive Behaviour: Descriptive Analysis

The pigeon's eating behaviour (Fig. 9.1A) requires the organization of several functionally distinct movement patterns into an adaptive behavioural sequence about 300 ms in duration. *Pecking* consists of a series of saccadic head movements made with the eyes closed, but punctuated by fixation pauses during which the eyes are open and the seed may be identified as food and localized in space (Zeigler et al. 1980; Goodale 1983). Eating also includes two prehensile jaw movement responses, grasping and mandibulation. *Grasping* is integrated into the pecking response and is divisible into opening and closing phases. During the opening phase, gape (i.e. interbeak distance) is scaled to the size of the food object, reaching its peak just prior to contact (LaMon and Zeigler 1984; Deich et al. 1985; Bermejo et al. 1989; see Sect. 8.3). The closing phase of grasping is initiated at contact and terminates with the seed held between the beak tips and the eyes fully open. The entire grasping phase takes between 30 and 60 ms. *Mandibulation* is divisible into two distinct phases, termed *stationing* and *intraoral transport*. Stationing consists of very rapid (10 to 15 ms) opening and closing movements of the jaw coordinated with "tossing" movements of the head, which are not transport movements, but serve to reposition the seed prior to transport. Intraoral transport of small seeds involves a lingual transport mechanism ("slide and glue") in which mucous secretions from intraoral glands cause the seed to adhere to the dorsal surface of the tongue, which conveys it from the beak tips to the caudal palate and pharynx. With larger seeds, an inertial ("catch and throw") mechanism is used, which combines a head jerk and a complete gape cycle, with the tongue retracted. During mandibulation, jaw movements are adjusted to both the size and the location of the seed. *Swallowing* is mediated primarily by the pharyngeal muscles, with little or no contribution from the jaw muscles. Dropping or failing to grasp the seed terminates the ingestive sequence, suggesting that mandibulation is under the sensory control of tactile stimuli from the seed (Zeigler et al. 1980; Zweers 1982a; Bout and Zeigler in revision a).

Drinking begins with the head elevated, the eyes open and the beak almost closed. Head descent is coupled with eye closure and both movements terminate on contact with the water (Fig. 9.1B). Water intake in the pigeon involves an active suction mechanism, in contrast to lingual transport of liquid in many mammals and to the passive gravity mechanism of some bird species (see also Sect. 10.4.2). Immersion of the beak is followed by a series of cyclical opening and closing movements of the jaw. During the jaw opening (lingual suction) phase, water enters the oral cavity and is transported in a caudal direction by a

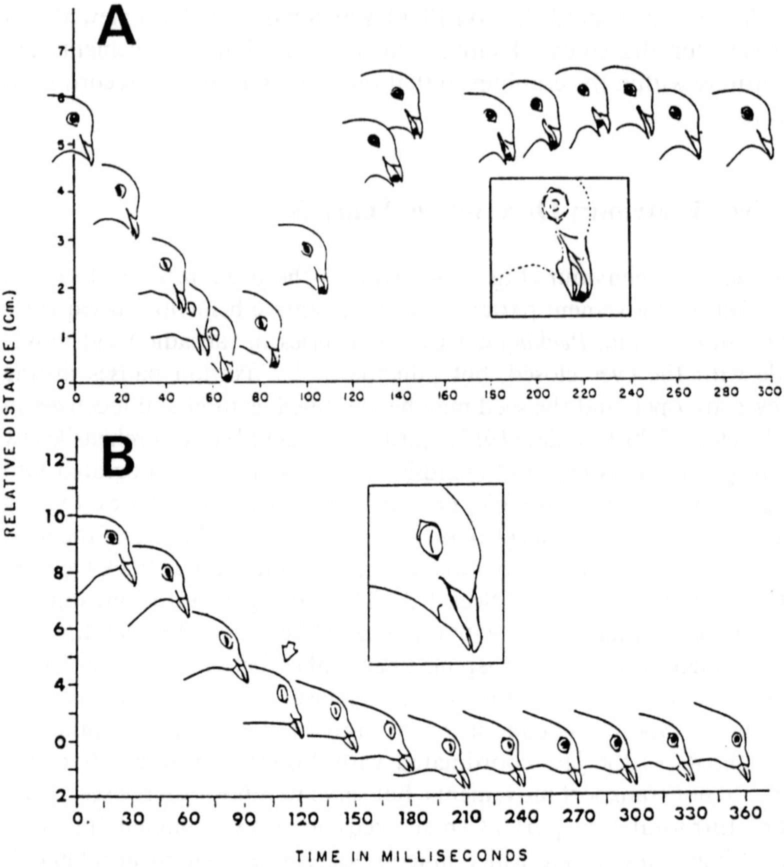

Fig. 9.1A, B. Ingestive behaviour in the pigeon. **A** Eating behaviour (*inset* shows the position of the tongue at the start of intraoral transport); **B** Drinking behaviour. *Arrow* indicates point of contact with the water surface (see *inset*). Based upon cinematographic records taken at 100 frames s^{-1} (Zeigler et al. 1980)

combination of capillary action, the piston-like action of the tongue and enlargement of the buccal cavity. During the jaw closing (Pharyngeal suction) phase, intraoral transport of the water is produced by simultaneous elevation of the floor of the buccal cavity and downward movement of the glottis and lingual base, creating an area of low pressure within the pharynx. This phase is accompanied by tongue protraction. The alternation of jaw opening and jaw closing movements produces cyclical variations in gape which are functionally related to each of these phases (Zweers 1982b; Klein et al. 1983).

Once initiated by contact with the water, drinking movements continue without any obvious phasic inputs. Indeed, "dry drinking" movements are sometimes made to the water container itself. In contrast, eating movements form a sensorimotor chain whose links are provided by phasic inputs from the food object. From a kinematic standpoint, however, the essential feature of jaw movement patterns during eating is the scaling of gape size to the size of the food object; for drinking it is the cyclicity and relative invariance of gape.

9.3 Functional Considerations

The jaw movement patterns generated during eating and drinking reflect the adaptive requirements of the pigeon's ingestive behaviour. During eating, the generation of jaw opening amplitudes approximately proportional to the size of the food object is a characteristic of all three phases. However, from a functional standpoint, each phase may be considered a different task, with unique requirements, eliciting stimuli and kinematic properties. Grasping is elicited and guided by the visual properties of the seed and the timing of the jaw opening movement must be coordinated with that of head descent so that it is completed before contact with the seed. Opening gape must be large enough to grasp the pellet, since undershooting of gape will lead to an unsuccessful grasp (Levine and Zeigler 1981). In stationing, the eliciting stimulus is probably tactile and the task is the appropriate positioning of the seed at the beak tips. Since the opening and closing of the jaws must be rapid enough to prevent loss of the object during the head toss, speed is critical. Finally, since pecking rates in excess of two or three per second have been reported (Zeigler et al. 1980), and ingestion of each new seed requires transport of the previous seed from the beak tips to the oesophagus, additional temporal constraints are present for intraoral transport.

A different, but no less stringent set of constraints operates during drinking. Water intake involves both a passive process of capillary action and the operation of an active suction mechanism for its transport within the buccal cavity. The cyclical alternation of jaw opening and closing movements and the relatively constant gape are critical to both processes. They allow, successively, intake of the water, formation of a tube for the generation of suction by the tongue, and isolation of a bolus of liquid within the oral cavity. Moreover, to the extent that drinking behaviour is sensitive to environmental constraints such as water depth, either cycle duration, jaw-opening amplitude or both may be modulated to control the volume of water ingested. Thus, both eating and drinking require the generation and modulation of characteristic and relatively constant jaw movement patterns.

9.4 Kinematic Analysis of Ingestive Jaw Movement Patterns

The first stage in clarifying the motor control mechanisms underlying the generation of ingestive jaw movement patterns is the quantitative characterization of these patterns. This may be achieved using a magnetosensitive transducer mounted on the upper beak and a small, rare-earth magnet mounted on the lower beak (Fig. 9.2A). The transducer generates a voltage proportional

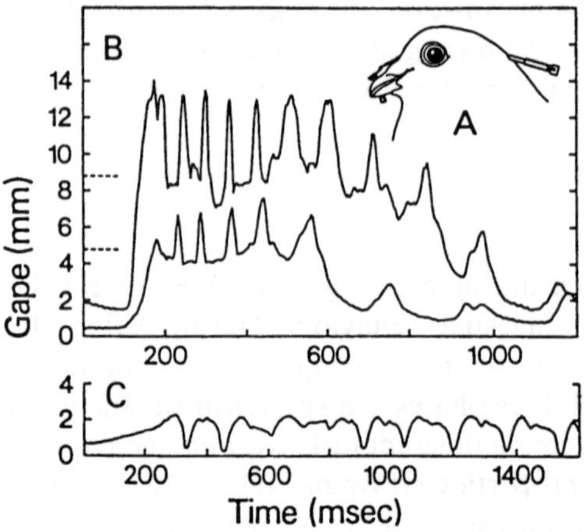

Fig. 9.2. Monitoring of jaw movements during eating and drinking in the pigeon. *A* A magnetosensitive (Hall effect) device is mounted on the upper beak and a rare-earth magnet on the lower beak; *B* jaw movements during eating. Pellet sizes (4.9, 8.7 mm) are indicated by the *dashed lines* on the *ordinate*. Upward deflections of the record indicate jaw opening; *C* jaw movements during drinking. Note the scaling of jaw opening for eating, and the cyclicity of drinking movements (Deich et al. 1985)

to the strength of the applied magnetic field (which varies with interbeak distance), generating a continuous record of variations in gape. Such records, obtained during drinking (Fig. 9.2C) and during the ingestion of food pellets of various sizes (Fig. 9.2B), provide kinematic variables (amplitude, velocity and duration) which are controlled to produce jaw movement patterns during eating and drinking. The results of such analyses may offer some insights into the motor control "strategies" underlying the movements (Bermejo and Zeigler 1989; Bout and Zeigler in revision a, b).

9.4.1 Kinematics of Drinking

Drinking topography varies with water level, and intake may involve either "rictus" (deep) or "tip" (shallow) drinking. In either condition, intake is mediated by cyclic opening and closing movements of the jaw. A continuous series of such cycles constitutes a drinking bout. Figure 9.3A illustrates variations in gape recorded during a single bout of "tip" drinking. The initial and terminal cycles are shown, together with a representative sample of cycles from the middle of the bout. Figure 9.3B shows a superimposed series of seven successive cycles from the same drinking bout, aligned with the start of the opening movement. Note that over the first few cycles gape amplitude is variable, and there is a gradual increase in the duration of each cycle and a transition to the waveform of the middle portion. Cycles in the middle of the bout exhibit one or two rapid, but transient, opening movements, a period of sustained opening of variable duration, and a closing movement which returns the gape record to baseline. Within a single drinking session, mean cycle durations do not change across the first, middle and final bouts of the session; that is, there are no systematic changes attributable to satiation. During either tip or rictus drinking, peak opening amplitude remains relatively constant. However, gape is larger during rictus

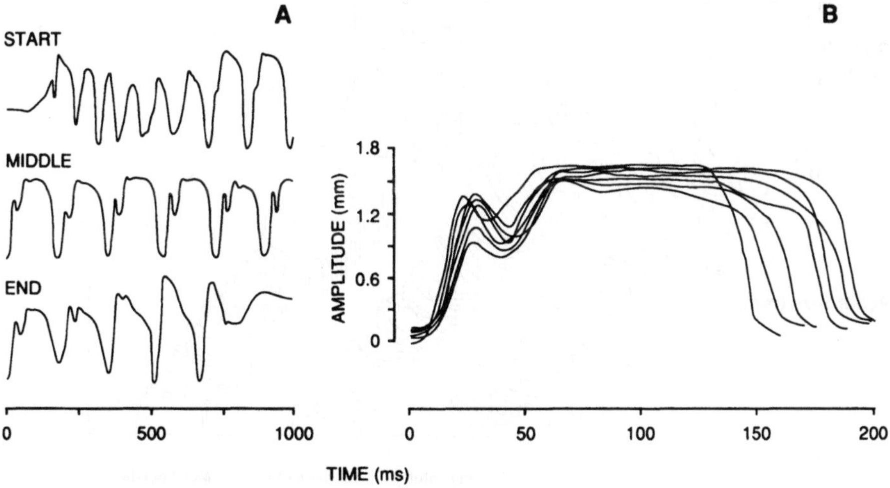

Fig. 9.3A, B. Kinematics of drinking in the pigeon. **A** Cyclic variations in gape during the initial, middle and end portions of a bout of "tip" drinking; **B** a superimposed series of seven successive jaw movement cycles from the middle portion of a single drinking bout, aligned with the start of the opening movement (Bout and Zeigler in press b)

drinking. This increase in opening amplitude is correlated with an increase in opening velocity rather than opening duration.

9.4.2 Kinematics of Eating

Figure 9.2B illustrates variations in gape recorded during the ingestion of pellets of two different sizes. The opening and closing phases of grasping appear as rapid upward and downward deflections in the gape record. After closing, the gape record does not return to baseline, but is maintained at a (reclosure) plateau whose value approximates the diameter of the food pellet (Deich et al. 1985, Table 1). Stationing is transduced as a second period of opening and closing, but with steeper slopes than grasping, while intraoral transport has both slow and fast opening phases.

With increasing pellet size, there is a proportional increase in the amplitude of the jaw opening response. The scaling of gape size to object size is particularly striking during grasping, because gape is minimal at the start of the response. Amplitude scaling of gape is shown in Fig. 9.4, which presents sets of jaw movement trajectories for grasping and stationing of a range of pellet sizes from 3.9 to 11.1 mm. Note that there may be some variation in peak gape within pellet size, but that across pellet sizes there is a systematic increase in the peak of the opening trajectory.

Our kinematic analysis indicates that, for the grasping response, scaling of gape amplitude to pellet size is produced by increasing either the velocity or

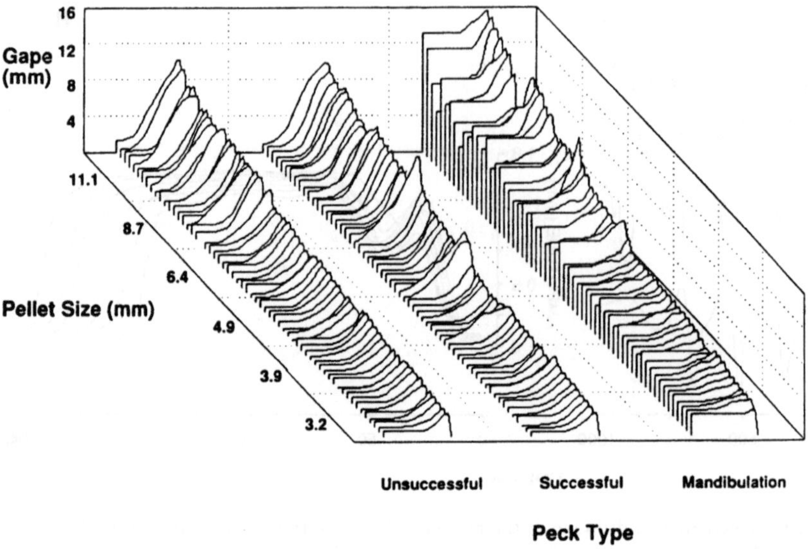

Fig. 9.4. Amplitude scaling of jaw movements in the pigeon. Jaw movement trajectories during eating as a function of food pellet size, for successful and unsuccessful grasping responses and for the stationing component of mandibulation (Bermejo et al. 1989)

the rise time of the jaw opening trajectory, or some combination of the two. Figure 9.5 presents examples of jaw opening trajectories selected to illustrate each of these strategies. In Fig. 9.5 (top) opening velocity, as reflected in the trajectory slope, remains relatively constant while rise time increases with pellet size. In the middle portion of Fig. 9.5, rise times are relatively constant and increased opening amplitudes are generated by increasing opening velocity. For the two subjects whose data are shown in Fig. 9.5 (top, middle), the two kinematic variables contribute about equally to amplitude scaling of gape, but their relative contribution may differ for individual birds.

The positive correlations between amplitude and rise time seen *across* pellet sizes are not present for populations of jaw opening trajectories *within* a single pellet size. Instead, there is a negative correlation between opening velocity and rise time within a single pellet size. An examination of individual trajectories suggests that rise time may function in a compensatory manner to "adjust" opening trajectories *during* their execution in order to generate peak amplitudes appropriate to the size of the target. Thus quite different initial velocities may generate very similar peak amplitudes depending upon rise time. Figure 9.5 (bottom) shows two trajectories made to the pellets of the same size, with initially different velocities but the same final amplitude. Analysis of the relation between several kinematic parameters indicates that initial accelerations predictive of gape undershooting are compensated for by increases in rise time; those predictive of gape undershooting are accompanied by decreased rise times (Bermejo and Zeigler 1989).

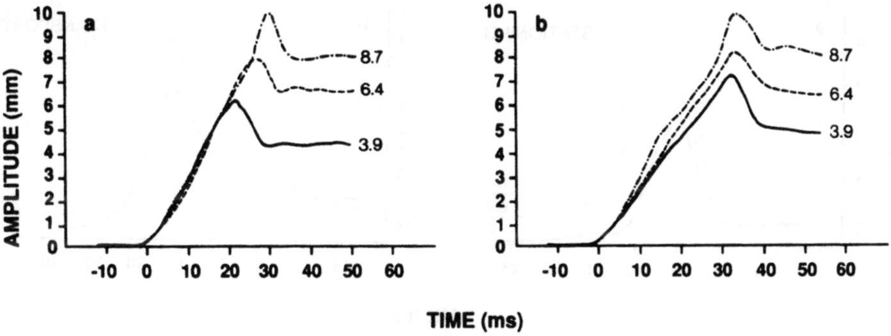

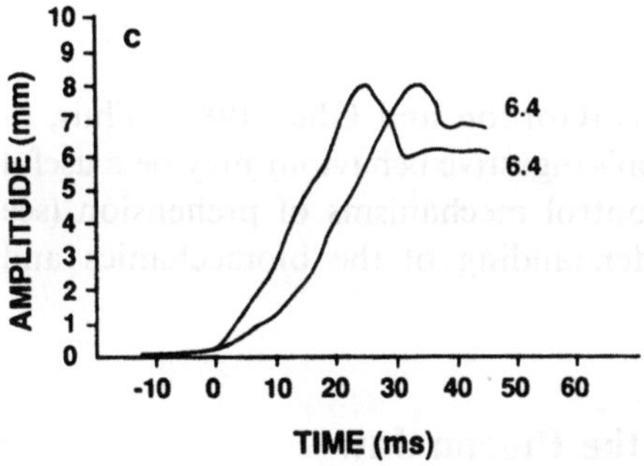

Fig. 9.5a–c. Jaw opening trajectories and the kinematics of amplitude scaling during eating. Amplitude adjustments across pellet sizes may be achieved either **a** by increasing rise time or **b** by increasing velocity. Amplitude adjustments within a single pellet size reflect compensatory adjustments in rise time. **c** Two trajectories made to the same (6.4 mm) pellet size with different initial velocities achieve the same final amplitude (Bout and Zeigler in press a)

Jaw opening trajectories for stationing and intraoral transport are presented in Fig. 9.6. Because the mouth is already open at the start of both movements, amplitude scaling of gape, though present, is less striking than for grasping. As in grasping, amplitude scaling involves increases in both velocity and rise time. In contrast with grasping, the variability of the scaling relationship is greater for these movements and individual birds often scale in a non-linear manner. Moreover, the contribution of rise time, although present, is noticeably reduced, perhaps reflecting the temporal constraints of the tasks.

Amplitude scaling (i.e. the accurate adjustment of movement amplitude to task requirements) is a critical function of motor control mechanisms and is particularly obvious in the case of prehensile behaviours. Comparison of the kinematics of reaching/grasping in humans and pecking/grasping in the pigeon reveals many similarities in both the topography of the response and the kinematic mechanisms mediating amplitude scaling (Jeannerod 1981; Klein et al. 1985; Bermejo and Zeigler 1989; Paulignan et al. 1990; Jakobson and Goodale 1991; Bout and Zeigler in press a). These include similar position-velocity profiles for the transport (reach, peck) components, and the adjustment of the prehensile effector organ (hand, jaw) to the stimulus properties of the target (size, shape). Moreover, "corrective adjustments" to movement trajectories during their execution have previously been reported in studies of the scaling

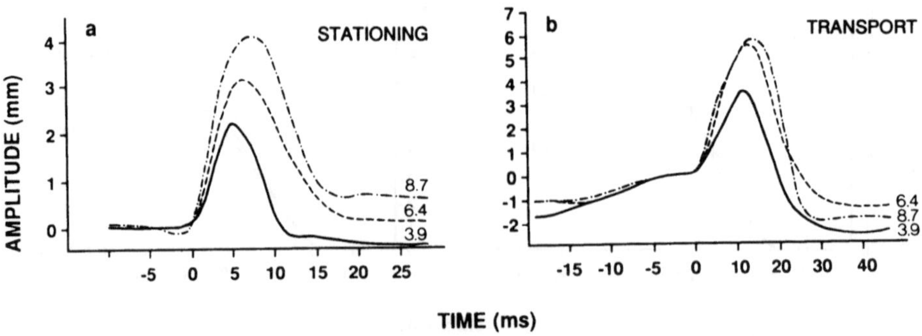

Fig. 9.6a, b. Kinematics of jaw opening trajectories across pellet size for **a** stationing, **b** intraoral transport (Bout and Zeigler in press a)

of isometric force trajectories in humans (Gordon and Ghez 1987). Thus, in addition to its intrinsic interest, the pigeon's ingestive behaviour may be a useful "model system" for studies of motor control mechanisms of prehension (see Sect. 8.8). Such studies require an understanding of the biomechanics and functional morphology of the jaw.

9.5 Morphology and Myology of the Pigeon Jaw

The pigeon is like other birds but unlike mammals in that both the upper and lower jaws are moveable. However, the operation of the lower jaw (mandible) is constrained by the presence of an inextensible (postorbital) ligament, which spans the quadrato-mandibular and the quadrato-cranial joints (Fig. 9.7A). In the resting position the action of this ligament prevents depression of the mandible. That is, at the start of eating or drinking, the lower jaw is effectively "locked" and must be "unlocked" to contribute to variations in gape (Bock 1964). Such "unlocking" may be achieved by a complex sequence of muscle contractions and quadrate movements, producing slackening of the ligament.

Jaw movements in pigeon are controlled by seven pairs of muscles (Fig. 9.7B), divisible into four functional groups: protractors, depressors, adductors and pterygoids (van Gennip 1986), of which all but the depressor are innervated by the trigeminal nerve (Wild and Zeigler 1980). However, none of the jaw muscles insert directly upon the maxilla, so that upper jaw opening must be indirectly produced by the contraction of its protractor muscle operating through a kinematic chain of bones, including the quadrates, pterygoids and palatines. Once the system is unlocked, opening of the lower jaw is produced by contraction of its depressor muscle.

Van Gennip (1988) proposed that "unlocking" may be achieved by mechanisms involving protractor contraction and quadrate movement, with or without elevation of the upper beak. Van Gennip's account predicts that,

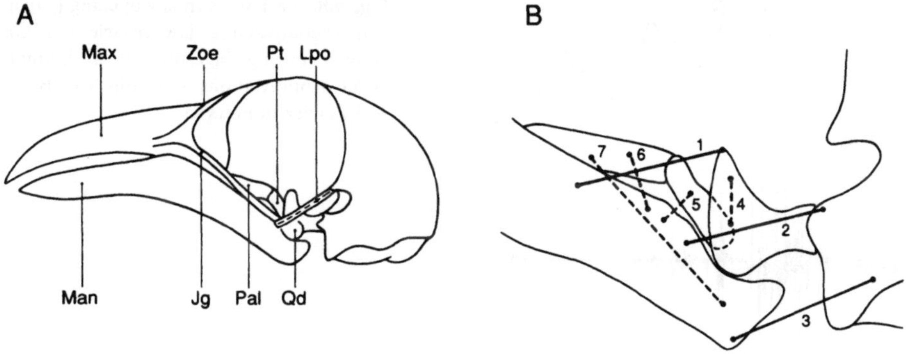

Fig. 9.7A, B. Morphology and myology of the pigeon jaw. **A** Lateral view of the pigeon skull, indicating its major components; **B** schematic diagram illustrating the origins and insertions of the pigeon's jaw muscles. *Jg* Jugal bar; *Lpo* postorbital ligament; *Man* mandible; *Max* Maxilla; *Pal* palatine bone; *Pt* pterygoid bone; *Qd* quadrate bone; *Zoe* zona elastica; *1* M. pseudotemporalis profundus (PTP); *2* M. adductor mandibulae externus pars medialis (AMEM); *3* M. depressor mandibulae (DM); *4* M. protractor quadrati et pterygoidei (PQP); *5* M. pterygoideus pars dorsalis caudalis (PDC); *6* pars dorsalis rostralis (PDR); *7* M. pterygoideus pars ventralis lateralis and pars ventralis medialis (PVL/PVM)

whichever of these mechanisms is used in a given movement, activity in the maxillary protractor muscle will always be the first event in the sequence of jaw muscle activity.

9.6 Electromyographic Analysis of Ingestive Jaw Movements

The jaw muscle activity patterns mediating ingestive behaviour in pigeons must reflect both the biomechanical constraints of the system and the functional requirements of the behaviour. Our analysis was designed to clarify the relation between muscle activity patterns during eating and drinking and the functional operation of the jaw muscle system during ingestion within the framework of those constraints (Bout and Zeigler in revision a, b).

Using conventional procedures, we recorded EMGs from muscles representing each of the pigeon's four functional groups, including the openers of the upper and lower jaws (M. protractor quadrati et pterygoidei: PQP; M. depressor mandibulae: DM), the skull and quadrate adductors (M. adductor mandibulae externus pars medialis: AMEM; M. pseudotemporalis profundus: PTP) and the pterygoid complex (M. pterygoideus pars ventralis lateralis: PVL).

9.6.1 Jaw Muscle Activity Patterns During Eating

The temporal organization of EMG activity during eating is illustrated in Fig. 9.8, which plots the activity of several jaw muscles during the ingestion of a

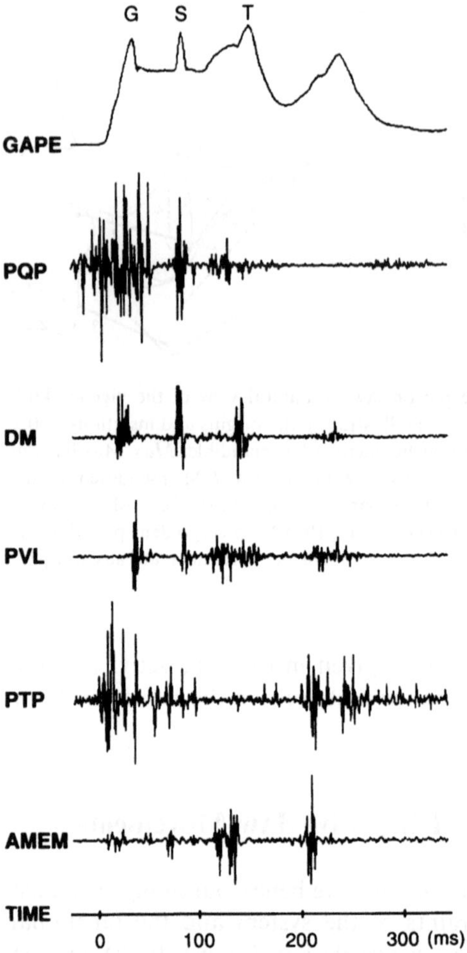

Fig. 9.8. Variations in jaw opening (gape) and accompanying jaw muscle (EMG) activity during ingestion of a 3.9-mm pellet. Abbreviations as in Fig. 9.7. (Bout and Zeigler in press a)

small (3.9 mm) food pellet. The onset of jaw opening during grasping is preceded by activity in the maxillary protractor (PQP) which continues throughout its rising phase and is accompanied by coactivation of the mandibular depressor (DM) and the skull and quadrate adductors (AMEM, PTP). Just prior to closing, the pterygoid antagonist (PVL), which has been relatively inactive, emits a brief burst of activity whose peak coincides with the peak in the opening amplitude record and declines rapidly. Jaw closing is followed by an amplitude plateau in the gape record accompanied by low level activity in the adductors. Stationing is preceded by activity in both the protractor and depressor muscles, which continues through the jaw opening phase and is followed by a PVL burst coincident with closing. During stationing, activity in the adductors is highly variable. During the slow phase of intraoral transport, jaw opening is not accompanied by activity in either opener muscle. During the rapid phase, opening is preceded by PQP activity, which declines as opening reaches a maximum and is followed by DM activity.

To examine the manner in which this basic pattern of muscle activity meets the functional requirements of eating, including the amplitude scaling of gape, we compared EMG and movement data obtained during the ingestion of food pellets of three different sizes (3.9, 6.4 and 8.7 mm). Figure 9.9 is a summary schematic diagram illustrating the correlated variations in jaw muscle activity and jaw movements accompanying the ingestion of each pellet size.

For all three pellet sizes, the onset of jaw opening during grasping is always preceded by PQP activity (consistent with its role is "unlocking" the system), but neither the magnitude nor the duration of the PQP burst varies with pellet size. Control of the opening trajectory involves coactivation of both opener (DM) and closer (adductor, pterygoid) muscles. The slope of the trajectory is controlled by the combined activity of the depressors and adductors. Its rise time is determined primarily by the duration of the DM burst, and the timing of its closing phase (the inflection point) reflects the onset of an activity burst in PVL. Amplitude scaling of the jaw opening trajectory during grasping is unrelated to activity in the upper jaw opener (PQP). It is produced by modulating the duration of activity in the lower jaw opener (DM) and the onset time of an activity burst in PVL, which serves as a brake on the activity of the openers. Although the average magnitude of the PVL burst is the same for all pellet sizes, its onset latency increases with pellet size. In other words, the "brake" is applied later in the movement, permitting a longer DM burst and consequently prolonging the rise time of the opening trajectory. The amplitude plateau following closing approximates the diameter of the pellet. Throughout this period "unlocking" may be maintained, passively, by the presence of the pellet between the jaws, while the active grasping required to hold the pellet in the place is probably produced by the low level activity of the adductor muscles.

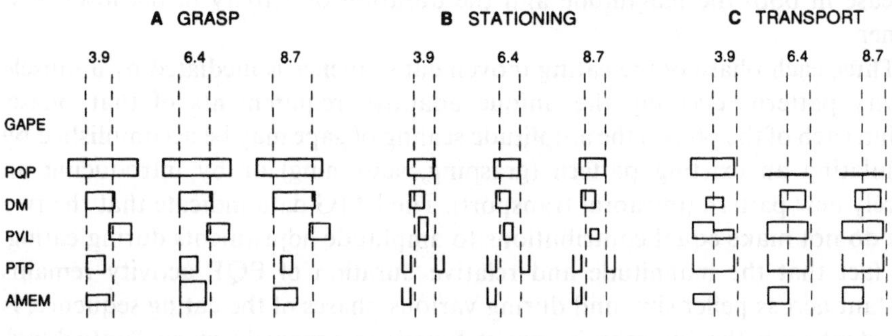

Fig. 9.9. Schematic summary diagram to illustrate the relation between the modulation of jaw muscle activity patterns and the adjustment of jaw opening. The width and height of each closed block are proportional, respectively, to the duration and magnitude of the EMG burst in the muscle represented. *Open blocks* indicate variable EMG activity. See text for details (Bout and Zeigler in press a)

The onset of stationing coincides with simultaneous bursts of activity in both opener muscles. Since the lower jaw is already "unlocked", both jaws contribute to the opening movement, accounting for its abrupt onset. The presence of increased DM activity together with the absence of antagonist (adductor) activity is consistent with its increased velocity. As in grasping, the timing of the closing phase involves the rapid onset of a brief activity burst in PVL, again serving as a brake on opener activity. Thus the jaw opening trajectory during scaling reflects primarily the activity of the two jaw openers (PQP, DM) and their pterygoid antagonist (PVL). Amplitude scaling of the trajectory with pellet size does not involve control of either the duration or magnitude of agonist activity nor of the latency of the PVL burst. However, there is a systematic (inverse) relation between pellet size and the magnitude of PVL activity, such that the larger the pellet the smaller the peak magnitude of the PVL burst and its accompanying force. Put another way, amplitude scaling during stationing is controlled by modulating the force with which the PVL "brake" is applied.

During the slow phase of intraoral transport, neither jaw opener muscle is active, suggesting that jaw opening in this phase must be produced by the action of the tongue, which is simultaneously protracting and being pushed under the pellet (Zweers 1982a). This movement is accompanied by adductor and pterygoid activity which could maintain active closing of the jaws against the forward movement of the pellet. For the smallest pellet, PQP activity rises above resting levels a few milliseconds prior to the onset of the rapid transport phase, but is almost absent for the two larger pellets. However, for all three pellet sizes, the rapid phase is preceded by a rapid burst of DM activity which parallels a trough in the pterygoid activity. A decline in DM activity and a small increase in the variable activity of PVL appears to time the onset of jaw closing. Amplitude scaling during intraoral transport involves different adjustments for the small and large pellets. For the smallest pellet, opening is produced by activity in both openers; for the larger pellets PQP activity is absent and the increased opening amplitude required for their transport is produced by a compensatory increase in both the magnitude and the duration of activity in the lower jaw opener.

Thus, each phase of the eating movement sequence is mediated by a muscle activity pattern reflecting the unique adaptive requirements of that phase. Within each of the phases the amplitude scaling of gape may be accomplished by modulating an existing pattern (grasping, stationing) or by introducing an entirely new pattern (intraoral transport). The EMG data indicate that the two jaws do not make equal contributions to amplitude adjustments during eating. The fact that the magnitude and relative duration of PQP activity remains constant across pellet size, and during various phases of the eating sequence, is consistent with the hypothesis that it functions primarily in an "unlocking" mechanism. The coactivation of agonist and antagonist (depressor and adductor) muscles to control opening trajectory has also been reported for jaw muscle activity during feeding in salamanders (Reilly and Lauder 1990).

9.6.2 Jaw Muscle Activity Patterns During Drinking

Jaw movements during drinking are cyclic, and a continuous series of individual cycles constitutes a drinking bout. Figure 9.10 illustrates, for a single subject, patterns of jaw movement and jaw muscle activity recorded during two successive cycles within a bout of "tip" drinking and aligned with the start of jaw opening. The EMG records shown are from jaw openers (PQP, DM), skull and quadrate adductors (AMEM, PTP), and pterygoids (PVL). Both openers are active during the initial period of rapid opening, but activity is seen only in the lower jaw opener during the period of sustained opening. Immediately prior to closing there is a gradual reduction in DM activity, accompanied by the abrupt onset of activity in the adductor and pterygoid muscles. The duration of jaw

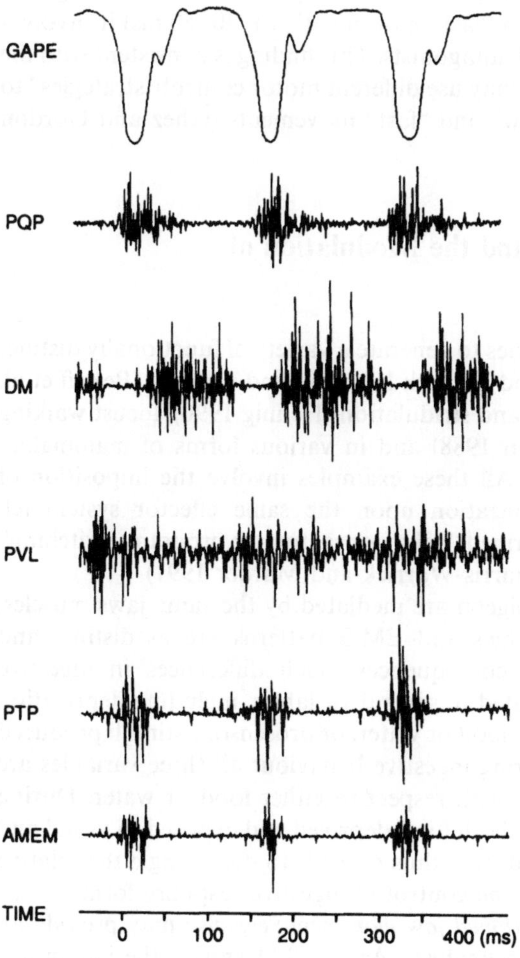

Fig. 9.10. EMG analysis of drinking in the pigeon. Variations in jaw opening (gape) and accompanying jaw muscle (EMG) activity during a bout of "tip" drinking. The records are aligned with the start of jaw opening. Abbreviations as in Fig. 9.7 (Bout and Zeigler in press b)

opening was significantly correlated with the duration (but not the amplitude) of DM activity, as well as with the onset latency of activity in AMEM and PVL which, as in eating, serves as a "brake" on opening.

Thus, both eating and drinking movements are mediated by characteristic and relatively constant patterns of jaw muscle activity. In both behaviours, variations in the pattern may occur under constant conditions (e.g. within a single drinking bout), or may be elicited by variations in sensory input (e.g. water depth, pellet size). There are a number of similarities and differences between the organization of jaw muscle activity patterns in eating and drinking. In both behaviours, PQP activity functions primarily in an unlocking mechanism, rather than in amplitude adjustment. In drinking, as in grasping and stationing, the duration of an opening movement is timed by the onset of activity in a jaw "closer" muscle, rather than the duration of activity in an "opener". However, during drinking (a "slow" movement, ca. 150–200 ms per cycle), the sustained portion of the jaw opening trajectory is controlled by the activity of the agonist (DM) muscle, while in grasping (a "fast" movement", ca. 30–50 ms) it involves coactivation of both agonists and antagonists. This finding is consistent with the observation that human subjects may use different motor control "strategies" to control force trajectories in "slow" and "fast" movements (Ghez and Gordon 1987).

9.7 Response Topography and the Modulation of Jaw Movement Patterns

Many animals use the same muscles to generate a variety of functionally distinct responses. Examples can be found in chick hatching and walking (Bekoff et al. 1987; see Chap. 6); cricket flight and stridulation (Hennig 1990); locust walking and flight (Ramirez and Pearson 1988) and in various forms of mammalian locomotion (Miller et al. 1975). All these examples involve the imposition of different forms of neural organization upon the same effector system (cf. Sect. 7.11), a process variously termed "behavioural (motor program) switching" or "network modulation" (see Harris-Warrick and Marder 1991).

Eating and drinking in the pigeon are mediated by the same jaws muscles, but their behavioural topographies and EMG patterns are as distinct and identifiable as their functional consequences. Such differences in ingestive topography may be causally related to several variables, including deprivation condition, visual stimuli from the food or water, or orosensory stimuli produced during eating and drinking. During ingestive behaviour all three variables are normally present and congruent with respect to either food or water. During eating, for example, the pigeon is both food-deprived and exposed to visual and orosensory stimuli from the food. It is thus difficult to disentangle the relative contribution of each variable to the control of ingestive response form.

However, the pigeon's *conditioned jaw movement response* may provide an appropriate preparation for such analysis. As Fig. 9.11 shows, the jaw move-

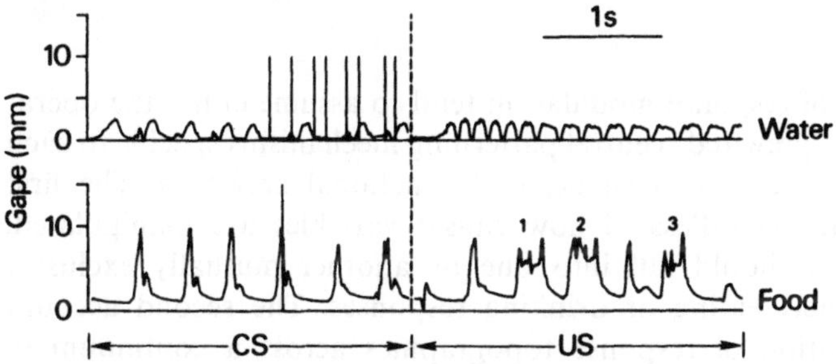

Fig. 9.11. Jaw movement topography during conditioned key pecking. (*Left*) Topography of conditioned jaw movements made to the key (*CS*) for water and food reinforcers. *Vertical lines* indicate key closures (pecks). *Right* Topography of ingestive jaw movements made to water and food (*US*). Five responses to food are shown; the three with multiple peaks (labelled *1, 2, 3*) were successful, while the two with single peaks were unsuccessful (Allan and Zeigler 1993)

ment component of the food- or water-reinforced conditioned key pecking response is similar in form to that of the unconditioned (eating, drinking) response. Put another way, the conditioned jaw movement response appears to preserve the essential topographic features of the unconditioned response (LaMon and Zeigler 1984, 1988; Allan and Zeigler 1993; Sect. 8.3). We have therefore begun to utilize the conditioned "gape" response to analyze the mechanisms underlying the modulation of ingestive response form in the pigeon.

As a starting point for such analyses we examined the topography of conditioned jaw movements under conditions in which both deprivation state and reinforcer type were systematically varied. Using a classical conditioning (autoshaping) paradigm, subjects initially acquired a key pecking response under either a food deprivation/food reinforcer or water deprivation/water reinforcer regimen (acquisition phase). In the transition phase, both groups were shifted to a regimen in which key-pecking was maintained under conditions of food and water deprivation with the two reinforcers having equal probabilities of presentation. In the reacquisition phase, the initial training conditions were reversed for each group. Peak gape amplitude was measured under the three conditions.

The results (LaMon and Zeigler 1988) showed that peak gape amplitude varied systematically and continuously across the experimental conditions from initial mean values of 5.7 mm (food) and 0.3 mm (water) during acquisition, through a mean of 2.25 mm for the transition session, concluding in the reacquisition condition with means of 4.3 mm (food) and 0.8 mm (water). Moreover, the gapes recorded in the transition phase did not simply reflect an average of the two dichotomous populations of "food" or "water" gapes. Instead, the transition manipulation produced an increase in the frequency of *intermediate* gape sizes (gapes between 1 and 3 mm); a class of responses which had been absent or extremely infrequent in either the initial food or water group.

9.8 Conclusions

Theoretical accounts of response modulation tend to assume either the operation of species-typical "prewired" central patterning mechanisms specific to each movement or the modulation of single, multifunctional networks. The first account suggests that, regardless of how causal variables are manipulated, response topographies should fall into one or another mutually exclusive response classes, namely eating or drinking responses. The second account envisages the modulation of response topographies across a continuum of response forms, from "eating" movements through movements of "intermediate" form to "drinking" movements. Tests of these alternative hypotheses involve kinematic analysis of the responses, identification of the relevant muscles, specification of their EMG patterns and, ultimately, correlations of these patterns with the activity of identified central neural elements.

Our behavioural (kinematic) results suggest that the jaw movement patterns mediating the pigeon's eating and drinking behaviour share a common, "multifunctional" premotor circuit. Perhaps more important, they illustrate the utility of conditioning paradigms for the study of mechanisms of response modulation in the pigeon. By combining EMG recording with the manipulation of deprivation and reinforcement variables, it should be possible to monitor the modulation of jaw muscle patterns during the transition from one movement form to another. Finally, because the neural circuits mediating jaw movements in pigeon have been delineated (Zeigler 1989), it may be feasible to explore, at cellular levels, the neural mechanisms mediating their motivational modulation.

Acknowledgements. The research on which this chapter is based was supported by grants from the National Science Foundation and the National Institute of Mental Health. Neuroanatomical contributions by Dr J.J.A. Arends and Dr J.M. Wild are gratefully acknowledged. Special thanks are due to Dr Garth Zweers and Dr Herman Berkhoudt (University of Leiden) for initiating us into the mysteries of functional morphology and electromyography.

References

Allan RW, Zeigler HP (1993) Conditioning of the jaw movement (gape) response during autoshaping of the pigeon's key peck. J Exp Anal Behav (in revision)

Bekoff A, Nussbaum MP, Sabichi A, Clifford M (1987) Neural control of limb coordination. I. Comparison of walking and hatching motor patterns in normal and deafferented chicks. J Neurosci 7:2320–2330

Bermejo RB, Zeigler HP (1989) Prehension in the pigeon II: Kinematic analysis. Exp Brain Res 75:577–585

Bermejo RB, Allan RW, Deich JD, Houben D, Zeigler HP (1989) Prehension in the pigeon I: Descriptive analysis. Exp Brain Res 75:569–576

Bermejo RB, Remy M, Zeigler HP (1992) Jaw movement kinematics and jaw muscle activity during drinking in pigeon. J Comp Physiol A 170:301–309

Bock W (1964) Kinetics of the avian skull. J Morphol 144:1–41

Bout R, Zeigler HP, Jaw muscle (EMG) activity and amplitude scaling of jaw movements during eating in the pigeon. J Comp Physiol (in press a)

Bout R, Zeigler HP, Drinking in the pigeon: an electromyographic analysis. J Comp Physiol (in press b)

Deich JD, Houben D, Allan RW, Zeigler HP (1985) A microcomputer-based system for the monitoring of jaw movements in the pigeon. Physiol Behav 35:307–311

Deich JD, Allan RW, Zeigler HP (1988) Orthogonal differentiation of gape during food-reinforced key pecking. Anim Learn Behav 16:268–276

Ghez C, Gordon J (1987) Trajectory control in targeted force impulses I: Role of opposing muscles. Exp Brain Res 67:225–240

Goodale M (1983) Visually guided pecking in the pigeon. Brain Behav Evol 22:22–41

Gordon J, Ghez C (1987) Trajectory control in targeted force impulses III. Compensatory adjustments for initial errors. Exp Brain Res 67:253–259

Harris-Warrick RM, Marder E (1991) Modulation of neural networks for behaviour. Annu Rev Neurosci 14:39–57

Hennig RM (1990) Neuronal control of the forewings in two different behaviours: stridulation and flight in the cricket, *Teleogryllus commodus*. J Comp Physiol 167:617–627

Jakobson LS, Goodale M (1991) Factors affecting higher order movement planning: a kinematic analysis of human prehension. Exp Brain Res 86:199–208

Jeannerod M (1981) Intersegmental coordination during reaching at natural objects. In: Long J, Baddely A (eds) Attention and performance IX. Erlbaum, Hillsdale, pp 153–168

Klein BG, LaMon B, Zeigler HP (1983) Drinking in the pigeon: response topography and spatiotemporal organization. J Comp Psychol 97:178–181

Klein BG, Deich JD, Zeigler HP (1985) Grasping in the pigeon: final common path mechanisms. Behav Brain Res 18:201–213

LaMon B, Zeigler HP (1984) Grasping in the pigeon: stimulus control during conditioned and consumatory responses. Anim Learn Behav 12:223–231

LaMon B, Zeigler HP (1988) Control of pecking response form in the pigeon: topography of ingestive behaviours and conditioned responses for food and water reinforcers. Anim Learn Behav 16:256–267

Levine RR, Zeigler HP (1981) Extratelencephalic pathways and feeding behaviour in the pigeon. Brain Behav Evol 19: 56–92

Mallin HD, Delius JD (1983) Inter- and intraocular transfer of color discrimination with mamdibulation as an operant in the fixed-head pigeon. Behav Anal Lett 3:297–309

Miller S, van der Burg J, van der Meche FGA (1975) Locomotion in the cat: basic programmes of movement. Brain Res 91:239–253

Paulignan Y, MacKenzie C, Marteniuk R, Jeannerod M (1990) The coupling of arm and finger movements during prehension. Exp Brain Res 79:431–436

Ramirez JM, Pearson K (1988) Generation of motor patterns for walking and flight in motoneurons supplying bifunctional muscles in the locust. J Neurobiol 19:257–282

Reilly SM, Lauder GV (1990) The strike of the tiger salamander: quantitative electromyography and muscle function during prey capture. J Comp Physiol 167:827–839

Remy M, Zeigler HP (1993) Classical conditioning of jaw movements in the pigeon: acquisition and response topography. Anim Learn Behav 21:131–137

Robinson D (1986) Is the oculomotor system a cartoon of motor control? In: Cohen B, Noth J (eds) The oculomotor and skeletomotor systems: differences and similarities. Progress in brain research. Elsevier, New York, pp 411–418

van Gennip EMSJ (1986) The osteology, arthology and myology of the jaw apparatus of the pigeon (*Columba livia*). Neth J Zool 36:1–46

van Gennip EMSJ (1988) A functional morphological study of the feeding system in pigeons. Thesis, University of Leiden, Leiden

Wild JM, Zeigler HP (1980) Central representation and somatotopic organization of the jaw muscles within the facial and trigeminal nuclei of the pigeon. J Comp Neurol 192:175–201

Zeigler HP (1989) Neural control of the jaw and ingestive behaviour: anatomical and neurobehavioural studies of a trigeminal sensorimotor circuit. In: Modulation of defined vertebrate neural circuits. Ann N Y Acad Sci 563:69–86
Zeigler HP, Levitt P, Levine RR (1980) Eating in the pigeon (*Columba livia*): response topography, stereotypy and stimulus control. J Comp Physiol Psychol 94:783–794
Zweers GA (1982a) Pecking of the pigeon (*Columba livia*). Behaviour 81:173–230
Zweers GA (1982b) Drinking in the pigeon (*Columba livia*). Behaviour 80:274–317

10 Motor Organization of the Avian Head-Neck System

G. ZWEERS, R. BOUT and J. HEIDWEILLER

10.1 Introduction

This chapter reviews progress in understanding the motor patterning and control of the avian cervical column, and its underlying anatomical and neuronal basis. The avian cervical column is a highly complex system which positions the head during all behavioural patterns. In many species over 20 highly mobile cervical vertebrae are found, and up to 200 muscles run along either side of the cervical column. Numerous modal action patterns occur, each serving one of many different functions which are primarily performed by the head, and these patterns appear very flexible when external conditions change (cf. Chap. 8). A versatile system is required in order to generate numerous specific modal action patterns and to achieve flexibility in each one of them. This is accomplished by a complex mechanical construction of the cervical column, and a flexible neuro-motor patterning.

Complex motor patterns are generated on the basis of information from all major exteroceptive senses in the head. They are tuned through feedback from the proprioceptive system, which provides information about the position and motion of all elements in the cervical column. The specific demands arising from different major activities, such as feeding, locomotion or vocalization, change during development, and these changes are reflected in head-neck motion. For proper operation, the head-neck system must keep up with these changes. In addition, scaling effects resulting from growth put specific demands on motor patterning.

The head-neck system has received only scattered attention in the literature on birds. One major study on the comparative anatomy is that of Boas (1929), and other anatomical studies are Zusi (1962), Burton (1974), Jenni (1981), Fritsch and Schuchmann (1987) and Vanden Berge and Zweers (1993). Some work on functional aspects has also been published (e.g. Virchow 1915; Duym 1951; Den Boer 1953; Popova 1972).

Recently, new anatomical descriptions have been carried out (e.g. Landolt and Zweers 1985; Zweers et al. 1987). These studies were followed by model studies on posture and motion patterning (e.g. Elshoud and Zweers 1987; Bout

Neurobehavioural Morphology, Institute for Evolutionary and Ecological Sciences, Leiden University, PO Box 9516, NL 2300 RA Leiden, The Netherlands

M.N.O. Davies and P.R. Green (Eds.)
Perception and Motor Control in Birds

et al. 1992), and analyses of the kinematic principles that control cervical motion (e.g. Heidweiller et al. 1992a). Other investigations of motor patterns also include an experimental approach (e.g. Cusick and Peters 1974; Bilo and Bilo 1983; Heidweiller et al. 1992b). Studies on anatomy and motor patterning during ontogeny are scarce (e.g. Murray and Drachman 1969; Heidweiller 1989; Heidweiller et al. 1992b). The latter authors explain that motor patterning in the head-neck system must change due to scaling effects during development.

Work on neurosensorial control of the head-neck system has been carried out in different areas (e.g. Friedman 1975; Goodale 1983; Martinoya et al. 1984; Deigh et al. 1985; Davies and Green 1989). These studies, combined with recent advances in modelling the control of motion (e.g. Peterson et al. 1989; Bout unpubl.), provide new insights into the control of the avian head-neck motor system.

10.2 Osteo-Muscular Design of the Avian Cervical Column

To understand motor patterning requires accurate knowledge of anatomy. The cervical columns of a variety of species have received attention from several authors (e.g. Boas 1929; Palmgren 1949; Davids 1952; Zusi 1962; Zusi and Storer 1969; Burton 1974; Jenni 1981; Johnson 1984; Zusi and Bentz 1984; Weisgram and Zweers 1988). However, the following anatomical description, selected from Baumel (1979), Komarek (1979), Vanden Berge (1979) and Zweers et al. (1987), will be restricted to the chicken, as it is only in this species that motor pattern analyses have been carried out.

10.2.1 Osteology

Chickens have 15 cervical vertebrae, all built according to the same design, although each has a specific size and shape, reflecting its specific function. Prominent processes allow the attachment of muscles and ligaments, or the support of articulation facets. The design (numbers refer to Fig. 10.1) comprises a corpus vertebrae (1) carrying the midsaggital major articulation facets (2, 3) at cranial and caudal ends, and strong processes for muscle attachment (e.g. 4, 5, 6). In addition, there is an arcus vertebrae (7), which is positioned dorsal to the corpus, enclosing the spinal cord. The arcus carries processes supporting the articulation facets of a second, bilateral joint (8, 9), while prominent processes for muscle attachment are present (e.g. 10, 11, 12).

10.2.2 Arthrology

The intercorporeal joints have saddle-shaped facies, allowing dorsoventral and lateral flexion. They each carry an intra-articular, cartilaginous meniscus. The

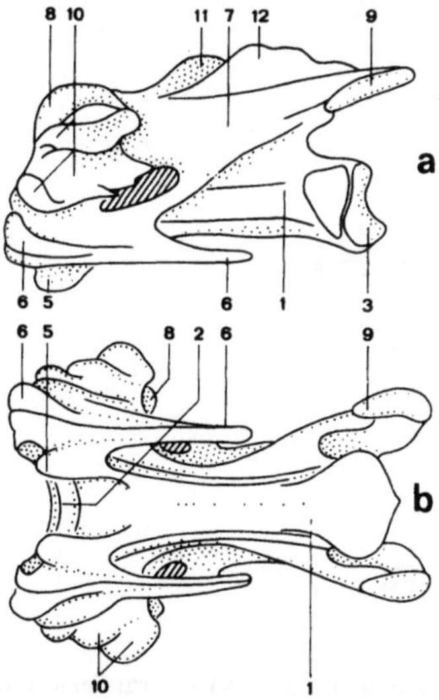

Fig. 10.1. Main elements of cervical vertebra 8 in the chicken (*Gallus gallus*): **a** Lateral view; **b** Ventral view. *1* Corpus vertebrae; *2* Facies articularis cranialis; *3* Facies articularis caudalis; *4* Proc. spinosus ventralis (not present in V8, but in more caudal vertebrae); *5* Proc. caroticus; *6* Proc. costalis; *7* Arcus vertebrae; *8* Proc. articularis cranialis; *9* Proc. articularis caudalis; *10* Proc. transversus; *11* Proc. spinosus dorsalis; *12* Proc. dorsalis (After Zweers et al. 1987)

interarcus joints have oval-shaped facies; cranially they face dorsomedially, and caudally joints face ventrolaterally. All joints carry capsules, while their facies and connecting ligaments restrict dorsoventral and lateral flexion. However, the degree of flexion possible in each joint varies greatly, and joints with similar ranges of flexion fall into neighbouring groups (Fig. 10.2). The cervical column may therefore be subdivided into several functional areas which strongly determine cervical motion.

Joints between cranium, axis and atlas are of a special nature. They are described by Boas (1929), Goedbloed (1958), Landolt and Zweers (1985), Zweers et al. (1987) and Weisgram and Zweers (1988). The combined action of these elements allows an extreme rotation, lateral and dorsoventral flexion of the head.

10.2.3 Myology

There are about 200 muscle slips on either side along the cervical column. Almost all slips are parallel-fibred and very long, spanning several articulations. For the latter reason their effects are very different from those of the mono-articular multipinnate muscles examined extensively in so many studies. Five major muscle groups (numbers refer to Fig. 10.3) are distinguished (Zweers et al. 1987; Heidweiller unpubl.). The first muscle group is the cranio-cervical system, which positions the head relative to the top of the cervical column. Five muscles are involved, connecting the head and first six vertebrae (1, 2, 3, 4, 5). The second group is the dorsal system, which elevates the head and neck and includes some very long muscles (e.g. 6, 7), which span nearly the full length of the cervical

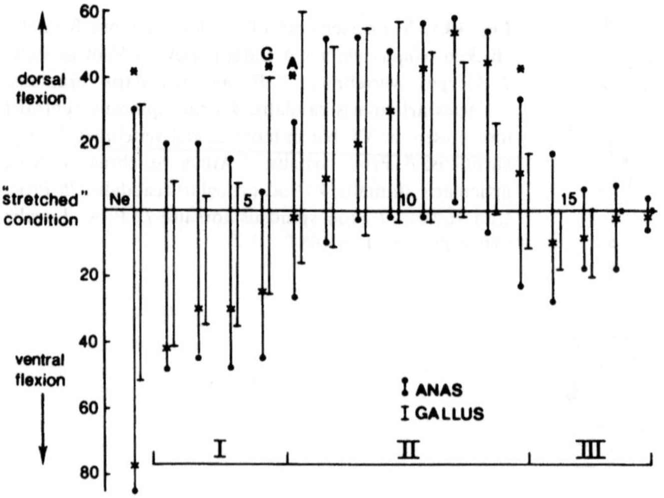

Fig. 10.2. Work envelope of head and neck represented by maximal dorsal and ventral flexion in subsequent cervical vertebrae of chicken (*Gallus gallus*) and mallard (*Anas platyrhynchos*). Three cervical regions have been indicated by horizontal *section I* (cranial end), *II* and *III* (caudal end). They are connected by pivot-like vertebral joints (*asterisk*). The angle pattern for the resting position in the mallard is indicated by *crosses*; *Ne* neurocranium (After Landolt et al. 1989)

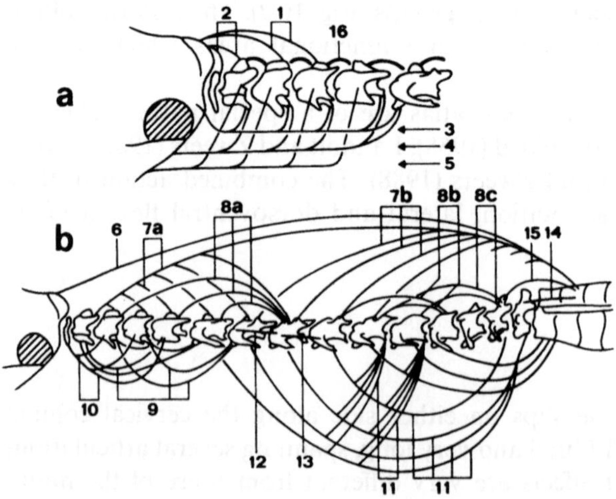

Fig. 10.3a, b. A selection of muscle slips of the main cervical muscles in a chicken (*Gallus gallus*). Muscles are shown as *lines* connecting origin and insertion **a** Cranio-cervical system; **b** cervical system. *1* M. complexus; *2* M. splenius capitis; *3, 4, 5* M. capitis lateralis, ventralis, dorsalis respectively; *6* M. biventer cervicis; *7* M. longus colli dorsalis; *8* M. cervicalis ascendens (3 bundles *a*, *b*, *c* from a series of 10); *9, 10* M. flexor colli medialis, lateralis, respectively; *11* M. longus colli ventralis; *12* M. intertransversarius; *13* M. inclusi; *14* M. longus colli dorsalis, pars thoracicus; *15* M. thoracicus ascendens, pars cranialis; *16* Lig. elasticum (After Zweers et al. 1987)

column. Branching slips from muscle (7) cover shorter parts of the cervical column. One muscle (8) forms a sequence of groups of long slips, which fan out cranially from one origin and span an increasing number of joints. Third, the ventral system depresses the head (e.g. 9, 10) and neck (e.g. 11). The muscle depressing the neck (11) is very long, with its number of slips increasing strongly towards the caudal end of the neck. The fourth group, the lateral system, serves to stabilize the cervical column and control lateral motion. The system consists of a long series of strongly pinnate mono-articular muscles (e.g. 12, 13). The final group, the thoraco-cervical system, comprises two strong muscles (14, 15) that elevate the cervical column.

10.3 Design Modifications of the Avian Cervical Column

Boas (1929) was the first to recognize that the avian cervical column comprises three regions, connected by transitional vertebrae, which differ in the relative dorsoventral flexion of the joints. Relative to a fully stretched neck, dorsal flexion dominates in some joints, and ventral flexion in others, while joints at the boundaries between regions are intermediate (Fig. 10.2; Landolt et al. 1989). The cranio-atlanto-axis joints are followed by the first region, in which ventral flexion predominates. Often at the level of vertebra 6, one or two joints allow dorsal as well as ventral flexion. The next seven joints allow only dorsal flexion. Often at about vertebra 13 again dorsal and ventral flexion is allowed, after which a third region is found comprising about three or four joints, allowing mainly ventral flexion, longitudinal rotation and large lateral flexion. The size of these regions, as well as the extent of flexion in each joint, may vary between taxa, and this variation can be related to the specific functions to which the cervical column is adapted. For example, Zusi (1962) explains the specific adaptations of the cervical column to the specialized feeding behaviour of the black skimmer (*Rhynchops nigra*), by correlating the bird's feeding technique to kinematic and morphological data.

10.3.1 Ligamentum Elasticum Cervicale

A long elastic ligament supports the weight of head and neck in many tetrapods. In birds it has a special shape. Separate elastic ligaments connect adjacent dorsal spinal processes of each vertebra (Fig. 10.3). However, the size of these ligaments varies along the cervical column. Boas (1929) distinguished three regions. Despite a large variation, related to specific functioning in different taxa, the following general picture emerges. First, a cranial region in which the ligaments are moderately developed usually runs as far as vertebra 6. This is followed by a middle region, which may run as far as vertebrae 10–12, in which the ligaments are small. Next, a third, proximal region containing highly extensible, very strong ligaments runs to the notarium. Bennett and Alexander (1987) studied the latter group of ligaments in turkeys (*Meleagris gallopavo*) and found experimentally that the proximal ligaments stretched to accommodate large strains. Even so, the ligaments were found to be incapable of developing enough force to support the extended head and neck without the aid of muscles. Hence, in the stretched posture, motor action is needed in the head-neck elevator muscles to maintain control (Heidweiller et al. 1992b).

In addition to postural control, elastic ligaments may also serve to save energy expended by the muscles. Birds keep their neck in an S-shape during most behavioural patterns, such as resting, walking and vocalization (Fig. 10.4). An S-shape reduces the energy needed to carry the head and neck, because the distance between the notarium and the centres of mass of head and neck is much shorter than in the stretched neck. In the S-shape these elastic ligaments make a large contribution to the conservation of energy expended by muscles, since the ventrally directed torques that result from gravity are clearly balanced by the ligaments in the caudal and cranial portions of the neck. In the mid-portion of the neck, however, the torques are directed dorsally so that the ligaments are of no use in balancing gravity. Indeed, in many birds ligaments are absent in the mid portion of the cervical column.

10.4 Patterning Head-Neck Movement and Motor Action

When moderately alert, and in the absence of visual cues, birds show characteristic head orientations (Duym 1951; Erichsen et al. 1989). For the purpose of modelling postures and motion, Elshoud and Zweers (1987) and Bout et al. (1992) reduced the head-neck system to a planar multi-element open chain, in which each vertebra has only one degree of freedom (Fig. 10.4). In this model, control of head-neck motion is similar to controlling the position of the end effector in a multi joint robot arm. As in robotics, the neural control system has to solve the problem of kinematic redundancy. The large number of combinations of joint angles (postures) by which a given point in a planar workspace can be reached by a multi-element chain must be reduced.

10.4.1 Postures: Minimal Flexion Model

One way to reduce redundancy is to minimize the amount of flexion between elements of the chain. Bout et al. (1992) developed a "minimal flexion" model in which a cost function is assigned to each joint. Given the length of each bar and the position of the end element of the chain, the model calculates the configuration of elements (posture) for which the sum of all cost functions is minimized. The model uses a modified simplex method (Bunday 1984) to calculate the angles. Best predictions are found when the cost function is represented by the squared angle of each element, so that increasing flexion becomes progressively more costly. This method leads to smoothly curved chain postures. Large angles

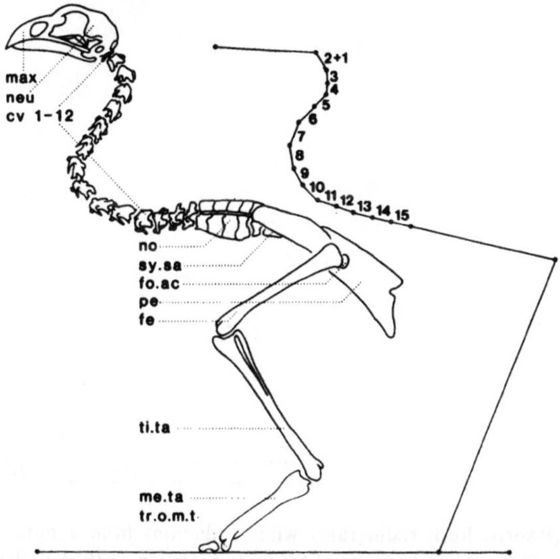

Fig. 10.4. The skeleton of the chicken as a planar, open kinematic chain *Left* Skeletal elements used in modelling head-neck patterns; *max* maxillare; *neu* neurocranium; *C1-12* cervical vertebrae 1-12; *no* notarium; *sy.sa* synsacrum; *fo.ac* foramen acetabulare; *pe* pelvis; *fe* femur; *ti.ta* tibiotarsus; *me.ta* tarsometatarsale; *tr.o.m.t.* trochlea ossis metatarsale III. *Right* The skeleton represented by 14 *bars*. *Bar lengths* are distances between articulation facets, which are considered as hinging points (After Elshoud and Zweers 1987)

are avoided and rotation is distributed over all the elements of the chain. The model predicts the S-shape of the neck when both position and orientation of the head with respect to the body are given.

Comparison of the model's predictions with measurements taken from radiograms of different postures in chickens show that the model predicts slightly "cheaper" curved neck postures than are observed in reality (Fig. 10.5a, c). In some simulations, predicted angles exceed biological flexion limits. Adding these maximal flexion limits as an extra constraint improves the model's predictions (Fig. 10.5c). The authors conclude that the minimal flexion model holds as a general description for static postures, but differences of several degrees in joint angles are present. Apparently other factors are superimposed on this general rule. This is due to the fact that the model ignores biomechanical aspects, and assumes that all joints move according to the same cost function. The accuracy of the model may be further improved when the difference in flexion limits of cervical vertebrae is taken into account (Bout in prep.).

10.4.2 Motion: Least Motion Model

Many modal action patterns in birds are shaped as a sequence of alternating static and dynamic phases (Zweers 1985). For example, pecking starts with a

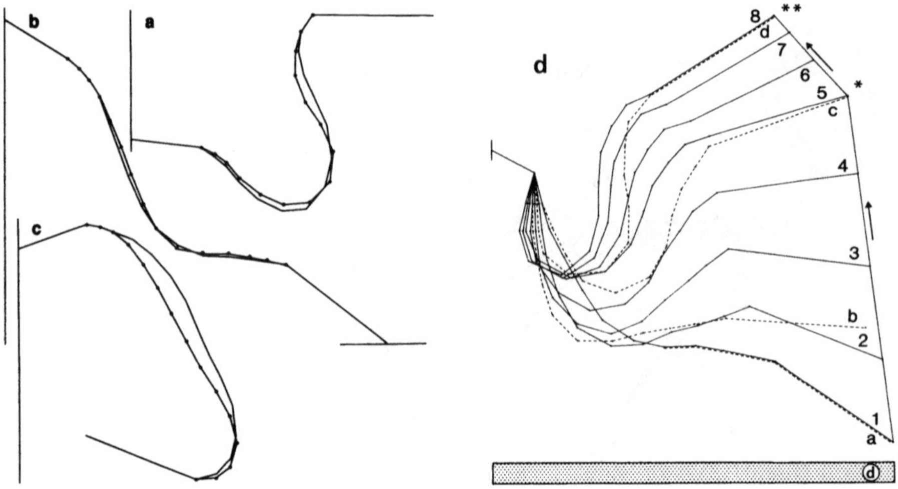

Fig. 10.5. Comparison of postures measured from radiograms with predictions from a model minimizing rotation between vertebrae. The most rostral part of the body is shown *on the left*, the head faces *right*. Joints are indicated by *dots* for the observed neck posture only; **a** shows a resting posture; **b** shows drinking, and **c** preening of breast feathers. In the preening posture some vertebrae reach their flexion limits; **d** predictions of a drinking upstroke pattern from a simulation using posture of the chain (*a,1*) and the position and orientation of the fully elevated head (*d, 8*) and the trajectory of the beak tips (to * and then to **) as input. See Sect. 10.4.2 for explanation (After Elshoud and Zweers 1987)

head fixation and is followed by a head depression. Alternatively, moving from one static position to another may be considered as a series of successive "pseudo-static" situations. In the pseudo-static approach, the minimal flexion model would predict cervical motion for a given trajectory. However, *change* in joint-angle rotation may be governed by its own mechano-economical rules.

Elshoud and Zweers (1987) developed a simulation model, the least motion model, based upon optimization principles similar to those used in the minimal flexion model. The model calculates, for a given starting posture and a new endpoint of the chain, the configuration of joints which differs the least from the starting posture (smallest sum of change in rotation). The model does not predict a trajectory, but a trajectory may be simulated by specifying beak tip points along a path. Straight line trajectories were taken from radiographic analyses of actions such as head upstroke during drinking, which runs from a beak tip down position during water intake, to a tip up position during swallowing. Radiographic testing of the predictions for the latter pattern shows that the least motion model clearly predicts the hooked appearance of the neck (see stages b and 2 in Fig. 10.5d) in the drinking upstroke (Elshoud and Zweers 1987). Differences between predictions and behaviour again make clear that other factors are superimposed on minimal angle rotation to govern cervical motion patterning.

Functional demands will also play a role. Notably, certain head trajectories and orientations will be required to fulfil certain functions. For example, during the upstroke of drinking, a bird may lose water from its beak. The main causes of such water loss come from gravity, if the beak opening faces downwards, and from the large centrifugal forces generated by curved head trajectories. To minimize the loss, a straight vertical trajectory of the beak tip and a fast upward rotation of the head are required (Heidweiller et al. 1992a). The sequence in Fig. 10.5d starts from the initial posture of the chain during beak tip down (a, 1). The trajectory is taken as two straight parts (to * and **; from radio-analyses), and only the final position and orientation of the head were used (8,d). Comparisons of predictions 2 and 5 to radiograms b and c (Fig. 10.5d) imply that if orientation of the head is determined by the need to keep water in the beak, better predictions will require the input of the complete change in orientation of the head as well as the complete trajectory.

10.4.3 Major Motion Principles

An empirical search for the principles underlying the control of cervical motion and an analysis of environmental factors influencing such motion was undertaken by Heidweiller et al. (1992a). A major problem was that despite modality in patterning, much variability also occurred. The authors selected the relatively stable, planar drinking pattern for measuring rotational changes of the head, cervical vertebrae and body. The position of these elements was digitized in a series of film frames of representative drinking scenes. Angles in all joints were calculated and plotted against time. This was done for hatchlings, birds from five developmental stages, and adult chickens (Fig. 10.6). Next, a search for motion principles was carried out by deduction of relationships between rotation and translation, and by developing mechanical characteristics for an open kinematic chain that must be operated by muscles. Predicted patterns were compared to observed patterns. This approach has led to recognition of five principles underlying the control of cervical motion.

The first two principles are the geometric movement and lever arm principles. Both principles minimize energy costs, since they maximize rotation efficiency, which is defined as the ratio of the sum of rotation in all joints and the resulting head translation. The geometric movement principle refers to the fact that horizontally positioned vertebrae are used to contribute primarily to vertical head translation, while vertically positioned vertebrae are used to contribute to horizontal translation (bars 1 and 2 in Fig. 10.7). The lever arm principle means that the length of a bar (several vertebrae in bar 1, Fig. 10.7) is maximized by positioning cranial vertebrae and the head in one straight line, at the same time as the bar rotates around the caudalmost joint.

The next two principles correct or tune a trajectory and are referred to as the curve development compensation principle and the overflexion compensation principle. The principle of curve development compensation means that a

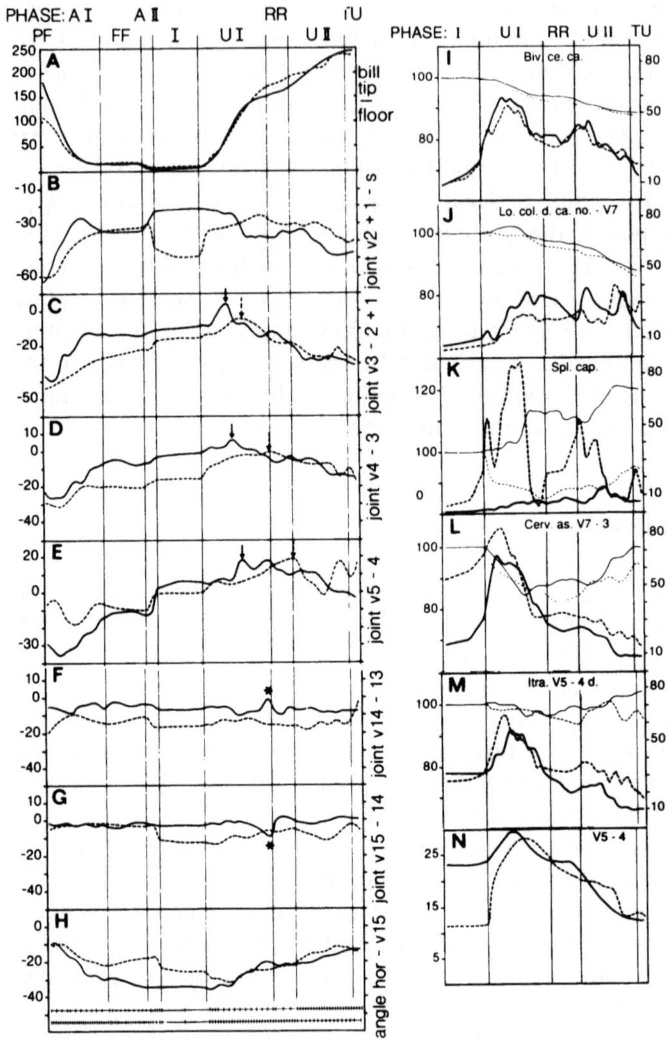

Fig. 10.6A–N. Kinematics, kinetics and motor patterns of the chicken's cervical system during drinking (selected from Heidweiller et al. 1992a, b). *Solid lines* represent drinking from a water box at a distance from the body; *dashed lines* represent drinking from a water box near the body. Time is scaled so that the boundaries (indicated by *vertical lines*) between subsequent phases correspond for the two positions of the water box. The film frame intervals of the two scenes are indicated at the *two bottom lines* of **H**. Phases are indicated in the *top row* by the following abbreviations: *PF* prefixation; *AI* approach I; *FF* final fixation; *AII* approach II; *I* immersion; *UI* upstroke I; *RR* relative rest; *UII* upstroke II; *TU* tip up; *v* vertebra. **A** Distance upper bill tip to floor (mm). **B–H** Change in angle of selected cervical joints as measured from combined film and X-ray analyses. The *arrows* in **C**, **D** and **E** illustrate the "bike-chain" pattern. The *asterisks* in **F** and **G** signify the occurrence of overflexion compensation during the upstroke from the more distant water box. **I–M** The relationship between change in relative muscle length and recorded muscle activity during the upstroke in drinking. The relative muscle length is plotted (*fine dashed and solid lines*) as the percentage of the length at the end of the immersion, and scaled along the left vertical axis of these plots. Muscle activity records are

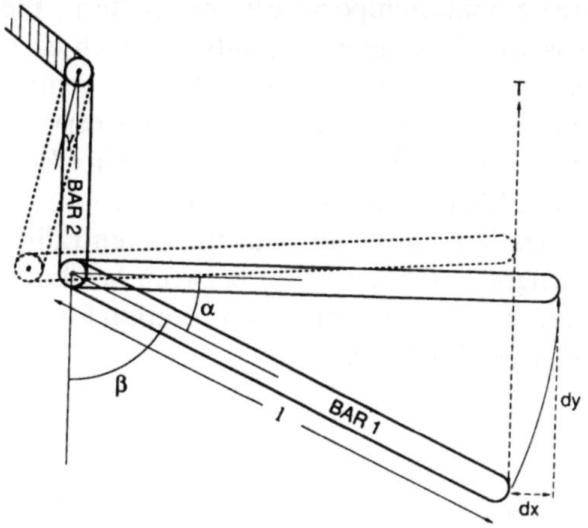

Fig. 10.7. A schematic illustration of the geometric relations acting within the craniocervical system of the adult chicken. *Bar 1* represents cranial vertebrae and head. The initial position of *bar 1* (angle *β*), the amount of rotation (*angle α*) and the length (*l*) determine the ratio of vertical (*dy*) and horizontal (*dx*) displacements of the cranial end of *bar 1*. The geometric movement principle is demonstrated by rotation of *bar 2*; *dx* is larger than *dy*, while the reverse holds for rotations in *bar 1*. The lever arm principle is demonstrated by a fixed change in α which causes a larger translation when *l* is greater. The curve development compensation principle, which serves to translate the cranial end of *bar 1* along the straight trajectory *T*, is shown as a rotation to ventral of *bar 2* (change over an *angle γ*) that must occur simultaneously with rotation to dorsal of *bar 1* (change over an *angle α*) (After Heidweiller et al. 1992a)

curved head trajectory is changed into a straight one by compensatory rotations in more caudally positioned joints (Fig. 10.7). Without such compensation, one cause of a curved trajectory would be operation of the lever arm principle. The principle of overflexion compensation states that if a joint rotates more than needed, a neighbouring joint rotates in the opposite direction.

If large forces tend to develop in the lever arm joint (see Fig. 10.7) as a result of the lever arm principle a fifth principle, the minimization of rotation force, sets to work. In this case, the lever arm is shortened through subsequent elevation from cranial to caudal of the lever arm vertebrae (Fig. 10.7). The operation of

---

Fig. 10.6A–N. *Contd.*
plotted (*bold lines*) as the percentage of the highest value observed for each electrode pair, and scaled along the right vertical axis of the plots running from 0 to 90%. Graphs **I–L** illustrate that Mm. biventer cervicis caudalis (*Bi.ce.ca*), longus colli dorsalis connecting notarium to V7 (*Lo.col.ca.d no-V7*), splenius capitis (*Spl. cap*) and cervicalis ascendens connecting V7 and V3 (*Cerv. as. V7-3*) are active when they shorten. Graph **M** shows that the dorsal portion of M. intertransversarius connecting V5 and V4 (*Itra. V5-4*) does not shorten, but is strongly active. **N** The relationship between torque and muscle action. A change in torque during the drinking upstroke is shown for the two water box positions in the cervical joint of vertebrae 4 and 5. Comparison of graphs **M** and **N** illustrates that the motor action of M. intertransversarius *V5-4d* is determined by the torque acting around the joint crossed by this muscle

this principle results in a characteristic spatiotemporal cervical pattern, the "*bike-chain*" pattern (Fig. 10.6). This is seen in a series of joints in which each joint shows a dorsal rotation, which immediately is followed by a ventral rotation. These rotations in cervical joints pass quickly down the cervical column from cranial to caudal. The same rotational pattern can be seen in a bike chain. If a chain lying on a flat surface is lifted by one end, a wave of rotations passes down the chain; each link rotates first "dorsally" and then "ventrally". The effect of this pattern is that the length of the lever arm between head and rotating cervical joint is kept to a minimum, so that a minimum rotation force is needed to elevate the head (Heidweiller et al. 1992a).

10.4.4 Motor Patterns

There have been few studies of motor patterning and the effects of muscle action in the avian cervical column (e.g. Den Boer 1953; Zusi 1962; Cusick and Peters 1974). Once the complex anatomy and motion principles are known, motor patterns may be investigated. However, the cervical column comprises 200 thin muscle bundles, most of which are very long, and which may shorten by centimetres. Recording complete electromyograms is hardly possible.

To gain insight into cervical motor patterning, Heidweiller et al. (1992b) applied a deductive procedure (cf. Zweers et al. 1981). These authors developed two kinds of models which predict muscle activity from changes in posture of the cervical column that occur during drinking in chickens. The first model is geometric, and calculates the distance between the origin and the insertion of a muscle for any set of joint angles in the cervical column. The second model calculates the gravity torques in each cervical joint for each posture during the upstroke. Motor patterns were assumed to be mainly determined by concentric contraction. Comparison of predicted motor patterns and recorded electromyograms (Fig. 10.6) indicated that muscle shortening and torque increase predict muscle activity well. The long dorsal muscles (M. biventer cervicis and M. longus colli dorsalis), the Mm. cervicales ascendentes and M. complexus cause elevation of head and neck. The unique action of the long, multi-articular dorsal muscles is responsible for the characteristic bike-chain pattern in cervical motion.

Applying to empirical rule that muscle shortening and torque increase predict motor patterning in the upstroke, Heidweiller et al. (1992b) deduced motor patterns for hatchlings and 4-week-old chicks also. They showed that motor patterns must have changed during ontogeny. This *motor pattern transition* is due to the fact that torques in the cervical joints increase more than the physiological cross-sectional area (representing maximal force) of the elevating muscles. For example, in hatchlings the long dorsal muscles alone can deliver sufficient force to elevate the head, but in adult chickens the Mm. cervicales ascendentes and M. complexus must contribute also.

10.5 Control of Head-Neck Movements

Head-neck motion serves mainly functions performed by the head. As a result most cervical postures and motions are primarily determined by head positions required for sensorial or operational reasons. Some examples of such head positions from recent research follow. First, when pigeons fixate a target at short distance, its image falls on the area dorsalis or "red field" of the retina (Goodale 1983). The near point of accommodation of this field is much closer than of the so-called lateral field (see Sect. 2.5). Also, in the final stage of a peck at grain a pigeon requires the grain to fall in the binocular field, in front of and directly below the beak (Martinoya et al. 1981; see Sect. 3.5.1). A third observation is that the distance between head and seed during pecking increases with increasing seed size (Klein et al. 1985; see Sect. 8.2). Fourth, in the absence of a target stimulus, pigeons adopt a characteristic head posture, which may be under vestibular control, since in a resting position the horizontal semicircular canal is kept slightly pitched up (Duym 1951; Vidal et al. 1986; Erichsen et al. 1989). As a flying pigeon approaches a landing perch, this posture changes, so that the perch falls 20°–25° above the beak in the visual field (Green et al. 1992). In the case of development, head extension and elevation can be elicited by both vestibular and visual stimuli in pigeon squabs (Davies and Green 1989). As a final example, the orientation of the head during drinking is critical for keeping water in the beak (Heidweiller et al. 1992b; see Sect. 10.4.2).

10.5.1 Comparator Model of Head-Neck Control

A behavioural model connecting functional demands and head-neck motion in the context of feeding in the pigeon has been formulated by Zweers (1985). Jaw operation and head-neck motion are highly tuned in feeding. At the start of a peck the head is fixated relative to the target seed at two stages F_1 and F_2 (see Goodale 1983, and Sects. 2.4 and 3.5.1). During fixation, information about colour, size, shape and position is gathered. Stage F_2 is the start of the ingestion phase and is followed by a head approach which may be swing-like, scoop-like or straight (see Sect. 8.2). After the seed has been grasped it is tested for size, hardness, taste and position within the mouth. The seed may be dropped, thrown away or retested during so-called stationing behaviour. Transport in the mouth also depends on the sensory information gathered. A catch and throw mechanism, which includes fast head jerks, occurs for large seeds; a lingual inertial transport mechanism without head movements is used for small seeds (Zweers 1982, 1985; Zeigler 1989; see Chap. 9).

The comparator model (Fig. 10.8) was developed to describe the static and dynamic phases of pecking. Three control units were proposed; a sensor selector, a comparator and a highest value passage (HVP) unit. The first selects an input set from a seed. This information is then compared to an internal representation of the input set based on experience. A decision is made concerning edibility of

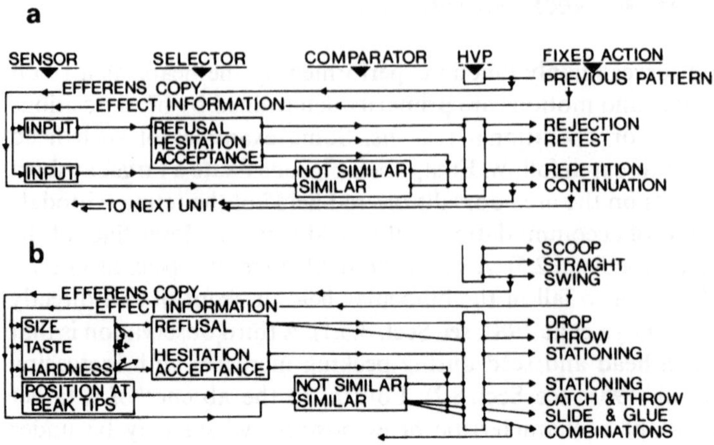

Fig. 10.8a, b. The comparator model for the control of pecking. **a** Control unit for a static phase and its subsequent dynamic phase; **b** control unit which describes head-fixation, head approach plus grasp, and subsequent transport of a seed in the mouth. *HVP* highest value passage; "*stationing*" a pattern of small "*catch and throw*" motions by which a seed is slightly repositioned at the beak tip; "*slide and glue*" lingual inertial transport of a seed that adheres to the tongue by saliva while the head is fixated (After Zweers 1985)

the seed and this information is sent to the HVP unit. The comparator unit compares afferent information with an expectancy based on immediately preceding motor commands (an efferent copy). This feedback serves to check whether the motor pattern executed produced the expected result. Again, the output is passed to the HVP unit. The sensor selector may gate the output of the comparator, but may also work in parallel. The HVP unit releases only the highest stimulated motor pattern initiating rejection, retesting or acceptance.

10.5.2 Connections in the Central Nervous System

Visual, tactile, vestibular and proprioceptive systems play a role in controlling head motion. The brain structures involved in the processing of sensory information and the execution of functions postulated in the comparator model are only partly known. We will briefly describe the central connections conveying tactile and visual information to telencephalon and brain stem (Fig. 10.9). At least two pathways carry information from the retina to the forebrain. The tectofugal pathway (R) projects to the ectostriatum (E), while the thalamofugal pathway projects to a thalamic nucleus (OPT) that sends its fibres to the caudal part of the hyperstriatum accessorium (HA). In the pigeon, stimulation of HA evokes coordinated movements of the neck, body and legs (Cohen and Pitts 1967). Information from mechanoreceptors in the beak reaches, via the mesencephalic nucleus princeps trigemini, the nucleus basalis in the forebrain. Intratelencephalic connections are very complex (e.g. Dubbeldam 1976, 1991; Ritchie 1979; Dubbeldam and Visser 1987) and will not be discussed here.

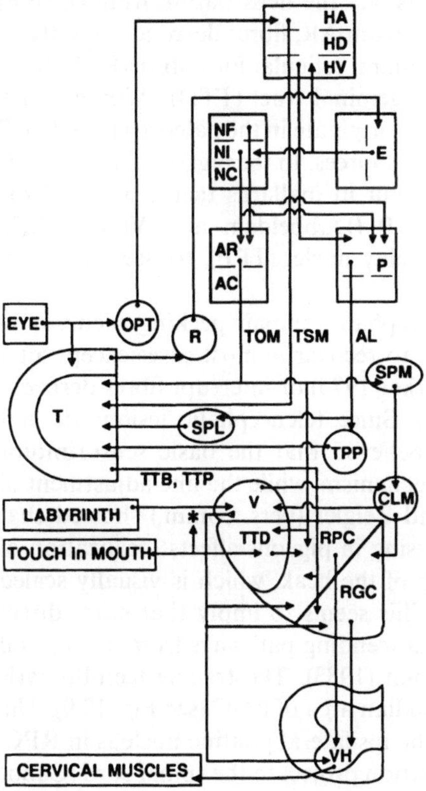

Fig. 10.9. Schematic illustration of the principal CNS pathways of the avian head-neck system; *AC* caudal part Archistriatum; *AL* Ansa lenticularis; *AR* rostral part Archistriatum; *CLM* cerebellum; *E* Ectostriatum; *HA* Hyperstriatum accessorium; *HD* Hyperstriatum dorsale; *HV* Hyperstriatum ventrale; *NC* Neostriatum caudale; *N* Neostriatum frontale; *NI* Neostriatum intermedia; *OPT* nucleus opticus thalamicus; *P* Paleostriatum; *R* nucleus rotundus; *RGC* nucleus reticularis giganto-cellularis (medial reticular area); *RPC* nucleus reticularis parvocellularis (lateral reticular area); *SPL* nucleus spiriformis lateralis; *SPM* nucleus spiriformis medialis; *T* tectum opticum; TOM, tractus occipito-mesencephalicus; *TPP* nucleus tegmenti pedunculo-pontinis; *TSM* tractus septo-mesencephalicus; *TTB* tractus tecto-bulbaris; *TTD* nucleus tractus descendens nervi trigemini; *TTP* tractus tectopontines; *V* vestibular complex; *VH* ventral horn. The *asterisk* represents the mouth-touch connection to the nucleus principalis nervi trigemini, nucleus basalis and neostriatum (Courtesy of J.L. Dubbeldam)

Three large extra-telencephalic motor systems may influence head motion. First, the caudal, visual part of HA projects (TSM) to various parts of the visual system (i.e. optic tectum; Reiner and Karten 1983). The rostral, somatosensory part of HA may descend in some species to the level of the hypoglossal nerve (Dubbeldam 1976). However, neither part of the HA seems to have a direct projection to the premotor areas in the brainstem. In the second system, E has indirect connections via the neostriatum intermedia (NI) with a second large source of telencephalic efferents (TOM): the rostral sensorimotor part of the archistriatum (AR). This area also receives, via the neostriatum frontale (NF) input from the trigeminal nucleus basalis. From AR, fibres descend not only to the tectum and cerebellum but also to the lateral reticular formation (RPC; Bout 1987) and the nuclei of the descending trigeminal tract (TTD). Afferents of a third extra-telencephalic motor system (AL) originate in the paleostriatum (P). P receives afferents from several telencephalic sources. In the pigeon, this system is considered part of a visuomotor system, but in mallards tactile areas of the forebrain also have massive input into P (Dubbeldam and Visser 1987). Descending fibres of P project, through relay nuclei (TPP), to deep layers of the tectum.

Lesions in different parts of the telencephalon impair pecking accuracy in chicks. However, pecking accuracy seems to recover in most cases except after lesions which involve AR (Salzen and Parker 1975) or interrupt fibres descending from AR (Levine and Zeigler 1981). Since telencephalic lesions do not abolish neck movements as such, it is believed that the basic sensorimotor coordination of pecking is a brainstem mechanism, while the fine adjustment of aiming involves the forebrain. Jaeger and Zeigler (pers. comm.) showed that obstruction of telencephalic visual processing in pigeons affects localization of pecking movements, but not the opening of the beak, which is visually scaled with seed size (see Sects. 8.3 and 9.4.2). This seems to imply that more direct connections to the brainstem exist. Two descending pathways from deep tectal layers are described by Reiner and Karten (1983). The tractus tectobulbaris (TTB) projects to the contralateral paramedian area of RGC (see Fig. 10.9). The tractus tecto-pontines (TTP) projects to the ipsilateral pontine nucleus in RPC. Both the lateral and medial reticular formation project to the spinal cord (Cabot et al. 1982).

Another important source of sensory input is the vestibular system. Vestibular efferents have direct connections with the motor neurons in the upper portion of the spinal cord (VH), RGC and RPC (Arends 1981) and CLM. It has been found that information from the mechano-receptors in the beak affects pecking in pigeons (Zeigler pers. comm.) and directs head motion during dabbling in mallards and probing in sandpipers (Zweers and Berkhoudt 1991). As soon as the beak tip touches the substrate the visual control of head motion is taken over by touch control (see Sect. 1.6.2.1). Although neck muscles in birds contain many muscle spindles (Saglam 1968) little is known about the proprioceptive part of the collimotor system.

10.5.3 Network Control

As discussed in Sect. 10.4.1, an algorithmic approach can be used to reduce the redundancy in a multi-element chain like the avian neck. The minimization of rotation in the chain produces the general features of different neck postures. However, one may wonder how the avian central nervous system can produce the different sets of joint angles which are described by the algorithm. A recent approach to this "implementation" problem is the use of neural network models.

To investigate the possibilities of such an approach we used a network model which consisted of three layers; an input layer of three units, a "hidden unit" layer of 5 units and an output layer of 14 units. The three input units represent the distance between the two ends of the chain (a measure of neck extension), the angle of this line with the body (a measure of head elevation relative to the body) and the angle between head and body. The output units represent the angles between vertebrae. Only feedforward connections between successive layers were present. The activity of a receiving unit is a function of the weighted sum of all units which have a connection with the receiving unit. The strengths of the connections (weights) which perform the transformation of a series of inputs (head position/orientation) into the matching output (angles between vertebrae) were calculated with the error back-propagation rule (Rumelhart et al. 1986). Given an initial set of weights this rule calculates a change in weights which will reduce the difference between actual and desired output. By repeatedly offering the network the set of inputs and adjusting the weights, the error converges to zero. To "train" the network (finding the right weights) a set of joint angles from seven different neck postures was used.

This network not only reproduces the training postures, but also predicts pecking postures (Fig. 10.10a) which were not among the training patterns. The rather elevated position of the head in the predicted posture is due to a deviation in the angle of a single vertebra (V11). The movement from the drinking posture to the "tip up" position (Fig. 10.10b) can be simulated as a series of interpolations between the two input sets. The network also offers the opportunity to explore neck movements which are difficult to produce in an experiment. For example, Fig. 10.10c shows two neck postures for which two input parameters (the extension of the neck and the orientation of the head) were constant but with a different elevation. The less elevated posture is reached by clockwise flexion of the vertebrae 12–8 and counter-clockwise flexion of vertebrae 7–4. The predicted solution, however, has an unexpected property. In spite of the fact that the input parameters which determine the position of the top of the neck were changed, the head is translated with respect to the body. This effect is more pronounced in the example shown in Fig. 10.10d. The only parameter varied was the orientation of the head. Both the minimization algorithm and the network predict the same change in posture: downward rotation of the head produces a more sharply curved S-shape of the neck. The network model also predicts retraction of the head. Such effects may be the result of a limited number of training patterns or may indicate that elevation, distance and orientation of the head are not controlled independently.

This type of network does not represent a particular part of the central nervous system. Any transformation like the one formulated here can be implemented by the type of network used (Barto 1990). However, it does show that "neural-like machines" can produce the kind of non-linear operations involved in the control of neck movements. Biologically more realistic models must distinguish between the problem of neural sensorimotor transformation and the transformation from motor signal to movement and incorporate knowledge of central connections.

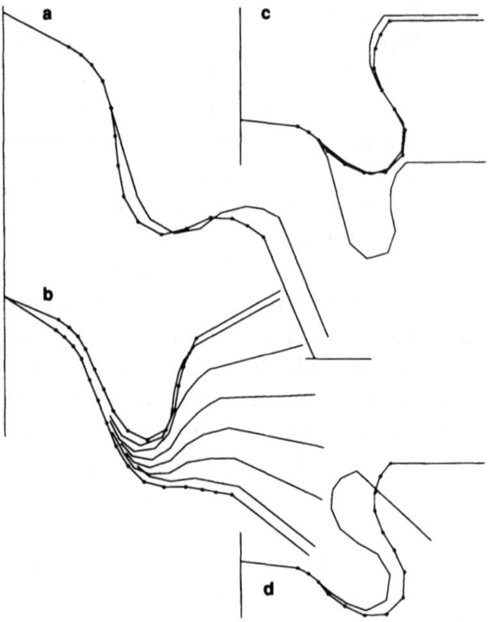

Fig. 10.10a–d. Four predicted head-neck patterns from a 3-layered neural network trained by back-propagation. *Dots* represent the positions of the vertebral joints, taken from radiograms. The postures predicted by the network are shown by *plain lines*. **a** The head-neck posture of a chicken grasping a seed; **b** a drinking posture and tip up posture, both part of the training set for the network. The postures produced by the network are shown, including a series of four intermediate postures simulating movement; **c** and **d** present the effect of translation and head rotation as predicted by the neural network; **c** shows the effect of the head lowered with respect to the resting position, without changing its orientation; **d** shows the effect of changing the orientation of the head by 40°, without changing its position (Courtesy of R. Bout)

Finally, a more analytical approach to constructing networks for neck motion was proposed by Pellionisz (1988) and Pellionisz and Peterson (1988). At the level of abstraction chosen (a chain of rigid bars) control of neck motion shares many problems with control of arm motion; Bullock and Grossberg (1988) and Cruse and Brüwer (1990) have developed neural network models for arm motion.

10.6 Conclusions

The main conclusions to be drawn from the research discussed in this chapter are as follows.

The avian head-neck system is a very complex, multi-element system, comprising over 15 vertebrae and over 200 muscle slips.

Kinematic patterns of the chain of cervical joints during drinking follow five mechanical principles. Search for further principles should be undertaken, particularly by working with a wider range of behaviour patterns, and by extending analytical methods to three dimensions.

Some static models have been developed which can predict head-neck postures. A semi-dynamic approach allows application of these models to motion patterning and leads to an understanding of the functional demands that are put on real head-neck operations. Subdivision of the very complex, multi-element system into functional components has not yet been achieved.

Motor patterning can be understood indirectly by applying a model of torque and muscle shortening. Changes in the patterning of motion and muscle activity during growth are known indirectly from scaling effects.

Despite slow progress in understanding neurosensorial control, promising results have been obtained from the use of network models of motion control. Such models help to understand head-neck motion patterning and may eventually be connected to neural design and to the demands to real, functional behaviour.

References

Arends JJA (1981) Sensory and motor aspects of the trigeminal system in the mallard (*Anas platyrhynchos* L.). Thesis, Univ Leiden

Barto AG (1990) Connectionist learning for control. In: Miller WT, Sutton RS, Werbos PJ (eds) Neural networks for control. MIT Press, Cambridge, pp 5–58

Baumel JJ (1979) Osteologia. Arthrologia. In: Baumel JJ, King AS, Lucas AM, Breazille JE, Evans HE (eds) Nomina anatomica avium. Academic Press, London pp 53–173

Bennett MB, Alexander RMcN (1987) Properties and function of extensible ligaments in the necks of turkeys (*Meleagris gallopavo*) and other birds. J Zool Lond 212:275–281

Bilo D, Bilo A (1983) Neck flexion related to activity of flight control muscles in the flow-stimulated pigeon. J Comp Physiol 153:111–122

Boas JEV (1929) Biologisch-anatomische Studien über den Hals der Vögel. K Dan Vidensk Selsk Skr Naturvidensk Math 9,1,3:101–222

Bout R (1987) Neuroanatomical circuits for proprioceptive and motor control of feeding movements in the mallard (*Anas platyrhynchos* L.). Thesis, Univ Leiden

Bout R, Postures of the avian cranio-cervical column. (in prep)

Bout RG, Kardong KV, Weisgram J, Zweers GA (1992) Modeling neck postures. Zool Jahrb Anat 122:167–169

Bullock D, Grossberg S (1988) Neural dynamics of planned arm movements. In: Grossberg S (ed) Neural networks and natural intelligence. MIT Press, Cambridge, pp 553–622

Bunday BD (1984) Basic optimization methods. Arnold, London

Burton PKJ (1974) Anatomy of head and neck in the Huia (*Heteralocha acutirostris*) with comparative notes on other Callaeidae. Bull Br Mus (Nat Hist) Zoo 127:1–48

Cabot JB, Reiner A, Bogan N (1982) Avian bulbospinal pathways: anterograde and retrograde studies of cells of origin, funicular trajectories and laminar terminations. In: Kuypers HGJM, Martin GF (eds) Anatomy of descending pathway to the spinal cord. Elsevier Biomedical Press, New York, pp 79–106

Cohen DH, Pitts LH (1967) The hyperstriatal region of the avian forebrain: somatic and autonomic responses to electrical stimulation. J Comp Neurol 131:323–335

Cruse H, Brüwer M (1990) A simple network controlling the movement of a three joint planar manipulator. In: Eckmiller R, Hartmann G, Hauske G (eds) Parallel processing in neural systems and computers. Elsevier, Amsterdam, pp 409–412

Cusick CJ, Peters JJ (1974) Electromyography, electroencephalographic and behavioural changes during the onset of erect postures in newly hatched chicks. Poul Sci 53:1456–1462

Davids JAG (1952) Etude sur les attachés au cranes des muscles de la tête et du cou chez *Anas platyrhynchos* L. 1, 2, 3. Proc K Ned Akad Wet C 55:81–94, 525–533, 534–540
Davies MNO, Green PR (1989) Visual head extension: transitional head coordination in the pigeon squab (*Columba livia*). Dev Psychobiol 22:477–488
Deigh JD, Klein BG, Zeigler HP (1985) Grasping in the pigeon: motor control mechanisms. Brain Res 337:362–367
Den Boer PJ (1953) On the correlation between the cervical muscles and the structure of the skull in *Phasianis colchicus* L. and *Perdix perdix* L. Proc K Ned Akad Wet C 56:335–345, 455–473
Dubbeldam JL (1976) The basal branch of the septo-mesencephalic tract in the mallard (*Anas platyrhynchos* L.). Acta Morphol Neerl-Scand 14:98
Dubbeldam JL (1991) The avian and mammalian forebrain. In: Andrew RJ (ed) Neural and behavioural plasticity. Oxford University Press, Oxford, pp 65–91
Dubbeldam JL, Visser A (1987) The organization of the nucleus basalis-neostriatum complex of the mallard (*Anas platyrhynchos* L.) and its connections with the archistriatum and the paleostriatum complex. Neuroscience 21:487–517
Duym M (1951) On the head posture in birds and its relation to some anatomical features, 1, 2. Proc K Ned Akad Wet 54:202–211, 260–271
Elshoud GCA, Zweers GA (1987) Avian cranio-cervical systems. Part 3: Robot kinematics for cervical systems. Acta Morphol Neerl-Scand 25:235–260
Erichsen JT, Hodos W, Evinger C, Bessette BB, Phillips SJ (1989) Head orientation in pigeons: postural, locomotor and visual determinants. Brain Behav Evol 33:268–278
Friedman MB (1975) Visual control of head movements during avian locomotion. Nature 255:67–69
Fritsch E, Schuchmann KL (1987) The Musculus splenius capitis of hummingbirds, Trochilidae. Ibis 130:124–134
Goedbloed E (1958) The condylus occipitalis in birds. Proc K Ned Akad Wet C 61:35–65
Goodale MA (1983) Visually guided pecking in the pigeon (*Columba livia*). Brain Behav Evol 22:22–41
Green PR, Davies MNO, Thorpe PH (1992) Head orientation in pigeons during landing flight. Vision Res 32:2229–2234
Heidweiller J (1989) Post natal development of the neck system in the chicken (*Gallus domesticus*). Am J Anat 186:258–270
Heidweiller J, van der Leeuw AHJ, Zweers GA (1992a) Cervical kinematics during drinking in developing chickens. J Exp Zool 262:135–153
Heidweiller J, Lendering B, Zweers GA (1992b) Development of motor patterns in cervical muscles of drinking chickens. Neth J Zool 42:1–22
Jenni L (1981) Das Skelettmuskelsystem des Halses von Buntspecht und Mittelspecht, *Dendrocopus major* und *medius*. J Ornithol 122:37–63
Johnson R (1984) The cranial and cervical osteology of the European oystercatcher *Haematopus ostralegus* L. J Morphol 182:227–224
Klein BG, Deich JD, Zeigler HP (1985) Grasping in the pigeon (*Columba livia*): final common path mechanisms. Behav Brain Res 18:201–213
Komarek VL (1979) Vertebra avia. Sci Agron Bohemoslovaca 2:35–49
Landolt R, Zweers GA (1985) Anatomy of the muscle-bone apparatus of the cervical system in the mallard (*Anas platyrhynchos* L.). Neth J Zool 35:611–670
Landolt R, Vanden Berge JC, Zweers GA (1989) The cervical column of mallard and chicken. Fortschr Zool 35:74–78
Levine RR, Zeigler HP (1981) Extratelencephalic pathways and feeding behaviour in the pigeon (*Columba livia*). Brain Behav Evol 19:56–92
Martinoya C, Rey J, Bloch S (1981) Limits of the pigeon's binocular fields and direction for best binocular viewing. Vision Res 21:1197–1200
Martionya C, Le Houezec J, Bloch S (1984) Pigeons' eyes converge during feeding: evidence for frontal binocular fixation in a lateral-eyed bird. Neurosci Lett 45:335–339

Murray PDF, Drachman DB (1969) The role of movements in the development of joints and related structures: the head and neck in chick embryos. J Embryol Exp Morphol 22:349–371
Palmgren P (1949) Zur biologischen Anatomie der Halsmuskulatur der Singvögel. In: Mayr E, Schuez E (eds) Ornithologie als biologische Wissenschaft. Winter, Heidelberg, pp 192–203
Pellionisz AJ (1988) Vistas from tensor network theory: a horizon from reductionist neurophilosophy to the geometry of multi-unit recordings. In: Cotterill RMJ (ed) Computer simulation in brain science. Cambridge University Press, Cambridge, pp 44–73
Pellionisz AJ, Peterson BW (1988) A tensorial model of neck motor activation. In: Peterson BW, Richmond B (eds) Control of head movement. Oxford University Press, New York, pp 178–186
Peterson BW, Pellionisz AJ, Baker JF, Keshner EA (1989) Functional morphology and neural control of neck muscles in mammals. Am Zool 29:139–149
Popova MF (1972) On Morpho-functional adaptations of the neck in swimming and diving birds. Vestn Zool 6:54–60
Reiner A, Karten HJ (1983) The laminar source of efferent projections from the avian Wulst. Brain Res 275:349–354
Ritchie TLC (1979) Intratelencephalic visual connections and their relationship to the archistriatum in the pigeon. Thesis, University of Virginia, Charlottesville, VA
Rumelhart DE, Hinton GE, Williams RJ (1986) Learning internal representations by error propagation. In: Rumelhart DE, McCelland JJ (eds) Parallel distributed processing, vol 1. MIT Press, Cambridge pp 318–364
Saglam M (1968) Morphologische und quantitative Untersuchungen über die Muskelspindeln in der Nackenmuskulatur des Bunt- und Blutspechtes. Acta Anat 69: 87–104
Salzen EA, Parker DM (1975) Arousal and orientation functions of the avian telencephalon. In: Wright P, Caryl PG, Vowles DM (eds) Neural and endocrine aspects of behaviour in birds. Elsevier, Amsterdam, pp 205–242
Vanden Berge JC (1979) Myologia. In: Baumel JJ, King AS, Lucas AM, Breazille JE, Evans HE (eds) Nomina Anatomica Avium. Academic Press, London, pp 175–219
Vanden Berge JC, Zweers GA (1993) Myologia. In: Baumel JJ (ed) Handbook of avian anatomy: Nomina anatomica avium. Nuttall Ornithological Club, Cambridge, pp 189–247
Vidal PP, Graf W, Berthoz A (1986) The orientation of cervical vertebral column in unrestrained awake animals. I. Resting position. Exp Brain Res 61:549–559
Virchow H (1915) Bewegungsmöglichkeiten in der Wirbelsäule des Flamingos. Arch Anat Physiol Anat: 244–254
Weisgram J, Zweers GA (1988) Avian cranio-cervical systems. Arthrology of the cranio-cervical system in the mallard (*Anas platyrhynchos* L.). Acta Morphol Neerl-Scand 25:157–166
Zeigler HP (1989) Neural control of the jaw and ingestive behaviour. Ann NY Acad Sci 563:69–86
Zusi RL (1962) Structural adaptations of the head and neck in the Black Skimmer, *Rynchops nigra* L. Publ Nuttal Ornithol Club 3:1–101
Zusi RL, Bentz GD (1984) Myology of the purple-throated carib (*Eulampis jugularis*) and other hummingbirds (Aves: Trochilidae). Smithson Contrib Zool 385:1–70
Zusi RL, Storer RW (1969) Osteology and myology of the head and neck of the pied-billed grebes (*Podilymbus*). Misc Publ Mus Zool Univ Mich 139:1–49
Zweers GA (1982) Pecking of the pigeon (*Columba livia*). Behaviour 81:174–230
Zweers GA (1985) Generalism and specialism in the avian mouth and pharynx. Fortschr Zool 30:189–201
Zweers GA, Berkhoudt H (1991) Recognition of food in pecking, probing and filter feeding birds. Acta XX Congr Int Ornithol, New Zealand Ornithological Congress Trust Board, Wellington, pp 897–902
Zweers GA, van Pelt HC, Beckers A (1981) Morphology and mechanics of the larynx of the pigeon (*Columba livia* L.). Zoomorphology 99:37–69
Zweers GA, Vanden Berge JC, Koppendraier R (1987) Avian cranio-cervical systems. Part 1. Anatomy of the cervical column in the chicken (*Gallus gallus*). Acta Morphol Neerl-Scand 25:131–155

Introduction to Section III

The chapters in the third section all address, in a variety of ways, the problem of linking perceptual processes and motor organization. Bringing together these two "sides" of more complex central nervous systems is a major challenge for neurobiology; some chapters describe direct physiological approaches to the problem, while others describe behavioural evidence leading to more abstract models.

The first three chapters are linked by a common concern with the control of gross locomotion, and particularly of flight manoeuvres. In Chapter 11, Bilo draws together a wide range of behavioural and physiological evidence to develop a cybernetic model of course control in pigeons. This specifies the roles of visual, vestibular and visceral input in controlling yawing turns during flight, and relies on an exact and thorough analysis of the aeromechanics of bird flight.

Bilo's analysis of flight control has important links with a number of themes developed in Section II. He uses EMG evidence, together with aeromechanical principles, to elucidate the role played by different muscles in adjusting flight course, as Zeigler et al. and Zweers et al. do in Chapters 9 and 10 for the adjustment of pecking and of head-neck posture. More specifically, control of the head-neck system by visual input turns out to play a key role in the control of flight muscle activity.

Some of the issues addressed by Bilo can also be linked to aspects of Frost, Wylie and Wang's review in Chapter 12 of their group's work on the responses of single cells in the pigeon visual system to different patterns of motion. Frost et al. begin their review with an analysis of the different patterns of optic flow which arise in a bird's natural environment, and the different environmental events and patterns of self-motion which they specify. Against the background of this ecological analysis, they describe neurophysiological results which demonstrate a basic two-fold division of motion processing between the tectofugal system and the accessory optic system (AOS).

In the tectofugal system, cells are selective for characteristics of *local* relative motion, and Frost et al. propose that the function of this system is to extract information about the motions of objects and animals in the bird's surroundings. One important recent discovery is the existence in this system of cells which respond to directly approaching objects and are selective for their time-to-contact. In the AOS, cells are selective for the direction of *whole-field* motion, and Frost et al. argue that the system is responsible for extracting information

M.N.O. Davies and P.R. Green (Eds.)
Perception and Motor Control in Birds

about the velocity of the bird's own motion. AOS cells share a common spatial frame of reference with the vestibular system, and are presumably closely involved in the control of the flight manoeuvres analyzed by Bilo in Chapter 11.

In Chapter 13, Lee also begins with a discussion of the information available in optic flow, but goes on to focus closely on the ways in which optic flow can provide time-to-contact information, through the optical parameter tau. This would be potentially useful to birds in a wide range of situations such as perching or timing strikes at prey, and Lee presents behavioural evidence that hummingbirds, gannets and pigeons use tau to control deceleration or to time actions. This evidence clearly parallels the physiological findings described by Frost et al. in Chapter 12. Lee also takes up a theme developed in Chapters 2 and 3, arguing that perception of distance or time-to-contact must be considered in a behavioural context, and he develops this point through the concept of scaling distance in terms of action cycles, such as wingbeat or head-bob cycles.

Lee goes on to demonstrate the scope of his analysis by showing that tau can also provide a strategy for echolocating bats to obtain time-to-contact information, providing that their hearing is directionally selective. This is a slight taxonomic digression, but a valuable one. Two species of birds are known to use echolocation (Griffin 1953; Medway 1959), and Lee's analysis has the interesting implication that the evolution of this ability may have required only a link between auditory input and pre-existing neural machinery for computing tau, without the need for any major new neural apparatus.

Directional selectivity in bird hearing is most dramatic in owls, some of which can hunt prey in complete darkness using hearing alone. In Chapter 14, Volman provides a thorough review of the elegant neurophysiological work which has, over the last 20 years, established how an auditory "map" of space is established in the inferior colliculus of barn owls. This is achieved through combining the outputs of two parallel pathways, one of which computes phase differences and the other intensity differences between the inputs from the two ears. Work is in progress to follow further the links from the collicular auditory map through the optic tectum to the control of motor output, and promises to provide a complete single-cell model of the sensorimotor integration underlying prey localization in these birds.

Volman goes on to put the neurophysiological analysis in a comparative context, by discussing findings from a variety of species in addition to the barn owl. One conclusion to emerge from these studies is that there is no major difference in the neural processing mechanisms between owls with symmetrically and asymmetrically placed ears, despite the ability of the latter to obtain more directional information from a single sound. Like Provine's "centripetal" hypothesis discussed in Chapter 7, Volman's conclusions are important for our understanding of the interactions between central and peripheral changes in neural and behavioural evolution.

In Chapter 15, Katzir is also concerned with the means by which birds orient towards prey targets, but now focuses on the use of vision in a specific situation, where a bird must strike accurately from air at underwater prey. Work on vision in amphibious birds has concentrated mainly on adaptations of optical design to cope with differences in refractive index between air and water (some examples are discussed in Chap. 1), but Katzir's work has highlighted a different problem; that of correcting for refractive error in striking at underwater targets. Experiments show that reef herons solve the problem by initiating strikes from points falling on a line along which real and apparent prey depth have a particular ratio. In effect, a highly constrained pattern of approach and strike allows a constant "rule of thumb" correction to be applied.

Comparative studies raise intriguing questions about the development of this strategy for correcting for refraction at the water surface. For example, terrestrial cattle egrets make errors when first striking at submerged prey, but learn to correct, whereas reef herons show no evidence of needing to learn the correction. It appears that the same strategy can be acquired in quite different ways.

In the final chapter, the editors discuss a general problem in the visual perception of distance; the existence in most contexts of multiple sources of distance information and the ability of most animals to detect more than one of them. Chapters 2, 3 and 13 discussed various ways in which each of accommodation, stereopsis and tau might be detected, and Chapter 16 considers how these and other parameters might be combined to provide the most accurate possible control of behaviour. Evidence from experiments on pigeon landing suggests a motivational influence on the distance cues used, and this finding leads to the proposal that the problem of depth cue integration should be treated in an "optimality" context. Animals would be expected to combine depth cues in different ways in different contexts depending upon the relative importance of accuracy and speed in the control of behaviour. This argument leads to a range of suggested factors which may influence the weights given to different cues in contexts such as the control of pecking.

References

Griffin DR (1953) Acoustic orientation in the oilbird *Steatornis*. Proc Natl Acad Sci USA 39:884–893

Medway L (1959) Echolocation among *Collocalia*. Nature 184:1352–1353

11 Course Control During Flight

D. Bilo

11.1 Introduction: The Avian Flight Control System

The avian flight control system has almost the same function as the technological flight control system in an aeroplane. Variables such as angular position, course, linear and angular velocities have to be controlled in both cases. In the biological flight control system the actual values of controlled variables are continuously measured by eye and mechanoreceptors (e.g. vestibular organ) and are compared in the central nervous system with the reference values (see Sect. 12.3.2). Control deviations caused by external disturbances such as gusts of wind modulate the activity of flight steering muscles. Such disturbance induces an adjustment of the final control elements, the wings and the tail, and thus a correction in spatial orientation and motion of the flying bird.

In general, the behaviour of a control system consists of responses to disturbances and to variations of the reference inputs. Under closed-loop conditions both kinds of responses are intermingled in such a way that it is often impossible to distinguish between them. For example, when analyzing the turning flight of birds of prey (Oehme 1976c), it is difficult to decide which of the tail and wing adjustments are intentional in order to initiate a turn and which of them are reflectoric responses to disturbances. The latter can be studied separately, if the feedback loops of the control system are opened. This is done simply by holding a bird such as the domestic pigeon with both hands, so that the movements of the wing and the tail cannot influence the angular position of the trunk. The wing and tail movements elicited by rotating the pigeon's trunk around its longitudinal axis (roll axis) transversal axis (pitch axis) or dorsoventral axis (yaw axis) are almost pure responses to the disturbances imposed on the angular position of the bird's trunk (the feedback loops of the head control system are closed in this situation).

An important result from these simple experiments is that the reflectoric responses of the wings and the tail to rotational stimuli are rational from an aeromechanical point of view. When, for example, a pigeon is tilted around its roll axis, the ipsilateral (i.e. depressed) wing will be abduced and the contralateral one will be adduced. During flight, this reflex would increase the lift

Universität des Saarlandes, Fachrichtung Zoologie 13.4, D-6600 Saarbrücken, FRG

M.N.O. Davies and P.R. Green (Eds.)
Perception and Motor Control in Birds

produced by the ipsilateral wing and decrease the lift generated by the contralateral one. Thus, the bird as a whole would roll backward into the horizontal position. When the pigeon is tilted around its pitch axis with its head down (pronation), the wings will be swept forward and the tail will be rotated upward, and vice versa when the pigeon is raised (supination) (Brown 1963). During flight, a forward sweep of the wings would shift the centres of pressure of the wings, so that they are in front of the centre of gravity of the whole bird, and vice versa when the wings are swept backwards. In both cases the torque generated would rotate the bird backward into a horizontal position. The fact that the wing and tail reflexes elicited in the fixed, non-flying bird can be interpreted in terms of aeromechanics *justifies studying the principles of neural flight control in the non-flying bird.* Thus we have the possibility, for example, of fixing a bird onto a turntable so that definite acceleration stimuli can be applied and the responses of flight steering muscles to these stimuli can be studied electromyographically.

The history of avian aeromechanics has been summarized by Oehme (1976b). In a theoretical investigation of the aeromechanical principles of avian flight control, he drew up a list of the moments (rolling, yawing and pitching moment) which were to be expected during gliding flight under the influence of various angles of attack of the two wings or the angular positions of the tail (Oehme 1976a). The aerodynamical interpretations of wing and tail reflexes which are to follow are mainly based on this study and on the contributions from Brown (1963), Jack (1953), Rüppell (1980) and Storer (1948).

The neural mechanisms of flight control, and especially the role of the vestibular system, have been most systematically studied by Groebbels in the years 1922 to 1929. He summarized his work in a paper with the remarkable title, *Der Vogel als sich automatisch steuerndes Flugzeug* – the bird as an automatically controlled aeroplane (Groebbels 1929). Parts of his concept of avian flight control still seem valid today, as will be shown below. On the basis of behavioural experiments, Mittelstaedt (1964, 1983) developed an ingenious hypothesis concerning the integration of vestibular and neck afferences in the pigeon's control of the roll angle. This hypothesis, however, neglects the existence of acceleration receptors which are located in the trunk. Although these receptors have not been directly established, their existence in the pigeon has been proved indirectly by the observation that rotation-compensating wing and tail reflexes persisted when both labyrinths had been destroyed (Trendelenburg 1906). Rotation-compensating tail reflexes persisted even when the spinal cord was transsected between the brachial plexus and the lumbar plexus (Singer 1884). The results obtained by Singer and Trendelenburg have been corroborated and refined by Biederman-Thorson and Thorson (1973). Moreover, Delius and Vollrath (1973) found that fibres within the dorsal roots of the pigeon's lumbosacral spinal cord respond to trunk rotations around the roll and pitch axes and less to rotation around a vertical axis. The spike rate of these fibres could be modulated by inflating a balloon which was implanted into the abdomen of the pigeon. This result suggests that the afferences originate from mechanoreceptors located in mesenteries. In thick body sections prepared by sawing frozen pigeons in various planes, Delius and Vollrath found:

"that the mesenteries have multiple, radial attachments to the peritoneal walls. This arrangement would seem to be adequate for the detection of motions of the viscera relative to the body as they occur during body rotations given that they are furnished with suitable stretch receptors. Such receptors are known to exist in mammals (Leek 1972) but have yet to be demonstrated in birds" (Delius and Vollrath 1973, pp 132–133).

11.2 Fundamentals of Avian Aeromechanics of Course Control

Course control, that is the control of the bird's flight direction in the horizontal plane, is an essential part of flight control. At first sight, course control during flight may seem to be a rather simple task, because it appears to be a purely two-dimensional problem which may be solved by adjusting the angular position of the bird in the horizontal plane (azimuth angle or heading), comparable to the course control of a ship. However, course control during flight is slightly more complicated: if a gliding bird (or the pilot of an aeroplane) wishes to fly a curve, it has not only to change its azimuth angle, but it has also to bank towards the inner side of the planned curve. In this way, the lift produced by the wings is tilted towards the curve centre and a centripetal force component is produced, which initiates the curve and compensates for the centrifugal force caused by the centripetal acceleration of the bird's mass (or of the mass of the plane). Thus, in flying birds (at least during gliding flight) and aeroplanes, the course is controlled by adjusting the bank angle and the azimuth angle simultaneously, or in other words, course control is accomplished via rotations of the bird or aeroplane not only in the horizontal plane, but in *space*.

For quantitative analysis, the instantaneous angular position of a bird or an aeroplane in space is decomposed into the three so-called Euler angles which are independent from each other: the azimuth angle (azimuth) or heading $\psi(t)$, the inclination angle (inclination) $\theta(t)$ and the bank angle (bank) $\phi(t)$. These angles as functions of time describe the angular movement of the body-fixed reference frame relative to the so-called vertical body-carried reference frame, whose angular orientation relative to the earth is constant (cf. textbooks on aeromechanics, e.g. Etkin 1972). Since the head of a bird is extremely mobile, not only a trunk-fixed and a vertical trunk-carried reference frame, $x_T y_T z_T$ and $x_g y_g z_g$, but also a head-fixed and a vertical head-carried reference frame, $x_H y_H z_H$ and $x_{g'} y_{g'} z_{g'}$, have to be defined. As shown in Fig. 11.1, the Euler angles $\psi_T(t)$, $\theta_T(t)$ and $\phi_T(t)$ describe the instantaneous angular position of the trunk in the horizontal $x_g y_g$ plane, in the vertical $x_T z_g$ plane and in the $y_T z_T$ (= transversal plane of the trunk), respectively. An analogous rule applies to the Euler angles of the head, $\psi_H(t)$, $\theta_H(t)$ and $\phi_H(t)$.

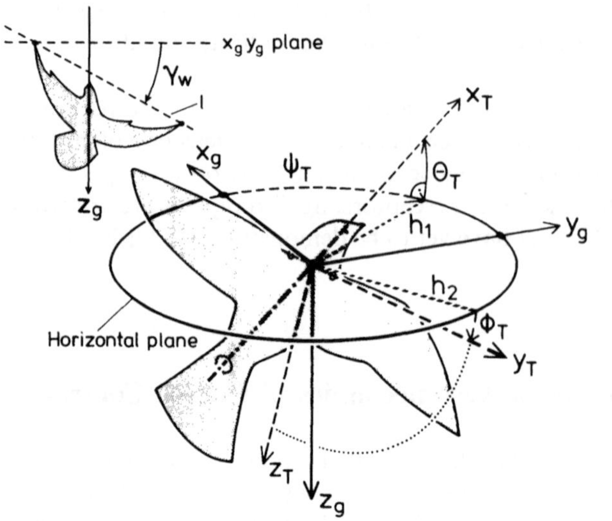

Fig. 11.1 Reference frames and angles; x_g; y_g and z_g indicate the axes of the trunk-carried vertical reference frame; x_T, y_T, and z_T indicate the axes of the trunk-fixed reference frame. The $x_T z_T$ plane and the $y_T z_T$ plane (indicated by the *dotted quater circle*) are parallel to the midsagittal and transversal plane of the trunk, respectively. Euler angles of the trunk: ψ_T azimuth angle (azimuth) = angle between the x_g axis and the projection h_1 of the x_T axis onto the $x_g y_g$ plane; θ_T inclination angle (inclination) = angle between the x_T axis and h_1; ϕ_T bank angle (bank) = angle between the y_T axis and the intersection line h_2 between the $y_T z_T$ and the $x_g y_g$ plane. The time series $\psi_T(t)$, $\theta_T(t)$ and $\phi_T(t)$ describe the angular movement of the trunk in the horizontal $x_g y_g$ plane, the vertical $x_T z_g$ plane and the $y_T z_T$ plane (transversal plane of the trunk), respectively. The head-fixed reference frame $x_H y_H z_H$ and the Euler angles of the head, ψ_H, θ_H and ϕ_H, relative to the $x_g y_g z_g$ system are defined by analogy (cf. Bilo 1992). The *inset* in the *left-hand top corner* defines the wing bank angle γ_W: it is the angle between the wingtip connection line (*l*) and its projection onto the $x_g y_g$ plane

In this connection two points are important:

1. As a rule, the azimuth of the trunk, ψ_T, differs in varying degrees from the bird's course (equal to horizontal flight direction or flight path azimuth), ψ_F. The difference between ψ_T and ψ_F is dependent, among other things, on the bird's course and flight speed relative to the air and on the wind speed and wind direction (Jack 1953). The azimuth angle of the trunk roughly equals that of the flight path only if the bird flies through calm air or if it flies straight upwind or downwind. *In the following we always assume that the bird flies through calm air and that $\psi_F(t) = \psi_T(t)$ at all times t.*

2. In birds, the *aerodynamically important* bank angle is seldom fulfilled by the angle $\phi_T(t)$ (which lies within the $y_T z_T$ plane) because the angle between the longitudinal axis of the trunk and the flight direction is often rather large (especially during slow turning flight) and the $y_T z_T$ plane is far from being perpendicular to the flight direction. An aerodynamically more suitable bank angle is achieved by the angle between the connecting line of the wing tips and

the horizontal plane, which I have called the wing bank angle γ_W later on (cf. Fig. 11.1, left-hand top corner).

In order to bank an aeroplane most effectively, two kinds of final control elements have to be activated simultaneously: the ailerons and the rudder. The ailerons are made by cutting out a section of the trailing edge of each wing and putting it on hinges. The pilot can increase or decrease lift by lowering or raising the aileron, thus increasing or decreasing the angle of attack for that part of the wing. The rudder is a mobile auxiliary airfoil hinged to the rear edge of the vertical fin, and like an aileron, it creates the sideward lift or push necessary for steering the plane. Birds do not have a vertical fin or specialized parts of the wings that act as ailerons. A gliding bird may use the whole wing or only its outer part as an aileron, rotating it to change the angle of attack (Storer 1948; Oehme 1976c). As pointed out in detail by Oehme (1976a), the angles of attack of the two wings can be adjusted in different ways, so that a pure rolling moment, a pure yawing moment or a combination of both will be produced. Moreover, the tail may be spread and twisted clockwise or counter-clockwise in order to produce an additional rolling moment which supports or counteracts the action of the wings. Oehme (1976c), for example, has filmed and analyzed a hen harrier's (*Circus cyaneus*) banking turn towards the right which was initiated by a counter-clockwise twist of the tail (as seen from behind the bird) and an adjustment of the wings in such a way that the angle of attack of the inner wing was larger than that of the outer one. This means that the tail produced a clockwise rolling moment (as seen from behind the bird), while the wings probably generated both a counter-clockwise rolling moment and a clockwise yawing moment (as seen from above). Obviously, the clockwise rolling moment produced by the tail was stronger than the counter-clockwise rolling moment produced by the wings, because the harrier banked clockwise. This rolling movement was stopped 250 ms after it began by a rapid clockwise twisting of the tail. The action of the tail was supported later on by an increase in the angle of attack of the right (inner) wing.

As pointed out by Brown (1963), the absence of a vertical fin implies that the gliding bird is most likely to be:

"essentially unstable, and that flight is maintained by the neuromuscular control of shape and position of wing and tail. This instability ... has certain advantages. It is a fact that stability and manoeuvrability are antagonistic. A highly stable system is unaffected by external forces and requires large control forces to change its direction of motion. An unstable system is easily upset but requires small control forces to correct deviations or change its direction. One may suggest that, while an engineer designs an aircraft with the maximum stability consistent with the manoeuvrability required, the animal evolves a system with the maximum manoeuvrability, and consequent instability, that the nervous system can control" (Brown 1963, p. 483 f).

While the aeromechanics of course control during *gliding* flight are not too difficult to grasp, at least in principle, course control during *flapping* flight is much more resistant to a thorough aeromechanical analysis. Stereo-cinematographic analysis of the domestic pigeon's slow turning flight has given us some preliminary ideas of the functional organization of this complex system (Bilo et al. in prep.). It became apparent that the pigeon initiates a narrow banking turn primarily by twisting each wing to a different degree during downstroke. The outer wing is much more twisted (i.e. the distal part of the wing is pronated) than the inner wing (cf. lower left-hand inset in Fig. 11.2). Thus, the distal part of the outer wing acts as a propeller and produces more thrust and vertical lift than the distal part of the inner wing. Moreover, the inner wing is more adduced during upstroke than the outer one, so that yawing and rolling moments are likely to be produced both during downstroke and during upstroke. The role of the spread-out tail in course control during slow flapping flight is not yet fully understood. It may function as a stabilizer, which dampens the rotatory oscillations of the trunk caused by the flapping wings. An extremely effective course control was observed by Oehme (1965) in the drongo (*Dicrurus macrocercus*). This species is capable of performing abrupt turns of up to 270° on the spot by stopping and supinating the inner wing immediately after the beginning of the downstroke, while the outer wing continues performing one or – in extreme turns – two "normal" downstrokes during a turn. In this way the drongo turns around its inner wing, which acts as a brake. The rolling moment produced by this asymmetrical movement of the wings is counteracted by the widely spread and twisted tail.

11.3 Head Stabilization and Head-Wing-Trunk Correlations During Slow Turning Flight

As already mentioned, a bird differs from an aeroplane in having, among other things, an extremely mobile linkage between head and trunk. This is advantageous for visual inspection of the surroundings during flight and on many other occasions in daily life. On the other hand, the integration of opto- and mechanosensory afferences is much more complex than with a rigid head-trunk linkage (cf. Sect. 10.5.2). In the following we will first take a look at the time courses of the azimuth angle of the head and of the azimuth and bank angle of the trunk during slow turning flight. Later on, we will turn to the neural mechanisms of the correlations between lateral head deflection and the adjustments of the wings and the tail.

Frame by frame analysis of a high-frequency film of a domestic pigeon (*Columba livia*) during a narrow banking turn has shown that the azimuth angle of the head changes stepwise while the trunk turns quite smoothly, its azimuth lagging behind that of the head (Fig. 11.2; cf. Bilo et al. 1985; Bilo 1991, 1992). The head azimuth remains more or less constant during the hold phases of the head. This intermittent stabilization of the head azimuth is assumed to be essential for visual orientation during turning flight in a narrow space.

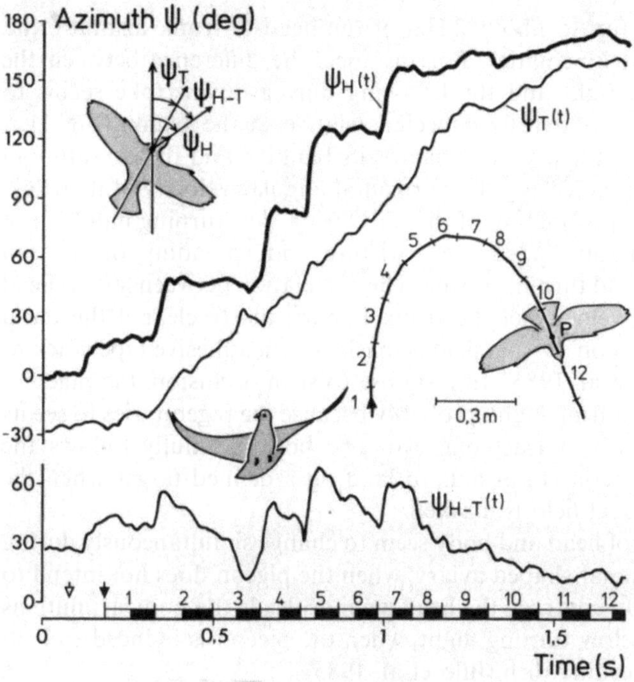

Fig. 11.2. Angular movement of the pigeon's head and trunk in the horizontal plane during slow turning flight (film speed 300 frames/s, top view). $\psi_H(t)$ and $\psi_T(t)$ indicate the time courses of the head azimuth and trunk azimuth (= horizontal angular position of the head and trunk respectively; cf. *inset* in the *upper left corner*) during a right-hand curve; $\psi_{H-T}(t)$ ($= \psi_H(t) - \psi_T(t)$) symbolizes the time course of the horizontal head deflection relative to the body (= head-to-body azimuth). The pigeon's horizontal flight path, that is the trajectory of the trunk point *P* in the horizontal plane, is shown on the *lower right inset*. The consecutive flapping cycles are marked and numbered above the time axis and along the flight path; one flapping-cycle is defined from the end of one downstroke to that of the next. Periods of upstrokes and downstrokes are plotted along the time axis as *white and black stripes*, respectively. The *open arrow* indicates the beginning of the take-off jump, when the pigeon begins to extend its legs. The *solid arrow* signifies when the pigeon's feet leave the ground. The *lower left-hand inset* is a frontal view of the pigeon during the first downstroke. The right wing points towards the curve centre. The angle of attack of the right distal wing region relative to the horizontal plane is clearly larger than that of the left one (Slightly modified from Bilo 1991, Fig. 1)

The pigeon's head performs a nystagmus-like movement relative to a trunk-fixed reference frame [Fig. 11.2, $\psi_{H-T}(t)$]. The fast nystagmus phases are directed towards the curve centre, the slow phases coincide with the hold phases relative to the earth. The slow phase velocity of the head relative to the trunk roughly equals the negative horizontal angular velocity of the trunk relative to the earth; it reaches more than 250° s^{-1}. The nystagmus amplitude varies between 20° and 40°, and the mean frequency amounts to 4–5 Hz (wing-beat frequency, 6–11 Hz). There are more or less clear correlations between the slow frequency components (up to about 2 Hz) of the head-to-trunk azimuth and some variables of flight kinematics. For instance, the difference between the angles of attack of the right and the left wing during downstroke seems to increase when the head is laterally deflected relative to the trunk. (Fig. 11.2, lowest inset.) In this way, the pigeon enhances its banking and the curvature of its flight path as well as increasing the horizontal angular velocity of its trunk. Although the aerodynamic function of the tail during *slow* turning flight is not clear in detail, presumably "adequate" twisting and spreading of the tail contributes to banking and turning as well. The correlation between lateral head deflection and angular velocity of the trunk is particularly clear if the nasal region ($\pm 90°$) of the pigeon's visual field is masked with adhesive tape stuck to the head feathers (Bilo et al. 1985). In response to such occlusion, the pigeon's head azimuth oscillates during flight, probably because the pigeon tries to see its frontal surroundings with at least one eye. The body faithfully follows the oscillating head. The pigeon is not able to land on a defined target when the frontal region of the visual field is masked.

The azimuth angles of head and body seem to change simultaneously during fast turning flight in a canal-shaped aviary, when the pigeon does not intend to land within the curve. By contrast, the head azimuth leads the body azimuth as has been described for slow turning flight, when the pigeon is induced to land within the curve on a feeding dish (Bilo et al. 1985).

11.4 Head Deflection and Activity of Flight Control Muscles in the Flow-Stimulated Pigeon

The time relationships between head and trunk azimuth during slow turning flight (the azimuth of the head leads that of the trunk) and the observed correlations between head-trunk azimuth and wing kinematics indicate that neck reflexes acting upon wing and tail muscles are very important in flight control. Groebbels (1929) reached this conclusion, although he did not succeed in establishing with any certainty the presence of these neck reflexes, because they are obviously switched on only during flight. They can, however, be elicited in the pigeon held in the hand or restrained in a support, when the bird is subjected to an airstream directed frontally or ventrally, and when its legs are hanging free. Shortly after switching on the blower, the animal moves its legs backwards into the flight position, whilst wing and tail muscles strongly react to passive or active lateral neck deflection (Bilo and Bilo 1983) and to galvanic stimulation of the labyrinth (Bilo and Bilo 1978). We conclude from these observations that the airflow stimulus, which is assumed to be perceived by mechanoreceptors attached to the bases of breast feathers (Gewecke and Woike 1978), produces an illusory state of flight, which enables us to study basic flight control reflexes in an immobilized bird.

Figure 11.3 illustrates the locations of those wing and tail muscles whose activities in relation to passive or active lateral neck deflection were investigated.

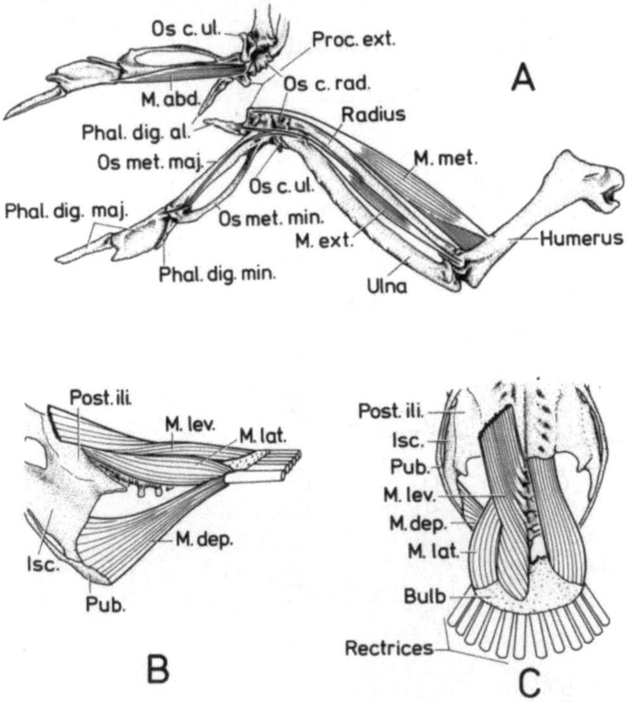

Fig. 11.3A–C. Locations of some wing and tail muscles of the domestic pigeon mentioned in the text. The muscles are named according to George and Berger (1966) and the new names introduced by Vanden Berge (1979) are given in parentheses. The skeletal elements are named according to Baumel (1979, 1988). **A** Dorsal (lower) and ventral (upper) views of the left wing skeleton. Skeletal elements: *Os c. rad.* os carpi radiale; *Os c. ul.* os carpi ulnare; *Os met. maj.* os metacarpale majus; *Os met. min.* os metacarpale minus; *Phal. dig. al.* phalanx digiti alulae; *Phal. dig. maj.* phalanges digiti majoris; *Phal. dig. min.* phalanx digiti minoris; *Proc. ext.* processus extensorius of the os metacarpale alulare. Muscles: *M. abd.* m. abductor indicis (m. abductor digiti majoris); *M. ext.* m. extensor digitorum communis; *M. met.* m. extensor metacarpi radialis. The *M. abd.* originates from the anterior and ventral side of the *Os met. maj.* and inserts on the ventral anterior edge of the proximal phalanx digiti majoris. The *M. ext.* originates from the dorsal head of the humerus and inserts, with its tendon dividing into two branches, on the dorsal side of the *Phal. dig. al.* and on the dorsal anterior edge of the proximal phalanx digiti majoris. The *M. met.* arises with a dorsal and a ventral head from the humerus and inserts on the *Proc. ext.* **B** lateral and **C** dorsal aspects of the caudal region of the pelvis (synsacrum) and of the proximal region of the tail. Skeletal elements: *Post. ili.* postacetabular ilium; *Isc.* ischium; *Pub.* pubis. Muscles: *M. lat.* m. lateralis caudae; *M. lev.* m. levator coccygis + m. levator caudae (M. levator caudae); *M. dep.* m. depressor caudae (M. pubocaudalis externus). The *M. lev.* and the *M. dep.* have been removed on the right side of the body in **C**. For further details of the origins and insertions of *M. lat.* and *M. dep.*, see Baumel (1988, pp. 22 and 24). (Original drawing, A. Gardezi)

Nomenclature of the muscles is also explained in Fig. 11.3, and their functions are described in Sect. 11.5.

Figure 11.4 shows the left-right activity pattern of some flight control muscles of the airflow-stimulated pigeon in response to stepwise lateral neck flexion, which was brought about by deflecting the head around the stationary

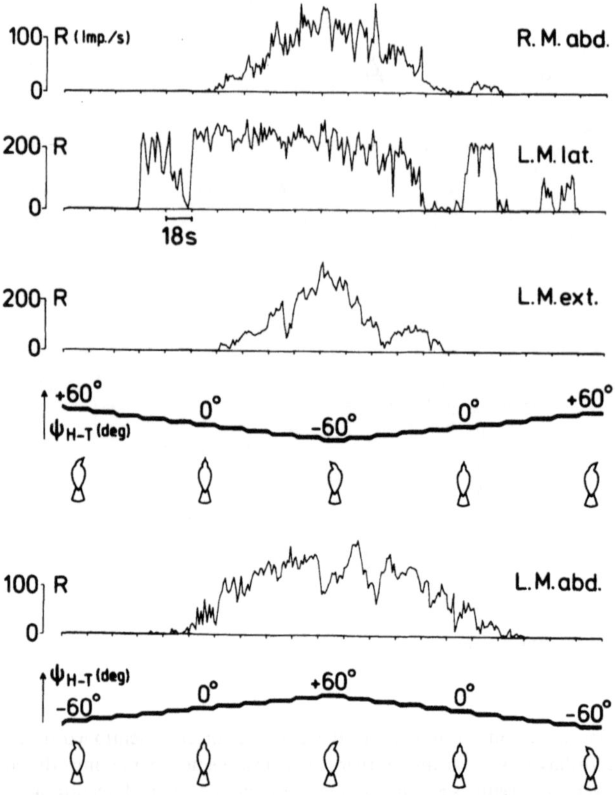

Fig. 11.4. Responses of the right (*R*) and left (*L*) m. abductor indicis (*M. abd.*), the left m. lateralis caudae (*M. lat.*) and the left *M.* extensor digitorum communis (*M. ext.*) of the airflow-stimulated pigeon to stepped lateral deflection of the head around the earth-fixed trunk. *R* symbolizes the mean impulse rate per 1.2 s, $\psi_{H-T}(t)$ the head-to-trunk azimuth. The responses of the right *M. abd.* and the left *M. lat.* were recorded simultaneously (After Bilo and Bilo 1983, Fig. 3)

trunk or deflecting the trunk around the stationary head. Whenever the neck is bent sideways, the impulse rates of the ipsilateral m. extensor digitorum communis (M. ext.) and m. lateralis caudae (M. lat.) and of the contralateral m. abductor indicis (M. abd.) and m. extensor metacarpi radialis (M. met.) increase, while the impulse rates of the opposing muscles decrease or remain zero. The same relationship is found between the head-trunk azimuth and the activities of flight control muscles in spontaneous horizontal head deflections (head saccades) (Bilo and Bilo 1983; Fig. 11.5). With spontaneous head deflections, the onset or drop of muscular activity follows the corresponding head saccade with a delay of less than 50 up to 500 ms. The increase in muscle activity following a head saccade is tonic or phasic-tonic. In some cases the phasic activity seems to be induced by a simultaneous change in both the azimuth and bank angle of the head.

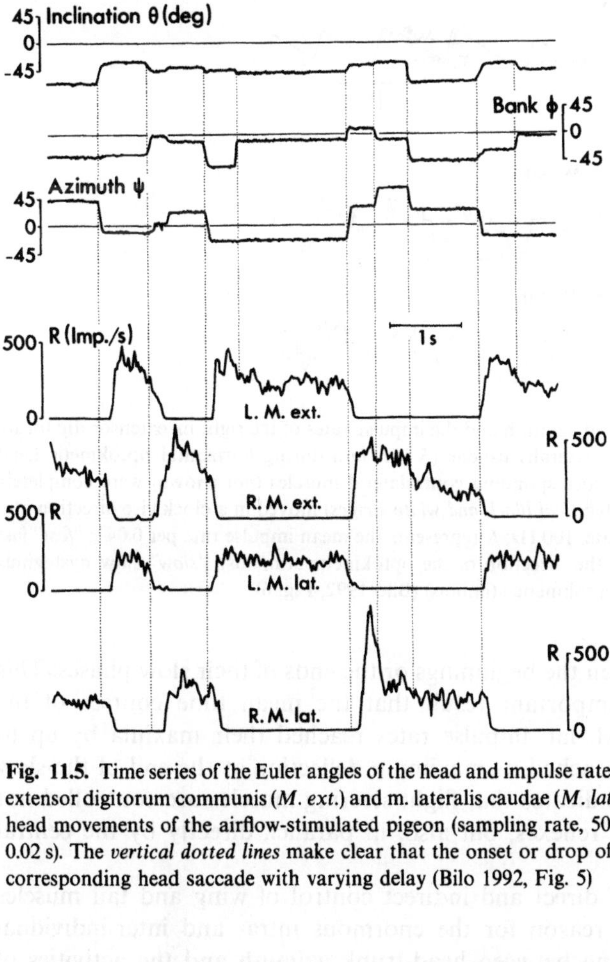

Fig. 11.5. Time series of the Euler angles of the head and impulse rates of the left (*L*) and right (*R*) m. extensor digitorum communis (*M. ext.*) and m. lateralis caudae (*M. lat.*) during spontaneous angular head movements of the airflow-stimulated pigeon (sampling rate, 50 Hz; *R* mean impulse rate per 0.02 s). The *vertical dotted lines* make clear that the onset or drop of muscular activity follows the corresponding head saccade with varying delay (Bilo 1992, Fig. 5)

During optokinetic head nystagmus (OHN) of the flow-stimulated pigeon, the activity of flight control muscles is frequently modulated in the rhythm of the OHN (Fig. 11.6). The muscular impulse rates of the ipsilateral M. ext. and M. lat., for example, increase in step with the lateral deflection of the head during each slow nystagmus phase. They pass a maximum towards the end of the slow phase and decrease to a minimum at the end of the fast phase.

The time courses of the ipsi- and contralateral muscle activities are mirror-inverted to each other if the range of head movement is centred around the midline, so that the beak crosses the midsagittal plane of the trunk during the fast and the slow nystagmus phase. The contralateral muscles are either inactive or their impulse rates run parallel to those of the ipsilateral muscles, if the range of head movements is confined to the side of the body which points in the direction of the visual pattern movement. The mean temporal relationships between head azimuth and muscular impulse rate were obtained by averaging the time courses of the muscular impulse rate over several nystagmus phases,

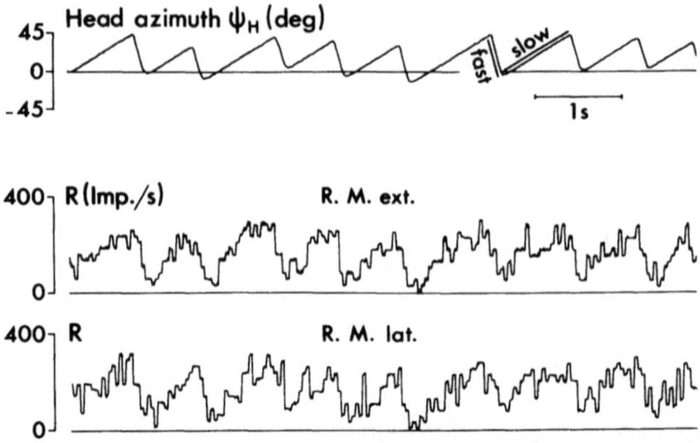

Fig. 11.6. Time series of the head azimuth and the impulse rates of the right m. extensor digitorum communis (*R. M. ext.*) and m. lateralis caudae (*R. M. lat.*) during horizontal optokinetic head nystagmus of the pigeon. The corresponding contralateral muscles (not shown) were completely inactive. The stimulus pattern (*vertical black and white stripes*) moved in a clockwise direction at a velocity of 69° s^{-1}. Sampling rate, 100 Hz; *R* represents the mean impulse rate per 0.04 s; "*fast*" fast nystagmus phase (contrary to the direction of the optokinetic stimulus); "*slow*" slow nystagmus phase (in the direction of the optokinetic stimulus) (Bilo 1992, Fig. 6)

which had been centred on the beginnings or the ends of their slow phases. This procedure yielded the important result that the mean time courses of the ipsilateral M. ext and M. lat. impulse rates reached their maxima by up to 200 ms *before* the head reached its maximum deflection at the end of the slow nystagmus phase. This indicates that flight steering muscles are controlled not only indirectly via neck reflexes, but also, in parallel, directly by the central nervous system.

This combination of direct and indirect control of wing and tail muscles could be an additional reason for the enormous intra- and inter-individual variability of the coupling between head-trunk azimuth and the activities of flight control muscles. This coupling with each individual control muscle can be strengthened, attenuated or completely switched off within minutes.

11.5 Effects of Control Muscle Activity During Flight

What effect could the muscular activity pattern coupled with lateral deflection of the head have in the flying bird? There is no definitive answer to this question for the time being, because there are no records of the activities of flight control muscles in flying birds. The functional anatomy of wing and tail and of the flight steering muscles (Sy 1936; Baumel 1988) leads to the following conclusions (cf. Fig. 11.3):

1. The M. met. will spread out the hand wing by abducing the metacarpus, if the elbow joint is locked by the biceps and triceps.

2. The M. abd. abduces and pronates the front primaries. Thereby the geometrical angle of attack of the outer wing region is reduced and its profile curvature is simultaneously increased.

3. The contraction of the M. ext. has the contrary effect: it supinates the front primaries thereby increasing the geometrical angle of attack and decreasing the profile curvature. In addition, the M. ext. raises the alular digit.

4. The M. lat. abduces and erects the lateral tail feathers, and if the contralateral m. depressor caudae (M. dep.), which abduces and depresses the lateral tail feathers, is active simultaneously, the entire tail is spread and twisted around its longitudinal axis (cf. Baumel 1988, pp. 74 and 75).

On the basis of these statements and the observed linkage between neck flexion and control muscle activity, the following hypotheses can be made: When a *gliding* pigeon is turning its head to the right for example, the activities of the right M. ext. and M. lat., and of the left M. met., M. abd. and, presumably M. dep. will be increased, whilst those of the opposite muscles will be decreased. According to statements (1) and (2), the right wing might produce more drag than the left one, thereby inducing a clockwise yawing moment. This will presumably be enhanced by the drag acting on the laterally deflected head and will rise further if the tail is spread unilaterally to the right. The twisted tail causes a rolling moment. If this is not compensated by an opposing rolling moment produced by the wings, the pigeon will roll and yaw clockwise, and thus fly towards the right. A muscular activity pattern symmetrical to that described can be expected if the pigeon turns its head to the left, in which case it will fly towards the left.

Similar correlations between head-to-trunk azimuth and flight control muscle activity may exist in flapping flight, and if so, at least during each *downstroke*. As far as the wing muscles are concerned, this hypothesis is supported by the observation that, halfway through downstroke in slow turning flight, the wing which is pointing away from the curve centre is often much more twisted (i.e. the distal part of the wing is pronated against the proximal one) than the wing pointing towards the curve centre (Fig. 11.2, lowest inset). Furthermore, this left-right difference in wing twisting seems to correlate with the low frequency component of the head-to-trunk azimuth (cf. Sect. 11.3). By contrast, the tail is more or less symmetrically spread during slow turning flight, indicating that the M. lat. and M. dep. are almost equally activated on both sides of the body.

11.6 Minimum Model of the Functional Organization of Course Control

The results and interpretations described above can be summarized in a minimum model of the functional organization of course control (Fig. 11.7, thick lines). In this model, a "loop" (either $1 \rightarrow 2 \rightarrow 3 \rightarrow 4 \rightarrow 5 \rightarrow 10 \rightarrow 1$ or

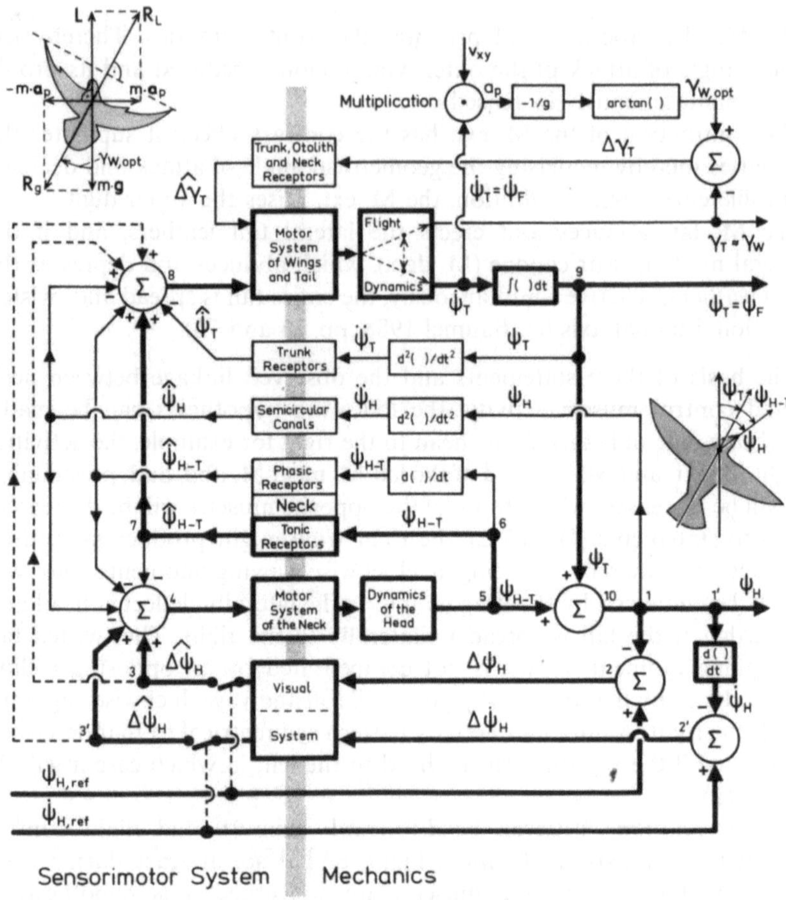

Fig. 11.7. Minimum model (*thick lines*) and extended model (*thick and thin lines*) of the functional organization of course control. All variables of the models are functions of time (e.g. $\psi_H = \psi_H(t)$). *Minimum model*: the model is a combination of a "loop" (either $1 \rightarrow 2 \rightarrow 3 \rightarrow 4 \rightarrow 5 \rightarrow 10 \rightarrow 1$ or $1' \rightarrow 2' \rightarrow 3' \rightarrow 4 \rightarrow 5 \rightarrow 10 \rightarrow 1'$) with a "unilateral mesh" ($5 \rightarrow 6 \rightarrow 7 \rightarrow 8 \rightarrow 9 \rightarrow 10$ *and* $5 \rightarrow 10$). *The visual system, which is the only exteroceptive system within this model, measures the deviations* $\Delta\psi_H$ and $\Delta\dot{\psi}_H$ from the reference values of the azimuth and the horizontal angular velocity of the head, $\psi_{H,ref}$ and $\dot{\psi}_{H,ref} = d\psi_{H,ref}/dt$, respectively. The neural outputs of the visual system, $\widehat{\Delta\psi_H}$ and $\widehat{\Delta\dot{\psi}_H}$, which correspond to $\Delta\psi_H$ and $\Delta\dot{\psi}_H$ are fed into the motor system of the neck. The output of this system, the head-to-body azimuth ψ_{H-T} (cf. the *right-hand inset* for definition), is measured by tonic neck receptors. Their total output $\hat{\psi}_{H-T}$ acts upon the motor system of the wings and the tail and causes the trunk azimuth ψ_T to follow the head azimuth ψ_H. The latter is obtained by adding ψ_T to ψ_{H-T}. It is assumed that $\psi_T(t)$ equals the course of the trunk (= flight path azimuth angle or horizontal flight direction), $\psi_F(t)$, at any time t. The same is assumed to be true for the rate of the flight path azimuth, $\dot{\psi}_F(t)$, and the horizontal angular velocity of the trunk, $\dot{\psi}_T(t)$. *Extended model*: (1) the horizontal angular acceleration of the head, $\ddot{\psi}_H$, stimulates the semicircular canal organs, especially the lateral ones. However, the primary semicircular canal afferences tend to reflect the instantaneous angular *velocity* of the head, $\dot{\psi}_H(t)$, more than its instantaneous acceleration $\ddot{\psi}_H(t)$. (Anastasio et al. 1985). Therefore the output of the box *Semicircular Canals* is symbolized by $\hat{\dot{\psi}}_H$ and

$1' \rightarrow 2' \rightarrow 3' \rightarrow 4 \rightarrow 5 \rightarrow 10 \rightarrow 1'$ in Fig. 11.7) is combined with a "unilateral mesh" ($5 \rightarrow 6 \rightarrow 7 \rightarrow 8 \rightarrow 9 \rightarrow 10$ and $5 \rightarrow 10$ in Fig. 11.7); for definition of the two terms, see Mittelstaedt (1964). The loop serves the purpose of controlling the azimuth $\psi_H(t)$ and/or the horizontal angular velocity $\psi_H(t)$ ($= d\psi_H(t)/dt$) of the head. This is exclusively achieved in the minimum model via the visual system (cf. Chap. 12), which measures the deviations of the two control variables, $\Delta\psi_H(t)$ and $\Delta\dot{\psi}_H(t)$, from the reference values $\psi_{H,ref}(t)$ and $\dot{\psi}_{H,ref}(t)$. If the switches in the feedback channels are closed, the neural equivalents of the control deviations, $\widehat{\Delta\psi}(t)$ and $\widehat{\Delta\dot{\psi}}(t)$, will be fed into the motor system of the neck and correction of the position and movement of the head will be initiated. When the head is moved in order to compensate for the control deviation, the head-trunk azimuth $\psi_{H-T}(t)$ will be changed. This angle is monitored by the neck receptors (only tonic neck receptors are considered in this minimum model). The afferences $\hat{\psi}_{H-T}(t)$ originating from these receptors are fed into the motor system of the wings and the tail and cause the trunk azimuth $\psi_T(t)$ to follow the head azimuth $\psi_H(t)$. The latter results from the addition of the head-trunk azimuth to the trunk azimuth. (The trunk azimuth may be considered as a disturbance input to the control system of the head azimuth.) Thus, not only the head azimuth but also the trunk azimuth can be stabilized by this minimum model of course control, although only the head azimuth is directly monitored by the visual system (cf. Chaps. 12 and 13).

Presumably, the head azimuth itself is continuously controlled only when the pigeon fixes its landing place with both eyes (upper switch in Fig. 11.7 closed). As long as the pigeon does not see or fix its goal (which frequently occurs during turning flight), the horizontal angular velocity of the head, not the

Fig. 11.7. *Contd.*
not by $\hat{\psi}_H$. The semicircular canal afferences $\hat{\dot{\psi}}_H$ help to stabilize both ψ_H and ψ_T. (2) The trunk azimuth ψ_T is additionally stabilized by afferences originating from visceral acceleration receptors (trunk receptors) which are assumed to have transfer characteristics similar to those of the semicircular canals, so that their output is symbolized by $\hat{\dot{\psi}}_T$ and not by $\hat{\psi}_T$. (3) Phasic neck receptors measure the angular velocity of the head relative to the trunk, $\dot{\psi}_{H-T}$. The phasic and tonic neck afferences, $\hat{\dot{\psi}}_{H-T}$ and $\hat{\psi}_{H-T}$, stabilize ψ_{H-T} at $\psi_{H-T} = 0$ and cause ψ_T to follow and catch up with ψ_H. (4) In the *uppermost part* of the model, the optimal wing bank angle $\gamma_{W,opt}$ (definition in the *left-hand top corner*) is calculated from $\dot{\psi}_T$ or ψ_F and from the horizontal flight-path speed v_{xy}. Multiplication of v_{xy} with $\dot{\psi}_T$ or $\dot{\psi}_F$ yields the centripetal acceleration a_p and, after division by the gravitational acceleration g, the tangent of $\gamma_{W,opt}$. Trunk, otolith and neck receptors measure the difference $\Delta\gamma_T$ between $\gamma_{W,opt}$ and the actual "bank" angle of the trunk, γ_T, which approximately equals the actual wing bank angle γ_W. The angles γ_T and γ_W are assumed to lie in parallel planes which are perpendicular to the bird's vector of horizontal flight path velocity, v_{xy}. The neural signal $\widehat{\Delta\gamma_T}$ corresponding to $\Delta\gamma_T$ is used to stabilize γ_T and subsequently γ_W at $\gamma_{W,opt}$. The physical interaction of γ_W (or γ_T) and ψ_T (or ψ_F) is indicated by $\leftrightarrow$ within the box *Flight Dynamics*. The *inset in the top left corner* defines the optimum wing bank angle $\gamma_{W,opt}$ (*m* body mass; **g** gravitational acceleration vector; $m \cdot \mathbf{g}$ vector of body weight; $\mathbf{a}_p$ centripetal acceleration vector; $m \cdot \mathbf{a}_p$ centripetal force vector; $-m \cdot \mathbf{a}_p$ centrifugal force vector; **L** vertical lift vector; $\mathbf{R}_g$ resultant of $m \cdot \mathbf{g}$ and $-m \cdot \mathbf{a}_p$; $\mathbf{R}_L$ resultant of **L** and $m \cdot \mathbf{a}_p$)

azimuth, seems to be controlled (lower switch in Fig. 11.7 closed, upper switch open). During the greater part of slow turning flight, the reference value of angular head velocity seems to be zero, as follows from the saltatoric change of the head azimuth. The supposition that the horizontal angular *velocity* and not the horizontal angular *position* is the control variable arises from the fact that the head azimuth shows a stochastic drift during the holding phases of the head (Fig. 11.1). This drift could originate from integration of the stochastic deviations from the zero reference value of head velocity.

In order to change the head azimuth, the bird switches over from velocity control to position control for a short time, so that the head is abruptly rotated towards a desired azimuth value. This is presumably triggered by visual inspection of the surroundings during the preceding holding phase. With each saccadic head rotation, the left-right asymmetry in the excitation of the cervical receptor system (muscle spindles, tendon organs, joint receptors) is suddenly increased. This asymmetry creates, via flight control muscles in the wings and tail, which act as final control elements, a comparatively smooth change in the trunk azimuth. The channels which are supposed to connect the visual system directly with the wing and tail muscles are symbolized by dashed lines in Fig. 11.7.

11.7 The Extended Model: The Influence of Visceral and Vestibular Afferences on the Activity of Flight Control Muscles

The extended model of the functional organization of course control shown in Fig. 11.7 (thick *and* thin lines) takes the following facts into consideration:

1. The head azimuth is stabilized not only by visual afferences but also by vestibular afferences originating from the semicircular canals (mainly the lateral ones) of the labyrinth (see Sect. 12.3).

2. Like cats (Kasper and Thoden 1981; Mergner et al. 1982), birds most likely possess phasic-tonic and phasic neck receptors as well as tonic receptors, and these measure the angular velocity of the head relative to the trunk, $\dot{\psi}_{H-T}(t)$. Together with the tonic neck receptors they stabilize the head-trunk azimuth and cause the trunk azimuth to follow the head azimuth.

3. The trunk azimuth is stabilized via flight control muscles both by afferences from visceral acceleration receptors (Biederman-Thorson and Thorson 1973; Delius and Vollrath 1973) and by afferences from the semicircular canals. The relative importance of the visceral and the semicircular canal afferences in controlling flight steering muscles was recently investigated by Schon (1987). Airflow-stimulated pigeons were submitted to horizontal rotary oscillations at different frequencies (0.05 to 1 Hz) at constant amplitude ($\pm 30°$). The blower was fixed to the turntable, so that the direction of air flow to the test animal was constant. The experiments were performed on the pigeon with its head fixed to the turntable, first in normal position and then upside

down (head turned by 180° around its longitudinal axis). The vertical axis of the turntable ran through the head between the two labyrinths.

At stimulus frequencies between 0.05 and 0.1 Hz, the muscle response was shifted by 150° to 180° in five of the eight birds tested when the head was turned upside down. In other words, the muscle response was inverted when the flow direction of the endolymph in the horizontal semicircular canals was inverted. At stimulus frequencies between 0.5 and 1 Hz the response phase was shifted by 0° to 90° (eight birds), while the amplitude remained unchanged, or was reduced by as much as 50% by inverting the head position. These results suggest that at low stimulus frequencies, which are coupled with low angular acceleration amplitudes, the influence of the semicircular canals on control muscle activity frequently exceeds that of the visceral acceleration receptors. However, the latter most often dominate when the stimulus frequency and therefore the angular acceleration amplitude are high. The influence of the otolithic receptors on muscle activity is considered to be negligible under the described experimental conditions, because the axis of rotation was between the two labyrinths and therefore the centrifugal force acting upon the otoliths should have been negligibly small.

For intact airflow-stimulated pigeons with their heads in normal position, the mean frequency characteristic of the M. ext. response (nine test animals) relative to the angular velocity of the turntable is similar to that of the horizontal vestibulo-ocular reflex in *non*-flow-stimulated pigeons (Anastasio and Correia 1988). Between 0.2 and 1 Hz, the activity of that M. ext. which is contralateral to the direction of rotation runs approximately parallel to the sinusoidal angular velocity of the turntable. When the stimulus frequency decreases below 0.2 Hz, the muscle activity increasingly leads the angular velocity and nearly follows the angular acceleration of the turntable at a stimulus frequency of 0.05 Hz (Schon 1987). The amplitude ratio (response amplitude/angular velocity amplitude) is almost constant between 0.05 and 0.5 Hz and decreases slightly towards 1 Hz.

4. As already mentioned, the bird has to bank its wing tip connecting line during turning flight by the wing bank angle $\gamma_W(t)$, in order to compensate for the centrifugal force caused by the centripetal acceleration of the bird's mass, $a_p(t)$. The wing bank angle $\gamma_W(t)$ approximately equals the analogous "bank" angle of the trunk, $\gamma_T(t)$, which lies in the plane perpendicular to the horizontal flight path velocity vector. Assuming that the turning flight is performed in the horizontal plane, the instantaneous optimum wing bank angle $\gamma_{W,opt}(t)$, is given by:

$$\gamma_{W,opt}(t) = \arctan\,(a_p(t)/g)\,, \tag{11.1}$$

where g is the gravitational acceleration and $a_p(t)$ the bird's instantaneous centripetal acceleration (cf. Fig. 11.6, left-hand top corner). One may expect that the instantaneous wing bank angle $\gamma_W(t)$ is automatically stabilized at its instantaneous optimal value $\gamma_{W,opt}(t)$, because this is identical with the direction of the instantaneous resultant of the bird's weight and the centrifugal force acting upon the birds mass and should therefore be the instantaneous "reference direction" for the bird's postural control system. Any deviation of the "bank" angle of the trunk, $\gamma_T(t)$ ($\approx \gamma_W(t)$) from $\gamma_{W,opt}(t)$ should elicit postural reflexes which operate on the wings and the tail, turning the bird into the optimal banking position.

The instantaneous centripetal acceleration $a_p(t)$ depends on the instantaneous rate of the flight path azimuth, $\dot{\psi}_F(t)$, and the instantaneous horizontal flight path speed $v_{xy}(t)$:

$$a_p(t) = \dot{\psi}_F(t) \cdot v_{xy}(t) . \tag{11.2}$$

[This equation is easily derived from $\dot{\psi}_F(t) = K_{xy}(t) \cdot v_{xy}(t)$ and $a_p(t) = K_{xy}(t) \cdot v_{xy}^2(t)$, where $K_{xy}(t)$ symbolizes the instantaneous curvature of the trajectory in the horizontal plane (cf. textbooks of mechanics, e.g. Spiegel, 1982)]. Substitution of $a_p(t)$ in Eq. (11.1) by Eq. (11.2) results in:

$$\gamma_{W,opt}(t) = \arctan\,(\dot{\psi}_F(t) \cdot v_{xy}(t)/g) . \tag{11.3}$$

With the assumption that $\dot{\psi}_F(t) = \dot{\psi}_T(t)$ (instantaneous horizontal angular velocity of the trunk), it follows from Eq. (11.3) that:

$$\gamma_{W,opt}(t) = \arctan\,(\dot{\psi}_T(t) \cdot v_{xy}(t)/g) . \tag{11.4}$$

Thus, the instantaneous optimum wing bank angle $\gamma_{W,opt}(t)$ is determined by $v_{xy}(t)$ and $\dot{\psi}_F(t)$ or $\dot{\psi}_T(t)$, respectively. In other words, a bird that wishes to change its flight path azimuth at a particular rate $\dot{\psi}_F(t)$ and at a particular horizontal flight path speed $v_{xy}(t)$, should bank its wing-tip connecting line by the angle $\gamma_{W,opt}(t)$, given by Eq. (11.4), in order to perform the optimum course change. As revealed by 3D high-speed cinematography of the pigeon's slow turning flight, $\gamma_W(t)$ roughly equals $\gamma_{W,opt}(t)$ halfway through each downstroke and each upstroke (Bilo et al. in prep.).

What are the receptors which participate in controlling γ_W? Since the head is not rigidly connected to the trunk, many birds, such as pigeons (Bilo pers. observ.) and ravens (Lorenz 1965, Fig. 35), hold their head upright during turning flight; the otolith organs *alone* are not sufficient for controlling γ_W. Visceral gravity receptors located within the trunk (Biederman-Thorson and Thorson 1973; Delius and Vollrath 1973) seem to be most suitable for an effective stabilization of the angular position of the trunk, because they are situated within that part of the body which bears the wings and the tail. In addition, however, γ_W could be stabilized by a control mechanism including both the otolith organs (utriculus and sacculus) and neck receptors. According to Mittelstaedt (1964, 1983) the afferences originating from these receptors are integrated by a "multiplicatively modulated" mesh, so that the postural limb and tail reflexes are independent from angular head position.

11.8 Improvement of Head Stabilization by Airflow Stimuli

Gioanni (1988a, b) quantitatively examined the relative contribution of the optocollic reflex (OCR) and the vestibulocollic reflex (VCR) to head and gaze stabilization in the pigeon. For the horizontal OCR he found that the horizontal angular head velocity during slow nystagmus phase, $\dot{\psi}_{HS}$, with constant pattern velocity $\dot{\psi}_P$ reached its upper limit at $\dot{\psi}_P \approx 40° \, s^{-1}$. Consequently, there was a drastic decrease of steady state gain ($= \dot{\psi}_{HS}/\dot{\psi}_P$) with $\dot{\psi}_P > 20\text{–}40° \, s^{-1}$. The free flying pigeon, on the other hand, is able to stabilize its head azimuth very well, even when the horizontal angular velocity of the trunk increases to more than $250° \, s^{-1}$ (cf. Fig. 11.2). The VCR, which is involved in the control of angular head position of the free flying bird, certainly improves the stabilization of the head in space (cf Sects. 12.3 and 13.2.1), nevertheless the performance of the OCR itself might be improved during flight. This idea was experimentally investigated as follows. The horizontal OCR was recorded in the earth-fixed pigeon with a stepwise increasing pattern velocity $\dot{\psi}_P(t)$ under three different conditions: (1) no airflow or acoustical stimulation; (2) acoustical stimulation without airflow stimulation (the airflow from the blower was not directed at the pigeon); (3) airflow stimulation, in which airflow was directed at the breast of the test animal. The chronological order was from (1) to (3). The test animals had never been used in previous experiments. Since the instantaneous angular head position was measured with the help of the search coil method (Koch 1980), the test animal's head was completely unrestrained.

The experiments yielded the following results (average of six subjects). For condition (1), the OCR gain decreased from 0.7 at $\dot{\psi}_P = 30° \, s^{-1}$ to 0.6 at $150° \, s^{-1}$ and dropped sharply with further increasing $\dot{\psi}_P$. In condition (2) there was improved performance of the OCR at pattern velocities up to $110° \, s^{-1}$, raising the gain to 0.9 at $30° \, s^{-1}$ and to 0.8 at $110° \, s^{-1}$. Finally, condition (3) yielded a high gain OCR over the whole $\dot{\psi}_P$ range. The gain decreased only from 0.9 at $\dot{\psi}_P = 30° \, s^{-1}$ to 0.8 to $230° \, s^{-1}$. In addition to an increase in OCR gain, airflow stimulation enhanced the rate of increase in $\dot{\psi}_{HS}$ following a $\dot{\psi}_P$-step (Bilo 1992) and at the beginning of each slow nystagmus phase with constant $\dot{\psi}_P$. We conclude from these results that airflow stimuli not only "link" the flight control muscles to the neck receptors, the labyrinth and the visual system, but also improve the performance of the OCR and presumably of the VCR, although this has not been verified so far. Thus, the gaze stabilization during the hold phases of the head and thereby the visual orientation during slow turning flight is optimized.

Acknowledgements. I would like to thank my wife A. Bilo and the editors for valuable comments on an earlier draft of this chapter, A. Gardezi for plotting the diagrams and W. Pattullo for improving the English. The report contains results of the doctoral theses of H.-M. Mörz and K. Saterdag and of the diploma theses of H. Schon, U. Reichl and J. Gzil. The investigations on which this report is based were supported by the Deutsche Forschungsgemeinschaft.

References

Anastasio TJ, Correia MJ (1988) A frequency and time domain study of the horizontal and vertical vestibuloocular reflex in the pigeon. J Neurophysiol 59:1143–1161

Anastasio TJ, Correia MJ, Perachio AA (1985) Spontaneous and driven responses of semicircular canal primary afferents in the unanesthetized pigeon. J Neurophysiol 54:335–347

Baumel JJ (1979) Osteologia. In: Baumel JJ, King AS, Lucas AM, Breazile JE, Evans HE (eds) Nomina Anatomica Avium. Academic Press, London, pp 53–121

Baumel JJ (1988) Functional morphology of the tail apparatus of the pigeon (*Columba livia*). In: Beck F, Hild W, Kriz W, Ortmann R, Pauly JE, Schiebler TH (eds) Advances in anatomy, embryology and cell biology 110. Springer, Berlin Heidelberg New York, pp 1–115

Biederman-Thorson M, Thorson J (1973) Rotation-compensating reflexes independent of the labyrinth and the eye: neuromuscular correlates in the pigeon. J Comp Physiol 83:103–122

Bilo D (1991) Integration opto- und mechanosensorischer Afferenzen bei der Flugsteuerung der Haustaube (*Columbia livia var. domestica*). Zool Jahrb Physiol 95:323–330

Bilo D (1992) Optocollic reflexes and neck flexion-related activity of flight control muscles in the airflow-stimulated pigeon. In: Berthoz A, Vidal PP, Graf WM (eds) Head-neck sensory motor system. Oxford University Press, Oxford, pp 96–100

Bilo D, Bilo A (1978) Wind stimuli control vestibular and optokinetic reflexes in the pigeon. Naturwissenschaften 65:161–162

Bilo D, Bilo A (1983) Neck flexion related activity of flight control muscles in the flow-stimulated pigeon. J Comp Physiol 153:111–122

Bilo D, Bilo A, Müller M, Theis B, Wedekind F (1985) Neurophysiological-cybernetic analysis of course control in the pigeon. In: Nachtigall W (ed) Biona report 3. Fischer, Stuttgart, pp 445–447

Bilo D, Reichl U, Gzil J, Scholl S, Lehmann R, Kinematics and control of slow turning flight in the domestic pigeon (*Columba livia* var. *domestica*) (in prep.)

Brown RHJ (1963) The flight of birds. Biol Rev 38:460–489

Delius JD, Vollrath FW (1973) Rotation compensating reflexes independent of the labyrinth: neurosensory correlates in pigeons. J Comp Physiol 83:123–134

Etkin B (1972) Dynamics of atmospheric flight. Wiley, New York

George JC, Berger AJ (1966) Avian myology. Academic Press, New York

Gewecke M, Woike M (1978) Breast feathers as an air-current sense organ for the control of flight behaviour in a songbird (*Carduelis spinus*). Z Tierpsychol 47:293–298

Gioanni H (1988a) Stabilizing gaze reflexes in the pigeon (*Columba livia*). I. Horizontal and vertical optokinetic eye (OKN) and head (OCR) reflexes. Exp Brain Res 69:567–582

Gioanni H (1988b) Stabilizing gaze reflexes in the pigeon (*Columba livia*). II. Vestibulo-ocular (VOR) and vestibulo-collic (closed-loop VCR) reflexes. Exp Brain Res 69:583–593

Groebbels F (1929) Der Vogel als automatisch sich steuerndes Flugzeug. Naturwissenschaften 17:890–893

Jack A (1953) Feathered wings. Methuen, London

Kasper J, Thoden U (1981) Effects of natural neck afferent stimulation on vestibulo-spinal neurons in the decerebrate cat. Exp Brain Res 44:401–408

Koch UT (1980) Analysis of cricket stridulation using miniature angle detectors. J Comp Physiol A 136:247–256

Leek BF (1972) Abdominal visceral receptors. In: Autrum H, Jung R, Loewenstein WR, MacKay DM, Teuber HL (eds) Handbook of sensory physiology, vol 3 (1). Springer, Berlin Heidelberg New York, pp 113–160

Lorenz K (1965) Der Vogelflug. Neske, Pfullingen

Mergner T, Anastasopoulos D, Becker W (1982) Neuronal responses to horizontal neck deflection in the group x region of the cat's medullary brainstem. Exp Brain Res 45:196–206

Mittelstaedt H (1964) Basic control patterns of orientational homeostasis. Symp Soc Exp Biol 18:365–385

Mittelstaedt H (1983) Einführung in die Kybernetik des Verhaltens am Beispiel der Orientierung im

Raum. In: Hoppe W, Lohmann W, Markl H, Ziegler H (eds) Biophysik, 2nd edn. Springer, Berlin Heidelberg New York, pp 822–830

Oehme H (1965) Der Flug des Fahnendrongos (*Dicrurus macrocercus*). J Ornithol 106:190–203

Oehme H (1976a) Die Flugsteuerung des Vogels. I. Über flugmechanische Grundlagen. Beitr Vogelkd 22:58–66

Oehme H (1976b) Die Flugsteuerung des Vogels. II. Kurzer Überblick über die Entwicklung der Flugsteuerungstheorien. Beitr Vogelkd 22:67–72

Oehme H (1976c) Die Flugsteuerung des Vogels. III. Flugmanöver der Kornweihe (*Circus cyaneus*). Beitr Vogelkd 22:73–82

Rüppell G (1980) Vogelflug. Rowohlt, Reinbek bei Hamburg

Schon H (1987) Drehreflektorische Modulation der Flugsteuermuskelaktivität bei der Haustaube (*Columba livia*): inter- und intra-individuelle Variabilität und relative Anteile vestibulärer, zervikaler und somatoviszeraler Afferenzen an der Muskelantwort. Diploma Thesis, Fachbereich Biologie der Universität des Saarlandes, Saarbrücken

Singer J (1884) Zur Kenntnis der motorischen Funktionen des Lendenmarks der Taube. S-B Akad Wiss Wien Math-Nat Kl 89(III):167–185

Spiegel MR (1982) Theory and problems of theoretical mechanics (Schaum's outline series, SI (metric) ed). McGraw-Hill, Singapore

Storer JH (1948) The flight of birds analyzed through slow-motion photography. Cranbrook Inst Sci Bull 28, Bloomfield Hills, pp 1–94

Sy MH (1936) Funktionell-anatomische Untersuchungen am Vogelflügel. J Ornithol 84:199–296

Trendelenburg W (1906) Über die Bewegung der Vögel nach Durchschneidung hinterer Rückenmarkswurzeln. Arch Anat Physiol 1906:1–126

Vanden Berge JC (1979) Myologia. In: Baumel JJ, King AS, Lucas AM, Breazile JE, Evans HE (eds) Nomina Anatomica Avium. Academic Press, London, pp 175–219

12 The Analysis of Motion in the Visual Systems of Birds

B.J. Frost, D.R. Wylie and Y.C. Wang

12.1 Introduction

Birds share with many other animal species a highly mobile life-style which produces, amongst other things, a very rich diversity of patterns of motion across their visual fields. As we have argued elsewhere (Frost 1982, 1985, 1993; Frost et al. 1990), these different patterns of visual motion fall generally into two broad classes, *object motion* and *self-induced motion*, the former usually resulting from some action by another animal, while the latter is usually produced by some action of the observing animal itself. Each category contains a large number of ecologically specific patterns of visual motion and, in a natural environment, patterns from one or both categories may occur concurrently. It is the task of the visual system to segregate or parse these various patterns of motion into the appropriate events that gave rise to them, and integrate them with other visual features, so that ultimately the bird may respond with behaviour that is appropriate for the situation.

In the last decade there has been an explosion of interest and research into both the mechanisms of motion analysis within the visual system, and theoretical and mathematical analyses of computational problems and potential solutions. For example, several groups have shown that all the information about figure-ground boundaries, object rigidity, object motion through the environment, and various parameters resulting from (and required for the control of) self-motion through the environment, is available from an animal's dynamic optic array (Nakayama and Loomis 1974; Longuet-Higgins and Prazdny 1980; Hildreth 1983; Reichardt et al. 1983; Koenderink 1985; Cutting 1986; Perrone 1992). In this chapter we will focus on empirical physiological work carried out in our own laboratory, which concentrates on motion analysis in various sites in the avian visual system, but the reader may wish to consult the above mentioned papers for the appropriate theoretical underpinnings of the area. Reviews by Nakayama (1985), and the collection of papers edited by Warren and Wertheim (1990), are also recommended for the reader interested in theoretical issues surrounding this topic.

Department of Psychology, Queen's University, Kingston, Ontario, Canada, K7L 3N6

M.N.O. Davies and P.R. Green (Eds.)
Perception and Motor Control in Birds

12.1.1 Local Motion, Figure-Ground Segregation and Camouflage

A few concrete examples of the different patterns of motion as they pertain to birds will now be given so that later we can relate these to neural mechanisms in the avian visual system. When an object (animal) within a bird's field of view moves, it results in either motion in a very local area of the visual field, if the observing bird's eye is stationary, or if the bird is moving, in a local patch of motion which is different from the motion in the surrounding area. This is illustrated schematically in Fig. 12.1a. Such patterns of "object" motion convey information to the observing bird that there is another animal in its vicinity, which may require further visual processing and integration for identification before an appropriate response can be selected. An important feature of the analysis of patterns of local motion is the presence of discontinuities in the motion domain (motion edges), which provide information about the boundaries of objects. They therefore convey important information about figure-ground relationships, and of course the shape or form of the object. Animal camouflage is generally thought of in terms of protective colouration and cryptic markings, but it also involves a motion component. When the background is motionless, freezing or remaining stationary is necessary to remove "motion edges" and avoid detection. When there is a moving background, such as foliage waving in the breeze, the animal wishing to hide must match its movements to the background to remove patterns of relative motion. In some species, such as Australian horned toads, chameleons, walking stick insects and some snakes, the animal may even produce motion patterns that simulate the motion patterns of wind-blown grasses and branches that surround them, and thus blend into the background to avoid detection by their predators or prey (Cott 1940; Portman 1959; Fleishman 1986).

12.1.2 Trajectory and Spin

The specific locus of motion in the visual field, and subsequent change in position of the area of local motion, can provide important information about the trajectory of the moving animal relative to the observing bird (see Fig. 12.1b), which it can use to intercept, avoid, or ignore the moving animal. Dynamic changes in size can also provide information about object trajectories in 3D space, as shown schematically in Fig. 12.1b, and can inform a bird if an animal or object is approaching or receding, or on a collision course. This latter event may portend "imminent disaster" for the observing bird if the approaching animal is a predator, or a "successful catch" if the observing bird itself is the predator.

In its simplest form, all the parts or features of an animal (object) may be moving together (coherently), but when some features move relative to other parts of a moving object, the observing bird will have access to information about the "action" of the observed animal, as well as its trajectory through

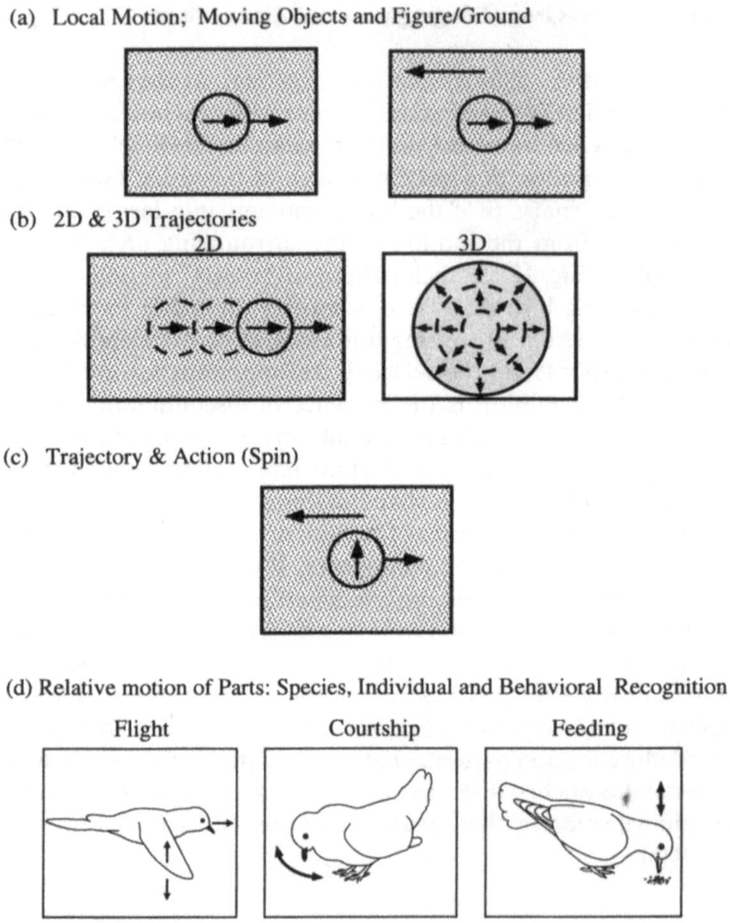

Fig. 12.1a–d. Examples of different forms of object motion, all of which are characterized by a localized region within the visual field where the motion vectors are the same or similar to each other (or have no sudden motion discontinuities), but differ from those in a surrounding zone (typically across a motion boundary or discontinuity). **a** Simple cases illustrating an object moving against a stationary background in the *left panel*, and the same object moving against a moving background in the *right panel*. These may occur when a stationary, fixating, observing animal views a moving object in the former case, or while translating itself in the latter case. These forms of motion may also contribute substantially to figure-ground segregation, where the coherent motion of the parts is seen as one unit. **b** Two examples of object motion illustrating how the 2D and 3D trajectories are available from patterns of local motion. **c** When there is a lack of coherence between the patch of local motion and the boundary of the object (motion discontinuity) then there is potential information available about the spin or action of the object. **d** Animate motion is also characterized by different patterns of relative motion of the parts: three simple examples are illustrated.

space. This is illustrated in Fig. 12.1c in a simple form as "spin" in a single dimension, and in more ethologically relevant form in Fig. 12.1d where it might represent flight, feeding, courtship, or some other behaviour. The integration of such "object-action" motion characteristics with other visual information such as shape, size and colour may be critical in the discrimination of predators from prey, conspecifics from non-conspecifics, males from females (e.g. Clark and Uetz 1992), and possibly even familiar individuals from strangers. The separation of complex forms of object motion into "trajectories" and "actions" is very similar to Johansson's findings that common motion and relative motion are separable in human visual motion perception, and that "biological motion" and actions are readily perceived when dots located on joints replace all other information about form (Johansson 1975; Cutting 1978). Whether such effects occur in birds and other species remains to be determined, but the importance of motion for eliciting imitation of feeding behaviour in chicks (Turner 1964), and many other forms of species typical behaviour is suggestive.

12.1.3 Self-Induced Motion

When an observing bird swims, walks, hops or flies through an environment filled with inanimate, stationary objects, or moves its eyes or head, it generates a rich variety of self-produced visual flow patterns. These dynamic patterns of visual transformations, some of which are illustrated schematically in Fig. 12.2, not only provide information about the bird's motion through space, but also help structure the world into its constituent parts and provide information about their 3D layout. Gibson (1950, 1966, 1979) in particular emphasized the important role such patterns and gradients of flow might have for both structuring the world and providing information necessary to guide locomotion (see also Chap. 13).

One of the simplest forms of self-induced motion is produced when a bird makes a saccadic eye movement, which results in similar patterns of displacement over the whole visual field (see Sect. 13.2.1). This can be understood by inspecting the left panel of Fig. 12.2c, where the darker arrows give a crude approximation of the flow that might be produced by a leftward horizontal saccade of the right eye. Head movements produce a somewhat similar pattern, but the flow in the binocular visual field would be similar to the total flow field shown in this panel, where left and right eye directions are opposite. Also, because the centre of rotation of the head is now displaced considerably from the nodal point of the eye, parallactic shear is produced in such a whole field flow pattern. Such gradients of motion parallax can specify the 3D structure of objects and provide information about their relative distances from the observing bird (Collett 1977; Rogers and Graham 1979, 1982; Sobel 1990). It is also likely that these gradients can specify an observer's heading (Regan and Beverley 1984; Warren and Hannon 1988; Cutting et al. 1992).

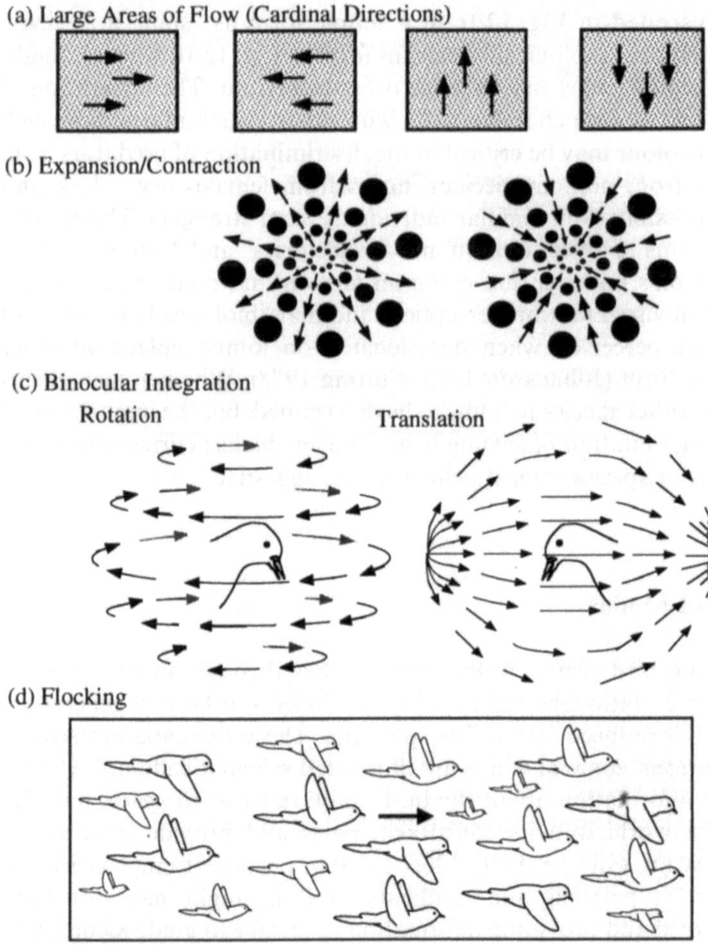

Fig. 12.2a–d. Examples of different forms of self-induced visual motion. These patterns typically occupy very large areas of an animal's visual field, and result in perceptions of self-motion or "vection" when viewed by human observers, or optokinetic responses of some form when viewed by animals. **a** Whole-field or large area flow patterns that may represent the cardinal directions preferred by cells at the lower levels of the accessory optic system of birds (see Sect. 12.3.1). These various patterns, originating from different areas of the binocular visual field, may be integrated with patterns of expansion and contraction like those shown in **b**, to result in selectivity for rotation and translation as shown in **c**. It is quite possible that the same or very similar mechanisms operate during flocking **d**, where other members of the group provide the essential visual stimuli to control optokinetic responses

When a bird is locomoting, its binocular visual flow fields can also provide information about its course in terms of both translational and rotational vectors. Later we will show how specific patterns of binocular integration of large flow field information can produce neuronal selectivity for roll, pitch and yaw information from forward/back, up/down and left/right translations. An oversimplified version of a translational flow field (see Sect. 13.2) is shown in the right panel of Fig. 12.2c, where it can be seen to consist of expansion and contraction zones similar to those illustrated in Fig. 12.2b, and relatively linear lateral flow zones like some of those illustrated in Fig. 12.2a. Also, just as rate of growth of images can provide information about the impending collision of objects and animals moving toward a bird, self-produced relative rates of expansion ("tau") can provide information about "time to contact" with stationary objects. This particular form of visual motion has been shown to be critical for the precise timing of a variety of visuomotor activities (see Sect. 13.2.2). Examples include landing in the milkweed bug and the housefly (Coggshall 1972; Wagner 1982), the aerodynamic folding of gannets' wings during diving (Lee and Reddish 1981), the controlled deceleration of hummingbirds as they dock at flowers (Lee et al. 1991), and the landing of hawks (Davies and Green 1990). However, Borst and Bahde (1988) have suggested that flies do not use tau to initiate the landing response, but rather employ an integrated spatio-temporal pooling from motion detectors, followed by a thresholding mechanism to trigger landing. All of these variables clearly provide a bird, or any other animal or machine for that matter, with information necessary to control and monitor its movements through its environment, and to precisely time its critical manoeuvres.

Finally, it is usual to consider visual flow fields as resulting from the motion of an animal relative to stationary inanimate objects that constitute its world. However, many species of bird are social and form flocks which may be considered to provide another form of flow field, like that illustrated in Fig. 12.2d. The synchronized turning, spacing and manoeuvring in such flocks could possibly be controlled by the same or similar mechanisms as those evolved to process self-induced flow fields.

Over the past decade, we have been exploring many of the varieties of "object" motion illustrated schematically in Fig. 12.1, and have contrasted these mechanisms with other motion detecting systems that seem to be specialized for processing self-induced motion or visual flow fields, like those shown in Fig. 12.2. Most of these studies have been performed on pigeons, using standard single neuron recording techniques, but a few have been carried out on kestrels and owls. The "object motion" studies have primarily involved the optic tectum and tectofugal pathway, where we have found that many motion and directionally specific neurons respond to local motion and are inhibited by large whole-field motion patterns. The "self-motion" studies have involved the accessory-optic system (AOS) where, conversely, large slowly moving flow fields have been shown to be the optimal stimuli. A schematic diagram of the three major visual pathways of birds is presented in Fig. 12.3. In the remainder of this chapter we will present some of the key observations, obtained from single cell recording techniques, that we feel throw some light on the different function these specialized motion detecting neurons and pathways may be playing in normal visual processing of birds. In most cases we have employed sophisticated graphics computers to present these complex moving visual patterns to the

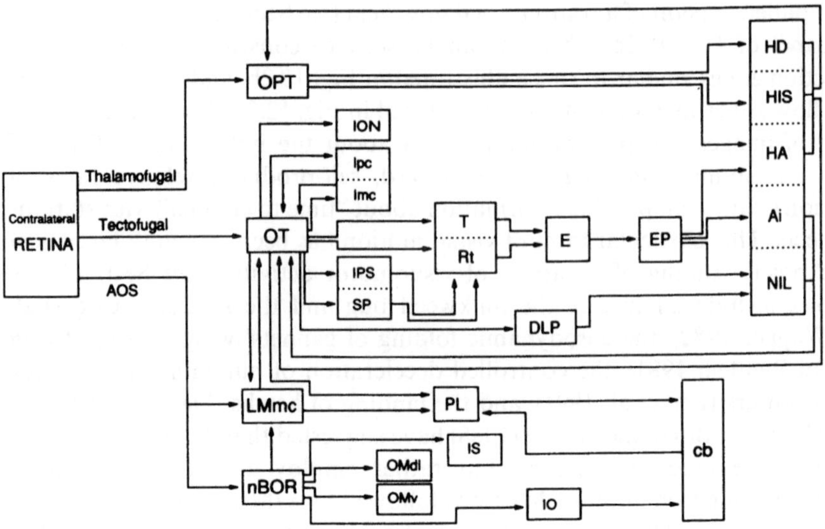

Fig. 12.3. Schematic projections of three visual pathways in the avian brain. *Ai* archistriatum intermedium dorsalis; *AOS* accessory optic system; *cb* cerebellum; *DLP* nucleus dorsolateralis posterior; *E* ectostriatum; *EP* periectostriatal belt; *HA* hyperstriatum accessorium; *HD* hypostriatum dorsale; *HIS* hypostriatum intercalatus superior; *Imc* nucleus isthmi, magnocellularis; *Ipc* nucleus isthmi, parvocellularis; *IO* nucleus olivaris inferior; *ION* nucleus isthmo-opticus; *IPS* nucleus interstitio-pretectalis-subpretectalis; *IS* nucleus interstitialis; *LMmc* nucleus lentiformis mesencephali, pars magnocellularis; *nBOR* nucleus of basal optic root; *NIL* neostriatum intermediale laterale; *OMdl* nucleus nervi oculomotori, pars dorsalis laterale; *OMv* nucleus nervi oculomotori, pars ventralis; *OPT* nucleus opticus principalis thalami; *OT* optic tectum; *PL* nucleus pontis lateralis; *Rt* nucleus rotundus; *SP* nucleus subpretectalis; *T* nucleus triangularis

birds. The specific experimental details of this work can be found in the original papers.

12.2 Object Motion in the Tectum and Tectofugal Pathway

In our earlier studies we found that nearly all visual neurons we encountered in the pigeon tectum below a depth of 400 μm from the tectal surface, through to the underlying ventricle, preferred relatively small stimuli, moving at moderate velocities (10–40° s^{-1}). Their directional tuning curves were very broad and often possessed a "backward" notch, or temporal to nasal null direction (Frost and DiFranco 1976). Moreover, these cells responded equally well to both directions of contrast, that is, black on white, and white on black, and to kinematograms (texture moved over texture) as shown in Fig. 12.4. From the response of this typical tectal neuron it can be seen that motion in the preferred direction resulted in an equally strong response to either a dark-on-light, light-on-dark, or kinematogram stimulus, while no response was elicited from motion in the reverse direction for these same three stimulus patterns.

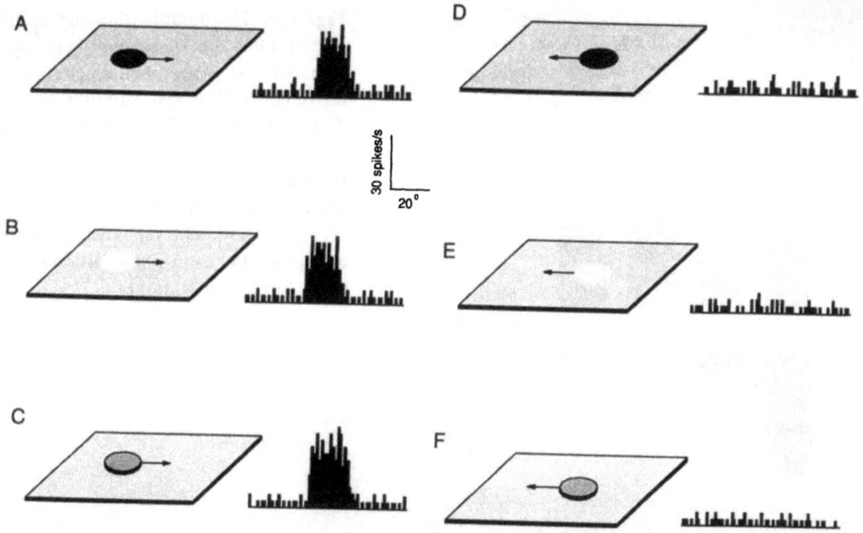

Fig. 12.4. Recording taken from a deep tectal neuron (OT-04-01) located in the stratum griseum et fibrosum superficiale (SGFS) (depth = 695 μm from tectal surface). Strong responses were produced by movement of a black **A**, or white spot **B**, 2° in diameter, moved horizontally across a stationary textured background in the preferred temporal to nasal direction (spot velocity = $20° s^{-1}$). However, no responses were obtained when the direction was reversed (null direction, nasal to temporal) (**D**, **E**). **C** and **F** illustrate kinematographic objects presented under the same conditions as pure luminance stimuli. The cell showed an identical response to moving kinematograms and to moving luminance spots. Stimulus presentation: 5 repetitions; bin width = 50 ms

12.2.1 Relative Motion

Perhaps the most important finding about these deep tectal neurons was that their response to a moving stimulus was dramatically modified by the motion of large background textured patterns presented simultaneously (Frost 1978b, 1982, 1985; Frost and Nakayama 1983; Frost et al. 1988, 1990). When a large textured background pattern was moved "in phase" with the moving target stimulus, these cells were completely inhibited, but when the background patterns were moved in "anti-phase", the directional response to the target stimulus was enhanced. This effect is shown in Fig. 12.5, where it can be seen that all velocities of "in-phase" motion inhibited these cells, while "anti-phase" motion of the backgrounds either enhanced their response or left it intact.

Other studies have shown that these same directionally specific deep tectal neurons have a very large inhibitory receptive field (IRF) surrounding their excitatory receptive field (ERF), and that they are best characterized as double opponent-process directional neurons (Frost et al. 1981; von Grunau and Frost 1983). They also respond best to opposed motion between a test stimulus and background pattern, independent of direction, and are inhibited when they both move in the same direction, independent of direction. These findings suggest

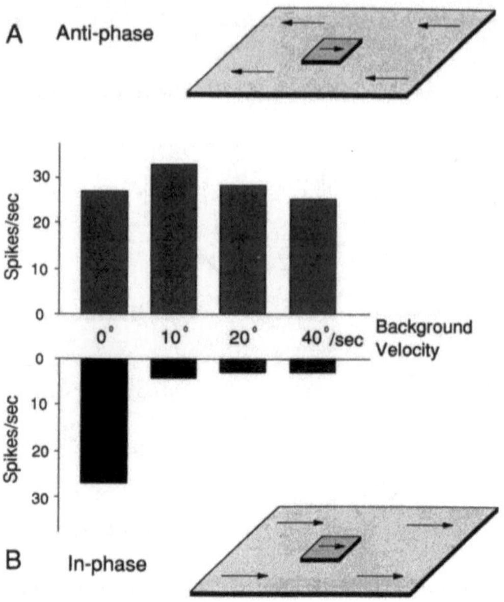

Fig. 12.5. Histograms showing firing rate as a function of different combinations of test object and background movements. The test stimulus was a 4° × 4° textured object swept in the preferred temporal to nasal direction through the ERF at 20° s^{-1}. The background texture is either stationary, or moving at different velocities in either anti-phase (**A**) or in-phase (**B**) relationships to the test stimulus

strongly that *relative* directions of motion, and not absolute directions, are important in controlling these cells' responses (Frost and Nakayama 1983).

All of these response characteristics suggest that these particular motion sensitive cells may play an important role in signalling local motion, while vetoing coherent motion over a large area of the visual field. As we illustrate it in Fig. 12.1a, local motion is most likely produced by an object or animal moving relative to its surroundings, and the deep tectal neurons would respond well to such a configuration. However, many if not most patterns of self-produced motion resulting from eye, head or body movements, such as those shown in Fig. 12.2, involve large areas of the visual field moving together in the same direction, and most deep tectal cells do *not* respond to these motion patterns.

12.2.2 Figure-Ground Segregation Through Motion

Not only deep tectal cells, but also neurons in other areas of the tectofugal pathway, such as the nuclei isthmi and the nucleus rotundus (Wang et al. 1993), respond as vigorously to kinematograms as they do to luminance stimuli. This fact suggests that coherent motion of features in a local area, and coherent motion of its boundary with these features, are the critical conditions to produce this selectivity (Frost et al. 1988). This in turn suggests that these visual neurons would be well suited to break camouflage, so long as a cryptically coloured animal moved relative to its background. In other words, they might play an important role in figure-ground segregation through motion, as well as in detecting motion of local visual areas already differentiated from their background by colour, luminance or texture. In this regard it should be remembered that camouflage when an animal freezes is achieved only if the surroundings are stationary, or moves in phase with them when they move (Fleishman 1986). Also, it should be noted that these "object" or animate motion detecting neurons prefer moderate to fast velocities and are relatively insensitive to slow motion. Thus the stalking behaviour of many predators may be considered a mechanism specifically selected so that these neurons are not stimulated. In other words, the slow stealthy motion of the predator might place their visual speed below the thresholds of the "object" motion neurons of their prey.

In order to tease apart some of the features of visual motion processing that might contribute to the response properties of these "object" motion neurons, we have recently carried out a number of investigations using kinematograms, because they represent "pure" motion stimuli. Kinematograms are produced by coherently moving a circumscribed region of random dots in one direction, while the remaining dots are moved with different motion characteristics (see Fig. 12.1a and Frost et al. 1988). Thus on any frame of such an animated sequence, there is no feature defining the "figure", but over successive frames the common motion of the elements produces a vivid appearance of the "object" in a manner that is quite analogous to random dot stereograms (Julesz 1975). Also, like random dot stereograms, kinematograms can be configured in two basic forms, one representing an "object" that appears to move over, or in front of, a similarly textured background, and another that appears as a window or "hole" in a textured surface that reveals a more distant similarly textured surface. These two kinematographic forms are differentiated by moving the boundary (separating the two directions of motion of elements) coherently with the patch of enclosed texture in the case of "objects", or coherently with the surrounding texture in the case of the "hole" (Frost et al. 1988).

Visually responsive neurons in the deep tectum, nucleus isthmi and nucleus rotundus all respond as vigorously to moving kinematographic objects as they do to moving light or dark spots (Frost et al. 1988; Wang and Frost 1992; Wang et al. 1993). However, these same cells specifically *do not* respond to moving holes of any configuration. A simple example of this is shown in Fig. 12.6 where it can be seen that a deep tectal neuron responded to a moving "object" configured kinematogram, but not to a kinematogram configured as a hole. By using much longer kinematographic bar-shaped objects and holes, where the leading and trailing edges are separated in space and time, we have recently been able to show that these deep tectal neurons specifically respond to occlusion edges and not to disocclusion edges (Jiang et al. 1989; Wang 1992). Although these studies have not been completed yet, it appears possible that such a specificity might also play a role in determining the depth relationships between the objects constituting a scene.

12.2.3 Motion in Depth and Time to Collision

Earlier qualitative single unit studies of the avian nucleus rotundus (Revzin 1970, 1979), a structure that receives substantial input in topological fashion

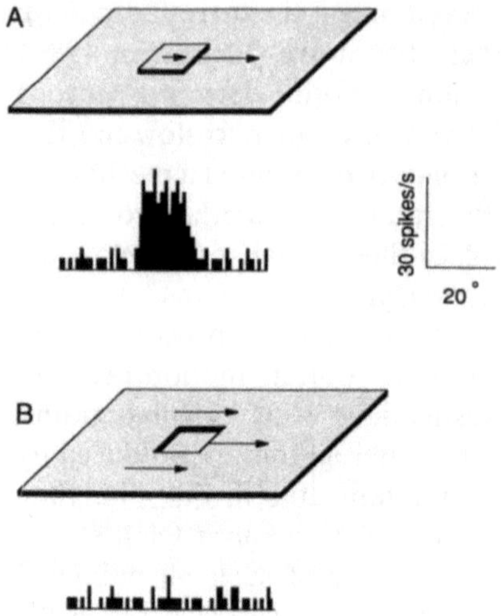

Fig. 12.6A, B. Responses of a single deep tectal cell to kinematograms in the "object" and "hole" configuration. This is typical of the response characteristics of all deep tectal neurons tested in this study. **A** "Object" moves over stationary textured background; **B** "Hole" moves with the background stationary. Both "object" and "hole" move in preferred temporal to nasal direction at optimal speed of 20° s^{-1}. Size of stimulus: 4° × 4° in both "object" and "hole" configurations. Stimulus presentation: 5 repetitions; bin width = 50 ms

from various laminae of the tectum (Benowitz and Karten 1976), had suggested that some rotundal neurons exhibited a selectivity for "motion in depth". Recently we have been able to isolate cells with this specificity for 3D motion in the dorsal posterior region of n. rotundus, and to obtain quantitative tuning curves by using computer generated stimuli. Specifically, we used a soccer ball-like stimulus, consisting of many black and white panels, which could be moved in any direction in spherical coordinates in simulated monocular 3D space. An example of the tuning curves for one of these cells is shown in Fig. 12.7, where it can be seen that this neuron responded vigorously only when the stimulus was on a direct collision course with the pigeon's head. All other directions of motion gave a response no different from the baseline firing rate. These neurons were found to constitute 16% of the cells studied in the dorsal posterior zone of the nucleus rotundus, while the majority of the others responded to more typical 2D directional characteristics. Not only were these cells very tightly tuned to the direct collision course, having half-width at half-height tuning curves of 3°–4°, but their peak firing rate occurred at almost exactly the same time (before contact), even when the size or the velocity of the stimulus was varied over a considerable range (Wang and Frost 1992). This suggests that these cells may be signalling "time to collision" (Lee 1980; see Sect. 13.2.2). Furthermore, their responses to a directly approaching soccer ball-like stimulus preceded increases in flight muscle activity and heart rate, which could possibly indicate their involvement with an escape or avoidance response system (Wang and Frost 1992; cf. Sect. 11.4). When tested with stimuli that simulated the bird's approach to a large stationary object (such as a "brick wall") these cells did not respond.

Other subdivisions in the nucleus rotundus respond to other parameters of visual stimulation such as colour, luminance, 2D motion, and occlusion edges (Wang et al. 1993). To date, we have not tested motion specific neurons in this

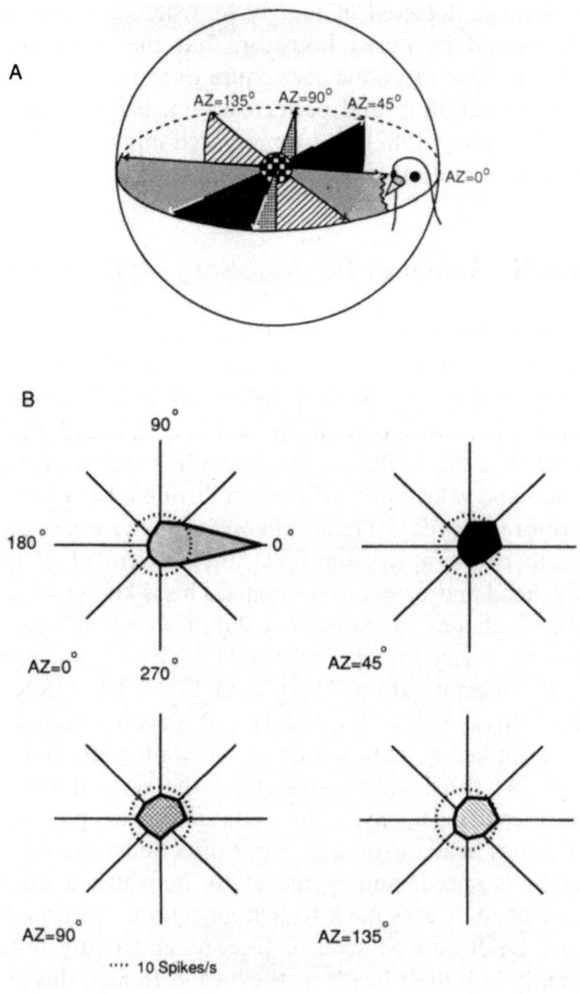

Fig. 12.7. A A soccer ball-like stimulus pattern consisting of *black and white panels*, was moved along simulated 3D trajectories 45° apart in spherical coordinates. The diagram illustrates the four planes along which stimuli were moved. **B** A typical single neuron from the nucleus rotundus of pigeons exhibiting clear selectivity for a looming visual stimulus. Firing rate is plotted for the different directions of motion of the soccer ball stimulus in 3D space. Each direction of motion was presented five times in a randomly interleaved sequence, and the values plotted represent the mean number of spikes for each 3D direction. Note that in the standard X–Y (tangent screen plane, or azimuth = 90°) plot, there is no indication of directional preference, and firing rate is quite low. However, for the 0° azimuthal plane (Z-axis) there is a strong preference for stimuli directly approaching the bird (0°). Polar tuning plots for directions specifying the azimuthal 45° and 135° planes likewise show no strong preference for any direction. Thus it is only the direct collision course or looming direction that produces an increased response in these neurons, and this pattern of activity was typical of the 27 neurons studied in the dorsal-posterior area of the nucleus

pathway with stimuli such as those depicted in Fig. 12.1c, where one class of object spin is portrayed. It should be noted however, that the anatomical subdivisions that occur in the nucleus rotundus also figure prominently in the patterns of projections to the ectostriatum and periectostriatal belt (Benowitz and Karten 1976; Nixdorf and Bischoff 1982). Future research might well look for even more complex patterns of motion processing in these structures.

12.3 Visual Analysis of Self-Motion by the Accessory Optic System

Whereas movement of external objects in an observer's visual environment results in localized motion in the optic array, self-motion of the observer through an environment cluttered with stationary objects results in global visual motion across the entire retina (cf. motion patterns in Figs. 12.1 and 12.2). Gibson (e.g. 1966) emphasized that these "flow fields" provide a rich source of information about the direction and velocity of self-motion through the environment (see Chap. 13). Furthermore, in order to maintain one's orientation within the environment, an organism can simply counteract any movement of the entire visual world with body, head and/or eye movements. This is known as the optokinetic response (OKR), which acts to maintain a stabilized retinal image. Retinal image stabilization is necessary for optimal visual acuity (Westheimer and McKee 1975) and velocity discrimination (Nakayama 1981). The OKR is ubiquitous in the animal kingdom (cf. Sects. 11.3, 13.2.1), and is demonstrated in the laboratory by placing a restrained animal in a drum containing alternating vertical black and white bars. As the drum rotates about the vertical axis, a nystagmus of the eyes or head results. This nystagmus consists of two phases; a *slow phase* in which the eyes and/or head move in the same direction as the drum motion, attempting to match its speed, and a *fast phase* in which a quick saccadic movement brings the head or eyes back to a more central position. In the natural environment, the OKR can be seen in pigeons and many other species of bird which apparently bob their heads as they walk. In fact, this is a visually driven OKR (Dunlap and Mowrer 1930; Friedman 1975; Frost 1978a; Davies and Green 1988). As the pigeon steps forward, the OKR keeps the head virtually stationary in space (slow phase), and then the head is thrust forward (fast phase). A more remarkable illustration of head stabilization by a pied kingfisher during flight is shown by Lee and Young (1986). As the bird hovers above its prey, despite the body position constantly changing as the wings flap, the vertical position of the bill does not change more than 0.5°!

An increasing body of evidence suggests that a separate visual pathway, the accessory optic system (AOS) (see Fig. 12.3), is involved in the analysis of whole-field motion and the control of the OKR (for review, see Simpson 1984). The avian AOS consists of two retinal recipient structures: the nucleus of the basal optic root (nBOR) and the pretectal nucleus lentiformis mesencephali (LM) (Karten et al. 1977; Reiner et al. 1979; Fite et al. 1981; Gamlin and Cohen 1988). Both the nBOR and LM in turn project to structures which receive input from the vestibular system and are implicated in postural and oculomotor control (Clarke 1977; Brecha et al. 1980; Bodnarenko and McKenna 1987; see Sect. 11.7). Lesions of the AOS markedly impair the OKR, yet massive lesions of the tectofugal and thalamofugal systems have comparatively little effect on it (Fite et al. 1979; Wallman et al. 1981; Gioanni et al. 1983a, b).

Electrophysiological studies have revealed that the response properties of AOS neurons contrast sharply with those of motion sensitive cells in the tectofugal pathways (Frost 1982, 1985; Frost et al. 1990). AOS neurons have large receptive fields (up to 100° in diameter) without inhibitory surrounds, whereas tectofugal pathway neurons tend to have small receptive fields with large inhibitory surrounds. AOS neurons respond best to large moving stimuli (flow fields), whereas tectofugal pathway cells do not respond to, or are strongly inhibited by, flow field stimuli. Tectofugal pathway neurons prefer small moving stimuli, which only weakly activate AOS neurons. Finally, AOS neurons do not adapt to repeated stimulation, whereas tectal neurons adapt quite quickly.

12.3.1 Cardinal Directions of Optic Flow

Figure 12.8 shows the directional tuning curves of three nBOR and one LM neuron in response to a large random dot stimulus ($90° \times 90°$) moving in the contralateral visual field. Typical of AOS neurons, these cells were excited by whole-field motion in a particular direction, and inhibited by whole-field motion in the opposite direction, although they are quite broadly tuned. Although there is some variation in preferred directions, nBOR neurons responding best to either upward, downward or backward (nasal to temporal) whole-field motion in the contralateral visual field are equally represented, but few cells prefer forward motion (Burns and Wallman 1981; Morgan and Frost 1981; Gioanni et al. 1984; Wylie and Frost 1990a). Most cells in the LM respond best to forward motion in the contralateral visual field (McKenna and Wallman 1981, 1985; Winterson and Brauth 1985). Thus, AOS neurons seem first to analyze visual flow fields into the four cardinal directions illustrated schematically in Fig. 12.2a.

12.3.2 Binocular Integration of Self-Induced Flow

These nBOR and LM neurons with monocular receptive fields cannot uniquely specify a given flow field. For example, backward whole-field motion in the left visual field results from forward translation of the bird as well as a rightward yaw rotation about a vertical axis. A neuron in the right nBOR with a contralateral receptive field preferring backward motion would respond equally well to these two flow fields. A simple solution to this problem would be to integrate flow-field information from the two eyes, such that a binocular

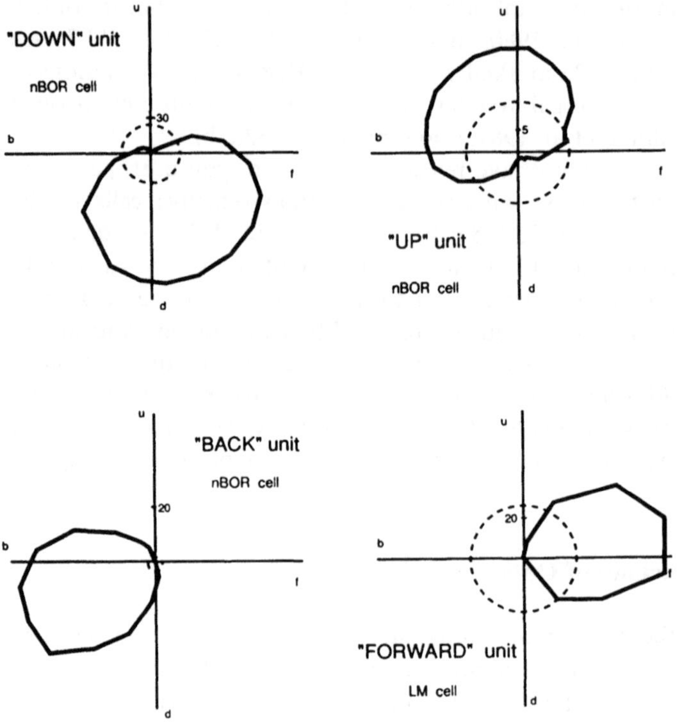

Fig. 12.8. Directional tuning curves of monocular neurons in the AOS of the pigeon. Firing rate is plotted in polar coordinates as a function of the direction of whole-field motion in the contralateral eye for three nBOR neurons and one LM neuron. The *broken circles* represent the cells' spontaneous firing rates. The horizontal is based on the measurements of the orientation of the pigeon head during walking and flying (Erichsen et al. 1989). *u* upward, *d* downward, *b* backward (nasal to temporal), and *f* forward whole-field motion (nBOR results adapted from Wylie and Frost 1990a)

receptive structure could uniquely specify either a translational or rotational flow field.

In our more recent studies of the AOS, we have shown that a small subpopulation of nBOR neurons, located in the posterior dorsolateral margin, has binocular receptive fields (see Fig. 12.9A), and specifies flow fields resulting from either self-translation or self-rotation (Wylie and Frost 1990b). In Fig. 12.9A the receptive field of a binocular nBOR neuron is shown. Note that there are two separate areas of the visual field, on opposite sides of the head, to which the neuron responds. In Fig. 12.9B and C, the responses of two binocular nBOR neurons to whole-field stimulation presented to the ipsilateral and contralateral eyes are shown. With monocular whole-field stimulation of the contralateral eye, the neuron in B was excited by backward-downward motion, and was inhibited by forward-upward motion. With whole-field stimulation of the ipsilateral eye, the neuron showed the opposite direction preference. Under

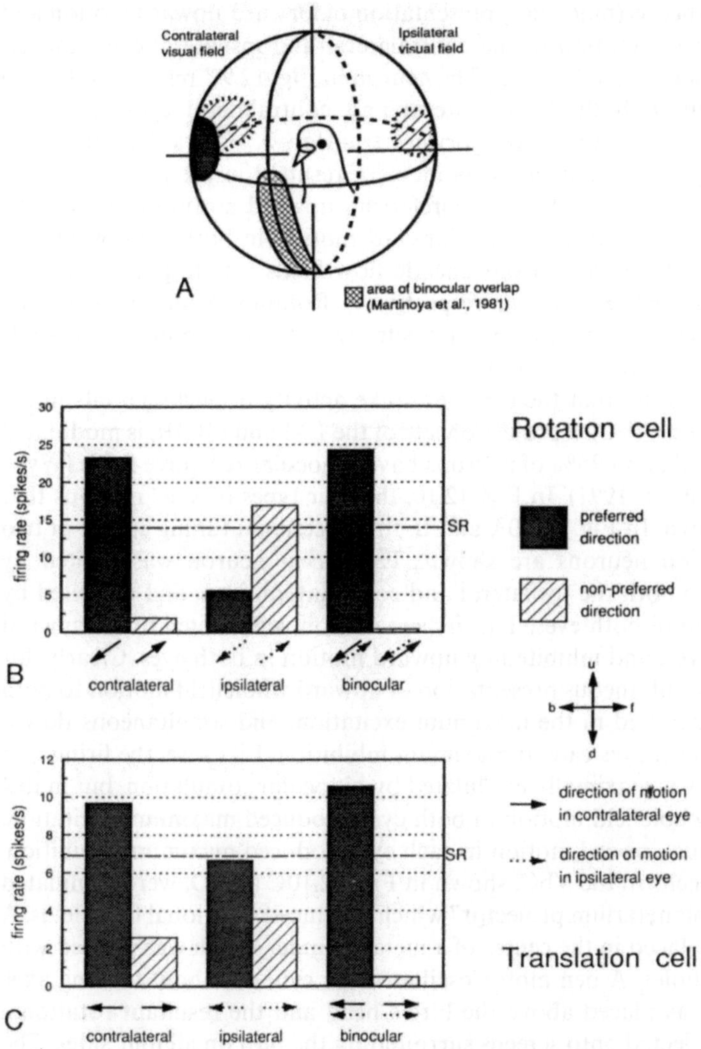

Fig. 12.9A–C. Specificity of binocular nBOR neurons for either translational or rotational flow-fields. **A** The ipsilateral and contralateral receptive field of a binocular nBOR neuron. The receptive fields are spatially separate and well outside the are a of binocular overlap. **B, C** Histograms showing the responses of a rotation cell **B** and a translation cell **C** to monocular whole-field stimulation in the preferred and non-preferred directions of either the contralateral (*left*) or ipsilateral eye (*middle*). The response to simultaneous (binocular) presentation is shown in the *right pair* of histograms of **B** and **C** (After Wylie and Frost 1990b)

natural binocular viewing conditions this neuron would respond best to a flow field resulting from a yaw rotation to the ipsilateral side, with a downward component. As shown in the far right of Fig. 12.9B, simultaneous presentation of backward-downward motion and forward-upward motion to the contralateral and ipsilateral eyes respectively resulted in maximal excitation of this cell.

Likewise, simultaneous (binocular) presentation of forward-upward motion and backward-downward motion to the contralateral and ipsilateral eyes respectively resulted in maximal inhibition. The neuron in Fig. 12.9C responded best to backward motion in both the ipsilateral and contralateral eyes, and thus specifies a flow field resulting from *forward translation.* Forward motion presented to both eyes simultaneously (binocular), resulted in the greatest excitation of the cell's firing rate. Neurons preferring upward motion in both eyes, downward motion in both eyes, and forward motion in both eyes, were also found in the nBOR. These neurons encode flow fields resulting from *descent*, *ascent*, and *backward translation*, respectively. Rotation neurons preferring upward and downward motion in opposite eyes, thus encoding flow fields resulting from *roll*, were also found.

We have also found that the complex spike activity of purkinje cells in the vestibulocerebellum (VbC), a projection site of the LM and nBOR, is modulated by flow-field stimuli, and 95% of neurons have binocular receptive fields (Wylie 1991; Wylie and Frost 1991). In Fig. 12.10, the four types of VbC neurons that we found are shown. In Fig. 12.10A and B, the directional tuning curves of two types of translation neurons are shown. The *ascent* neuron was excited by upward motion in both the ipsilateral and contralateral eyes, and inhibited by downward motion in both eyes. The *descent* neuron was excited by downward motion in both eyes, and inhibited by upward motion in both eyes. Clearly, for the *descent* cell, simultaneous presentation of upward wholefield motion to both eyes (binocular) resulted in the maximum excitation, and simultaneous downward motion in both eyes caused maximum inhibition. Likewise, the firing rate of the *ascent* cell was maximally modulated by binocular stimulation, but in this case downward wholefield motion in both eyes produced maximum excitation, while simultaneous upward motion in each eye produced maximum inhibition.

The rotation cells in the VbC, shown in Fig. 12.10C and D, were stimulated with a rotating "planetarium projector" which produced rotational flow fields. A light source was placed in the centre of a metal cylinder which was pierced with numerous small holes. A pen motor oscillated the cylinder about its long axis. This apparatus was placed above the bird's head and the resultant rotational flow field was projected onto screens surrounding the bird on all four sides. The axis of rotation of the planetarium could be oriented to any position in three-dimensional space. These experiments revealed that the directional preferences of rotation cells in the VbC were organized in vestibular coordinates. That is, rotation neurons in the VbC respond best to a particular rotational flow field which would result from a head rotation that maximally stimulated one of the three pairs of vestibular semicircular canals. Figures 12.10C and D show the responses of two VbC rotation neurons to rotation of the planetarium about the axes of three semicircular canals. *Yaw* (or *vertical axis*) neurons (see Fig. 12.10C) were maximally excited by rotation of the planetarium about the vertical axis in the direction producing forward and backward motion in the ipsilateral and contralateral eyes, respectively. Rotation about the vertical axis in the opposite direction maximally inhibited these neurons, and rotation about the axes of the

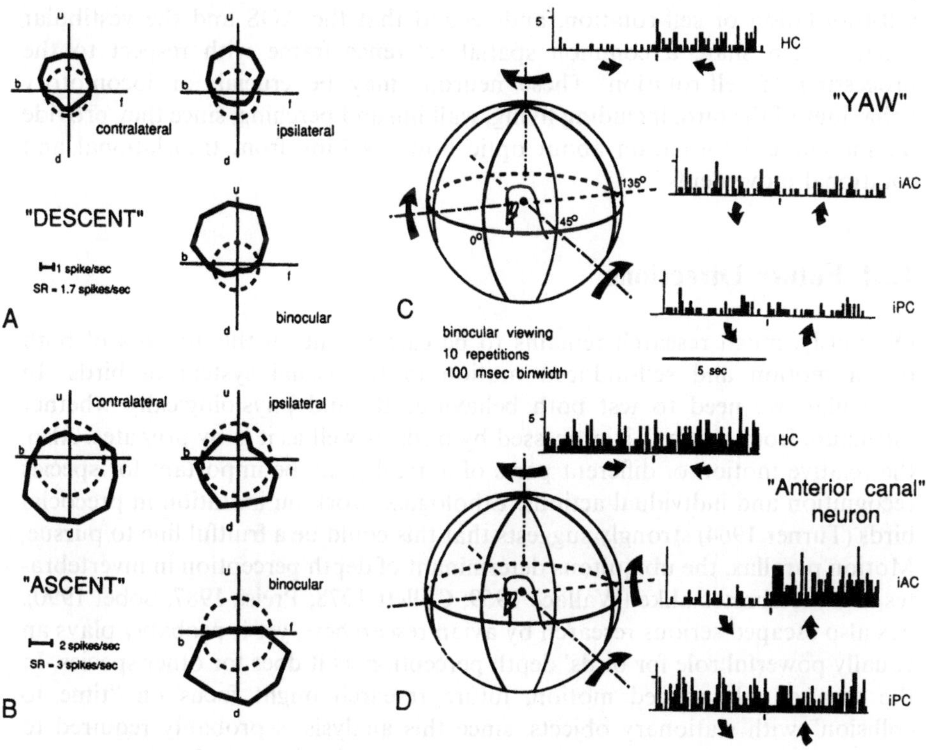

Fig. 12.10A–D. Specificity of binocular purkinje cells in the pigeon VbC for either translational or rotational flowfields. **A, B** Ipsilateral, contralateral and binocular directional tuning curves of the complex spike activity of *descent* **A** and *ascent* **B** Purkinje cells in the VbC. **C, D** PSTHs of the complex spike activity of a yaw **C** Purkinje cell and an *anterior canal* **D** Purkinje cell in response to rotation of the planetarium about the axes of the three semicircular vestibular canals. The planetarium rotated at 1.2° s^{-1} for 5 s in each direction. The PSTHs are averaged over ten sweeps in each direction. The *large arrows* indicate the direction of motion for the second half of the sweep. *HC* horizontal canal; *iAC* ipsilateral anterior canal; *iPC* ipsilateral posterior canal. (After Wylie 1991 and Wylie and Frost 1991)

ipsilateral anterior and posterior canals resulted in very little modulation of the cells' firing rates. Thus, *yaw* neurons respond best to optic flow resulting from a head rotation maximally stimulating the horizontal canals. The neuron in Fig. 12.10D was maximally modulated by rotation of the planetarium about a horizontal axis 135° from the midline on the ipsilateral side, which corresponds to an axis which maximally stimulates the ipsilateral anterior canal and contralateral posterior canal. Rotation about axes orthogonal to this (i.e. the vertical axis or the ipsilateral posterior canal axis) did not affect the neurons' firing rates.

Collectively, these experiments show first that higher-order neurons in the AOS of the pigeon encode specific patterns of optic flow resulting from either self-translation or self-rotation, and second that the AOS and the vestibular canal system share a common spatial reference frame with respect to the processing of self-rotation. These neurons may be crucial for locomotory behaviour of the bird, including flying, walking and perching, since they provide unambiguous information about optic flow resulting from translational and rotational movements.

12.4 Future Directions

Obviously, much research remains to be carried out on the analysis of both object motion and self-induced motion in the visual system of birds. In particular, we need to test both behaviourally and physiologically whether "structure from motion" is processed by birds as well as it is by primates. Also, the relative motion of different parts of animals may be important for species recognition and individual actions. Ethological work on imitation in precocial birds (Turner 1964) strongly suggests that this could be a fruitful line to pursue. Motion parallax, the ubiquitous determinant of depth perception in invertebrates and vertebrates alike (Wallace 1959; Collett 1978; Preiss 1987; Sobel 1990), has also escaped serious research by avian researchers, yet it probably plays an equally powerful role for birds' depth perception as it does for other species. In the area of self-induced motion, future research might focus on "time to collision" with stationary objects, since this analysis is probably required to control both steering through the environment and such critical manoeuvres as landing or striking prey. Also, the problem of heading determination, with all its complexities, is surely in need of attention.

References

Benowitz LI, Karten HJ (1976) Afferentation of the nucleus rotundus and ectostriatum of the pigeon: a retrograde transport analysis. J Comp Neurol 167:503–520

Bodnarenko SR, McKenna OC (1987) Efferent control of the lentiform nucleus of the mesencephalon in chicken. Soc Neurosci Abstr 13:864

Borst A, Bahde S (1988) Spatio-temporal integration of motion. A simple strategy for safe landing in flies. Naturwissenschaften 75:265–267

Brecha N, Karten HJ, Hunt SP (1980) Projections of the nucleus of basal optic root in the pigeon: an autoradiographic and horseradish peroxidase study. J Comp Neurol 189:615–670

Burns S, Wallman J (1981) Relation of single unit properties to the oculomotor function of the nucleus of the basal optic root (AOS) in chickens. Exp Brain Res 42:171–180

Clark DL, Uetz GW (1992) Morph-independent mate selection in a dimorphic jumping spider: demonstration of movement bias in female choice using video-controlled courtship behaviour. Anim Behav 43:247–254

Clarke PGH (1977) Some visual and other connections to the cerebellum of the pigeon. J Comp Neurol 174:535–552

Coggshall JC (1972) The landing response and visual processing in the milkweed bug, *Oncopeltus fasciatus*. J Exp Biol 57:401–413

Collett TS (1977) Stereopsis in toads. Nature 267:349–351

Collett TS (1978) Peering – a locust behaviour pattern for obtaining motion parallax information. J Exp Biol 76:237–241
Cott HB (1940) Adaptive colouration in animals. Methuen, London
Cutting JE (1978) Generation of synthetic male and female walkers through manipulation of a biomechanical invariant. Perception. 7:393–405
Cutting JE (1986) Perception with an eye for motion. MIT Press, Cambridge
Cutting JE, Springer K, Braren PA, Johnson SH (1992) Wayfinding on foot from information in retinal, not optical, flow. J Exp Psychol Gen 121:41–72
Davies MNO, Green PR (1988) Head-bobbing during walking, running and flying: relative motion perception in the pigeon. J Exp Biol 138:71–91
Davies MNO, Green PR (1990) Optic flow-field variables trigger landing in hawk but not in pigeon. Naturwissenschaften 77:142–144
Dunlap K, Mowrer OH (1930) Head movements and eye functions in pigeons. J Comp Psychol 11:99–112
Erichsen JT, Hodos W, Evinger C, Bessette BB, Phillips SJ (1989) Head orientation in pigeons: postural, locomotor and visual determinants. Brain Behav Evol 33:268–278
Fite KV, Reiner T, Hunt S (1979) Optokinetic nystagmus and the accessory optic system of pigeon and turtle. Brain Behav Evol 16:192–202
Fite KV, Brecha N, Karten HJ, Hunt SP (1981) Displaced ganglion cells and the accessory optic system of the pigeon. J Comp Neurol 195:279–288
Fleishman LJ (1986) Motion detection in the presence and absence of background motion in an Anolis lizard. J Comp Physiol 159:711–720
Friedman MB (1975) Visual control of head movements during avian locomotion. Nature 225:67–69
Frost BJ (1978a) The optokinetic basis of head-bobbing in the pigeon. J Exp Biol 74:187–195
Frost BJ (1978b) Moving background patterns alter directionally specific responses of pigeon tectal neurons. Brain Res 151:599–603
Frost BJ (1982) Mechanisms for discriminating object motion from self-induced motion in the pigeon. In: Ingle DJ, Goodale MA, Mansfield JW (eds) Analysis of visual behavior. MIT Press, Cambridge, pp 177–196
Frost BJ (1985) Neural mechanisms for detecting object motion and figure-ground boundaries contrasted with self-motion detecting systems. In: Ingle D, Jeannerod M, Lee D (eds) Brain mechanisms of spatial vision. Nijhoff, Dordrecht, pp 415–449
Frost BJ (1993) Subcortical analysis of visual motion: relative motion, figure-ground discrimination and self-induced optic flow. In: Miles FA, Wallman J (eds) Visual motion and its role in the stabilization of gaze. Elsevier, Amsterdam, pp 159–175
Frost BJ, DiFranco DE (1976) Motion specific units in the pigeon optic tectum. Vision Res 16:1229–1234
Frost BJ, Nakayama K (1983) Single visual neurons code opposing motion independent of direction. Science 220:744–745
Frost BJ, Scilley PL, Wong SCP (1981) Moving background patterns reveal double opponency of directionally specific pigeon tectal neurons. Exp Brain Res 43:173–185
Frost BJ, Cavanagh P, Morgan B (1988) Deep tectal cells in pigeon respond to kinematograms. J Comp Physiol 162:639–647
Frost BJ, Wylie DR, Wang Y-C (1990) The processing of object and self-motion in the tectofugal and accessory optic pathways of birds. Vision Res 30:1677–1688
Gamlin PDR, Cohen DH (1988) Retinal projections to the pretectum in the pigeon (*Columba livia*). J Comp Neurol 269:1–17
Gibson JJ (1950) The perception of the visual world. Houghton Mifflin, Boston
Gibson JJ (1966) The senses considered as perceptual systems. Houghton Mifflin, Boston
Gibson JJ (1979) The ecological approach to visual perception. Houghton Mifflin, Boston
Gioanni H, Rey J, Villalobos J, Richard D, Dalbera A (1983a) Optokinetic nystagmus in the pigeon (*Columba livia*). II. Role of the pretectal nucleus of the accessory optic system (AOS). Exp Brain Res 50:237–247

Gioanni H, Rey J, Villalobos J, Dalbera A (1983b) Optokinetic nystagmus in the pigeon (*Columba livia*). III. Role of the nucleus ectomamillaris (nEM): interactions in the accessory optic system (AOS). Exp Brain Res 50:248–258
Gioanni H, Rey J, Villalobos J, Dalbera A (1984) Single unit activity in the nucleus of the basal optic root (nBOR) during optokinetic, vestibular and visuo-vestibular stimulations in the alert pigeon (*Columba livia*). Exp Brain Res 57:49–60
Hildreth EC (1983) The measurement of visual motion. MIT Press, Cambridge
Jiang S-Y, Wang Y-C, Frost BJ (1989) Response properties of nucleus rotundus cells in the pigeon. Soc Neurosci Abstr 15:460
Johansson G (1975) Visual motion perception. Sci Am 232:76–89
Julesz B (1975) Experiments in the visual perception of texture. Sci Am 232:34–43
Karten HJ, Fite KV, Brecha N (1977) Specific projection of displaced retinal ganglion cells upon the accessory optic system in the pigeon (*Columba livia*). Proc Natl Acad Sci 74:1752–1756
Koenderink JJ (1985) Space, form and optical deformations. In: Ingle D, Jeannerod M, Lee D (eds) Brain mechanisms of spatial vision. Nijhoff, Dordrecht, pp 31–58
Lee DN (1980) The optic flow field: The foundation of vision. Philos Trans R Soc Lond B 290:169–179
Lee DN, Reddish PE (1981) Plummeting gannets: a paradigm of ecological optics. Nature 293:293–294
Lee DN, Young DS (1986) Gearing action to the environment. Exp Brain Res (Suppl) 15:217–230
Lee DN, Reddish PE, Rand DT (1991) Aerial docking by hummingbirds. Naturwissenschaften 78:526–527
Longuet-Higgins HC, Prazdny K (1980) The interpretation of a moving retinal image. Proc R Soc Lond B 208:385–397
McKenna O, Wallman J (1981) Identification of avian brain regions responsive to retinal slip using 2-deoxyglucose. Brain Res 210:455–460
McKenna O, Wallman J (1985) Accessory optic system and pretectum of birds: comparisons with those of other vertebrates. Brain Behav Evol 26:91–116
Morgan B, Frost B (1981) Visual response properties of neurons in the nucleus of the basal optic root of pigeons. Exp Brain Res 42:184–188
Nakayama K (1981) Differential motion hyperacuity under conditions of common image motion. Vision Res 21:1475–1482
Nakayama K (1985) Biological image motion processing: a review. Vision Res 25:625–660
Nakayama K, Loomis JM (1974) Optical velocity patterns, velocity sensitive neurons and space perception: a hypothesis. Perception 3:63–80
Nixdorf BE, Bischoff HJ (1982) Afferent connections of the ectostriatum and visual wulst in the zebra finch (*Taeniopygia guttata castanotica* Gould) – an HRP study. Brain Res 248:9–17
Perrone JA (1992) Model for the computation of self-motion in biological systems. J Opt Soc Am A 9:177–194
Portman A (1959) Animal camouflage. University of Michigan Press, Ann Arbor
Preiss R (1987) Motion parallax and figural properties of depth control flight speed in an insect. Biol Cybern 57:1–9
Regan D, Beverley KL (1984) Psychophysics of visual flow patterns and motion in depth. In: Spillman L, Wooten BR (eds) Sensory experience, adaption and perception. Erlbaum, Hillsdale, NJ, pp 215–240
Reichardt W, Poggio T, Hausen K (1983) Figure-ground discrimination by relative movement in the visual system of the fly. Part II: Towards the neural circuitry. Biol Cybern 46:1–30
Reiner A, Brecha N, Karten HJ (1979) A specific projection of retinal displaced ganglion cells to the nucleus of the basal optic root in the chicken. Neuroscience 4:1679–1688
Revzin AM (1970) Some characteristics of wide-field units in the brain of the pigeon. Brain Behav Evol 3:195–204
Revzin AM (1979) Functional localization in the nucleus rotundus. In: Granda AM, Maxwell JH (eds) Neural mechanisms of behaviour in the pigeon. Plenum Press, New York, pp 165–176

Rogers B, Graham M (1979) Motion parallax as an independent cue for depth perception. Perception 8:125–134

Rogers B, Graham M (1982) Similarities between motion parallax and stereopsis in human depth perception. Vision Res 22:261–270

Simpson JI (1984) The accessory optic system. Annu Rev Neurosci 7:13–41

Sobel EC (1990) The locust's use of motion parallax to measure distance. J Comp Physiol 176:579–588

Turner ERA (1964) Social feeding in birds. Behaviour 24:1–46

von Grunau MW, Frost BJ (1983) Double opponent-process mechanism underlying receptive field structure of directionally specific cells of cat lateral suprasylvian visual area. Exp Brain Res 49:84–92

Wagner H (1982) Flow-field variables trigger landing in flies. Nature 297:147–148

Wallace GK (1959) Visual scanning in the desert locust *Schistocerca gregaria.* J Exp Biol 36:512–525

Wallman J, McKenna OC, Burns S, Velez J, Weinstein B (1981) Relation of the accessory optic system and pretectum to optokinetic responses in chickens. In: Fuchs AF, Becker W (eds) Progress in oculomotor research developmental neuroscience, vol 12. Elsevier, Amsterdam, pp 435–442

Wang Y-C (1992) The processing of luminance, colour, motion and looming stimuli in neurons in the tectofugal pathway of pigeon. Thesis, Queen's University, Kingston, Canada

Wang Y-C, Frost BJ (1992) Time to collision is signalled by neurons in the nucleus rotundus of pigeons. Nature 356:236–238

Wang Y-C, Jiang S-Y, Frost BJ (1993) Visual processing in pigeon nucleus rotundus: luminance, colour, motion and looming subdivisions. Visual Neurosci 10:21–30

Warren R, Wertheim AH (1990) Perception and control of self-motion. Erlbaum, Hillsdale, NJ

Warren WH Jr, Hannon DJ (1988) Direction of self-motion is perceived from optical flow. Nature 336:162–163

Westheimer G, McKee SP (1975) Visual acuity in the presence of retinal image motion. J Opt Soc Am 65:847–850

Winterson BJ, Brauth SE (1985) Direction-selective single units in the nucleus lentiformis mesencephali of the pigeon (*Columba livia*). Exp Brain Res 60:215–226

Wylie DR (1991) Neural mechanisms for distinguishing self-translation and self-rotation in the pigeon. Thesis, Queen's University, Kingston, Canada

Wylie DR, Frost BJ (1990a) Visual response properties of neurons in the nucleus of the basal optic root of the pigeon: a quantitative analysis. Exp Brain Res 82:327–336

Wylie DR, Frost BJ (1990b) Binocular neurons in the nucleus of the basal optic root (nBOR) of the pigeon are selective for either translational or rotational visual flow. Visual Neurosci 5:489–495

Wylie DR, Frost BJ (1991) Purkinje cells in the vestibulocerebellum of the pigeon respond best to either translational or rotational wholefield visual motion. Exp Brain Res 86:229–232

13 An Eye or Ear for Flying

D.N. LEE

13.1 Introduction

A bird, like any animal, has to control its actions through perception and thereby couple itself to the environment. Furthermore, it has to control its actions prospectively, which means that it requires predictive perceptual information to guide its current movement for future ends.

Consider, for example, a bird flying through foliage to alight on its nest. To avoid colliding with branches en route, it has to perceive what would happen if it were to continue its present course of action (e.g. flying a certain flight path or slowing down with a certain deceleration) and, when necessary, take timely evasive action. As it approaches its nest, it has to steer an accurate course, slow itself down adequately and prepare for landing by, for example, extending its feet at the right time. In all aspects of this everyday manoeuvre, birds exhibit exquisitely precise prospective control (see Chap. 11).

Prospective control of action in the environment requires information about the position, orientation and movement of the animal as a whole and of its parts relative to surfaces in the environment. Since control requires making contact with some surfaces while at the same time avoiding others, information must be available simultaneously about surfaces lying in different directions. Furthermore, since the animal has to control *movement*, the information must have a temporal component as well as a spatial component. The information therefore must constitute a spatio-temporal flow field.

In this chapter, a theory will be given of optic information a bird could use to guide itself in flight, and relevant empirical evidence will be reviewed. Flight guidance is not, however, the prerogative or vision. Bats, using echolocation in place of vision, show similarly precise control in flight. The acoustic information available to bats will also be analyzed.

The central theme will be *control of approach*. This is a fundamental problem for all animals and appears in several guises; controlling *linear* approach to a destination as when landing on a perch, controlling *rotary* approach to a desired direction as when orienting the head to the perch when in flight, and controlling linear and rotary approach conjointly as when flying a curved path to a perch.

Psychology Department, University of Edinburgh, 7 George Square, Edinburgh EH8 9JZ, Scotland

M.N.O. Davies and P.R. Green (Eds.)
Perception and Motor Control in Birds

Here we will concentrate on the analysis of control of linear approach but will indicate how other types of control of approach can also be explained in terms of a particular function – the *tau* function – of the optic or other sensory input.

13.2 Flying by Eye

The starting point is the optic flow field. Visual perception of the world is made possible by light reflected from surfaces covered with texture elements – facets and patches of different pigmentation – that have different light reflecting properties from their neighbours. The light reflected from a texture element normally radiates over a wide angle and so at a bird's point of observation in the environment light will be incident from many different texture elements. The light from each element to the point of observation forms a thin optic cone, with the base of the cone on the element and the apex at the point of observation. The set of cones constitutes the optic array at the point of observation (Gibson 1961).

At each point of observation there is a unique array, and so when the bird is translating relative to the environment the optic array at its eye is continuously changing, giving rise to an *optic flow field.* (see e.g. Lee 1980; Koenderink 1986.) The optic flow field is of fundamental importance in understanding visual perception because it stands outside particular visual systems. The bird's chambered eye with its lens and retina and the insect's compound eye with its radial bundle of light guides (ommatidia) both pick up the information in the optic flow field.

When a bird is flying straight through (stationary) foliage the *linear* optic flow field at its eye is as illustrated in Fig. 13.1. One invariant property of the linear optic flow field is that the radial bundle of optic cones reflecting from texture elements on surfaces ahead opens up about the central optic cone, which corresponds to the line of movement of the point of observation. A second invariant property is that when a faster moving optic cone (e.g. from point B on branch) catches up a slower one (e.g. from point L on leaf) it temporarily envelopes and replaces it. This is because the faster moving cone corresponds to a nearer surface, which occludes the further one.

These two invariant properties are important because they are the *defining* properties of the linear optic flow field (Lee 1974). An optic flow field at a bird's eye with these properties *specifies* that the eye is moving in a straight line through a rigid environment. If the flow field lacks either or both properties, then the eye is not moving straight and/or there are moving objects in the field of view. Thus, if a linear optic flow field is presented to a subject (e.g. by moving the visible surroundings as a unit), this visually specifies movement of the self relative to rigid surroundings. The head bobbing of pigeons (Friedman 1975), the hovering position of a bee (Kelber and Zeil 1990) or a hawk moth (Pfaff and Varju 1991) and a person's balance (Lee and Lishman 1975) can all be visually driven by moving the visible surroundings, showing that visual information for

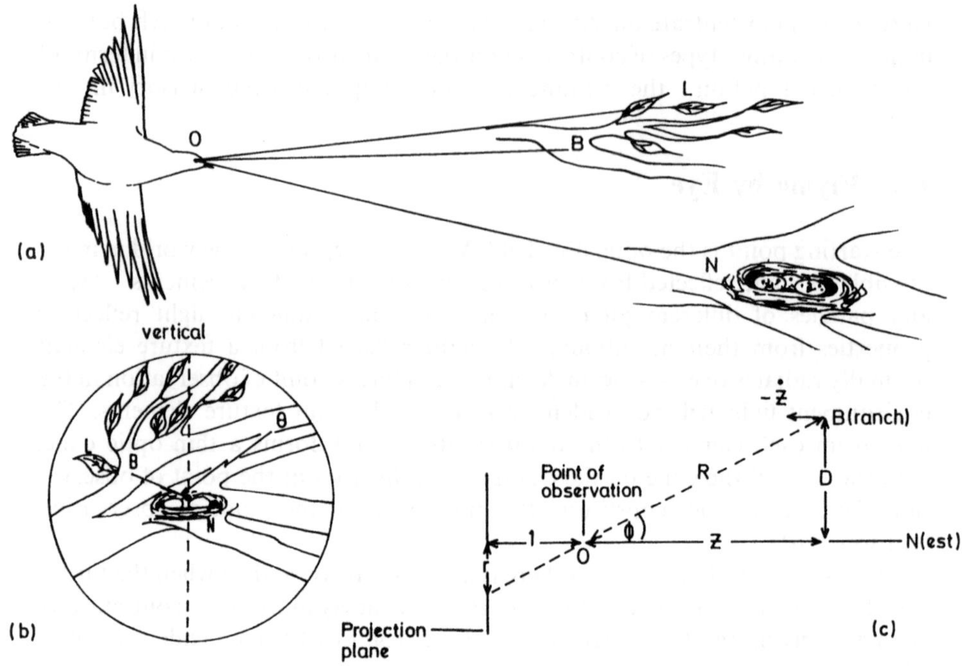

Fig. 13.1a–c. Linear optic flow field for a bird. **a** from each small surface texture element facing the point of observation O, light is reflected to O forming a narrow optic cone. Only three optic cones are illustrated, represented by lines ON, OB and OL, but in general a densely packed radial bundle of optic cones converge on O. As the point of observation O moves in the direction of the surface element N on the nest, a radial bundle of cones opens out about the central cone ON, which lies along the path of locomotion, so as to form a *linear optic flow field*. The flow field has two invariant properties: (1) each optic cone, approximated as a line, fans outward from ON, remaining in a particular plane through ON; (2) when two optic cones moving in the same plane coincide, the faster one envelopes and replaces the slower one. These two properties are illustrated by optic cones OB and OL, which are fanning out in same plane through ON. As point B on the branch is nearer to the bird than point L on the leaf, optic cone OB is fanning out faster than OL. As OB catches OL it envelopes it, because B is then occluding L. **b** Another way of visualizing a linear optic flow field, as a section through the optic flow field perpendicular to line ON. This is equivalent to a movie taken from O with the optical axis of the camera pointing at N. Note that the images of B and L on a projection plane or retina are moving outward from the image of N along the same radial flow line at angle θ to the vertical. **c** Section through the linear optic flow field containing line ON and surface element B, approximated as a point. The projection onto a plane is also shown

self-movement can over-rule mechanical information. In the case of flight, the need for optical specification of self-movement is paramount since mechanical information is subject to the vagaries of the wind (see Sect. 12.1.3).

13.2.1 Stabilizing Vision

In the present account we will concentrate on straight flight. This is not an over simplification since flight paths are approximately straight in sections. However, before describing the detailed information available in the linear optic flow field, another fundamental optic flow field should be mentioned. If, for example, a bird were to turn its head when flying straight, then a *rotary optic flow field* would be superimposed on the linear optic flow field at its eye (considering the frame of reference of the flow field fixed in the head). The rotary flow field is normally kept at low amplitude intermittently by well-controlled head and/or eye movements. There is evidence of this, for example, in pigeons (see Chap. 11, Fig. 11.2), pied kingfishers (Lee and Young 1985), hoverflies (Collett 1980), waterstriders (Junger and Dahmen 1991), as well as humans (Berthoz and Pozzo 1988; Daniel and Lee 1990). Keeping the rotary optic flow field at low amplitude makes sense because it provides little information (only about rotation), and if the rotary flow is too large it is likely to impede the pick-up of information from the much richer linear optic flow field in the ways discussed in the remainder of this section. If the rotary amplitude is sufficiently low, the linear and rotary flow fields are perceptually separable by humans and information can be obtained from the linear flow (Warren and Hannon 1990). All this is not to say that the rotary component of an optic flow field always should or can be eliminated entirely. The composite linear plus rotary optic flow field, in fact, provides vital information for prospective control of steering on a curve (Lee and Lishman 1977; Warren et al. 1991).

13.2.2 The Tau Function

A bird flying to its nest requires detailed information from the linear optic flow field at its eye. We will first consider the *temporal* information and later show how spatial information is derivable from the temporal information.

Consider a surface texture element B (on a branch) in the environment depicted in Fig. 13.1a, b. The cone of light from B to the point of observation O and the line of movement of O define a plane represented in Fig. 13.1c. Since O is moving, the distances Z and R are decreasing over time, while angle ϕ is increasing. D remains constant. Denoting rate of change over time of Z, R and ϕ by $\dot{Z}$, $\dot{R}$ and $\dot{\phi}$, $\dot{R} = \dot{Z}\cos\phi$ and $Z = R\cos\phi$, hence:

$$Z/\dot{Z} = R/\dot{R}\cos^2\phi \,. \tag{13.1}$$

We now introduce the *tau function*, which will play a central role in the following. Tau (τ) of any quantity X is defined to be X divided by the rate of change of X; i.e.

$$\tau(X) = X/\dot{X} \,. \tag{13.2}$$

Thus Eq. (13.1) can be written

$$\tau(Z) = \tau(R)\cos^2\phi \,. \tag{13.3}$$

Considering the optic cone from a surface texture element B to O, if α is the (small) angle subtended at O by B in the direction perpendicular to the plane defined by points B, O and N, then α is inversely proportional to R. Therefore

$\tau(R) = R/\dot{R} = -\alpha/\dot{\alpha} = -\tau(\alpha)$, and so from Eq. (13.3)

$$\tau(Z) = -\tau(\alpha)\cos^2\phi\ . \qquad (13.4)$$

The negative of $\tau(Z)$ ($= -Z/\dot{Z}$) is the time it would take for the point of observation to reach the nearest position to a surface texture element under constant velocity. This time has been termed the *tau margin* (Lee and Young 1985), but in order to develop the more general theory we will here use the notation $\tau(Z)$. $\tau(Z)$ would be valuable information for a bird in a task such as timing its foot extension when landing on a perch. There is evidence that $\tau(Z)$, measured eye to perch, is in fact used for such timing by hawks (Davies and Green 1990) and, measured foot to perch, by pigeons (Lee et al. 1993). $\tau(Z)$ would also be valuable information for a bird to time wing closure to get past a branch. It is also the information needed for timing footfalls over irregular ground (Warren et al. 1986) and for catching moving objects, an ability that infants develop remarkably early (von Hofsten 1980).

It follows from the foregoing that, in principle, it is not necessary to pick up information about distance or speed in order to register $\tau(Z)$, because $\tau(Z)$ is directly specified in the optic flow field [Eq. (13.4)]. Experiments in which time to contact is judged from simulations of approaching objects have shown that $\tau(Z)$ can indeed be perceived without perceiving either distance or speed of the object (Schiff and Detwiler 1979; Todd 1981). Furthermore, many species of animal have been shown to be sensitive to the visual information about impending collision given in such simulations (Schiff 1965). Recently, Wang and Frost (1992; see also Sect. 12.2.3) have found neurons in the nucleus rotundus of the pigeon that signal when $\tau(Z)$ for a simulated approaching object reaches a particular value.

13.2.3 Other Optical Specifications of $\tau(Z)$

$\tau(Z)$ is separately specified by several different optic variables. Depending on the circumstances in which the animal finds itself, one optic variable may be more easily registered than another. The following are two other ways in which $\tau(Z)$ is optically specified.

13.2.3.1 Specification by Direction Relative to Locomotor Axis

$\tau(Z)$ is optically specified in terms of the angle formed at the eye by the line to an object point and the line of locomotion. Referring to Fig. 13.1c, $Z/D = \cot\phi$. Differentiating gives $\dot{Z}/D = \mathrm{d}(\cot\phi)/\mathrm{d}t$. Hence,

$$\tau(Z) = \tau(\cot\phi)\ . \qquad (13.5)$$

This optical specification of $\tau(Z)$ could be useful in situations where the surface being approached has no fine grain texture, so there are no thin optic cones reflecting from it to the point of observation, or where the surface contains an aperture through which the animal is aiming to pass. All that is required for the specification of Eq. (13.5) to apply is the presence of "points" associated with surfaces, such as corners or other aspects of edges. Equation (13.5) is equivalent to $\tau(Z) = \tau(r)$, where r is the projection of NB on a flat projection surface orthogonal to ON and unit distance behind O (Fig. 13.1c); this description of the optic flow field has been used previously (Lee 1980).

13.2.3.2 Movement in Frontal Plane

If the relative motion between B and O is in a direction perpendicular to ON rather than parallel to ON, as in Fig. 13.1c, then Z will be constant and D will be changing. Since $\tan\phi = D/Z = r$, it follows in this case that

$$\tau(D) = \tau(\tan\phi) = \tau(r) . \tag{13.6}$$

This optical specification could be useful in tasks such as catching an insect moving in a frontal plane or avoiding collision when cutting across the flight path of other birds.

13.2.4 More General Tau

So far we have dealt only with the tau function of the distance (denoted Z or D in the above) of a surface element from the point of nearest approach to the point of observation. While this is relevant for many behaviours, it does not apply to cases where the movement of a point other than the eye is being controlled – for instance a hummingbird moving its bill tip to a flower or a person reaching for something with the hand. Figure 13.2a shows the nodal point of the eye above the plane in which a point H on the hand is moving toward a point T on an object. H is assumed to be moving in a straight line toward T (it can be shown that linear approach is optically specified). At a certain time t, H is distance X from T and moving towards T with velocity $-\dot{X}$. We take a visual frame of reference, defined by a nodal point unit distance from a projection plane, such that the projection plane is parallel to the trajectory of the hand. (Other projection surfaces, such as a sphere centred on the nodal point, may be used to describe the optic flow field, but these result in mathematically equivalent descriptions. The present projection plane is chosen for clarity of exposition because it enables the optic-environmental relationships to be expressed in simple mathematical form.) The images of H and T on the projection plane are H' and T' respectively. At the time t, $H'T' = x$ and H' is moving toward T' with velocity $-\dot{x}$. From similar triangles, $x/1 = X/Z$, where Z is the *fixed* distance of the nodal point from the trajectory of H. Differentiating this equation with respect to time gives $\dot{x}/1 = \dot{X}/Z$, and dividing the two equations we obtain $x/\dot{x} = X/\dot{X}$, that is

$$\tau(x) = x/\dot{x} = X/\dot{X} = \tau(X) . \tag{13.7}$$

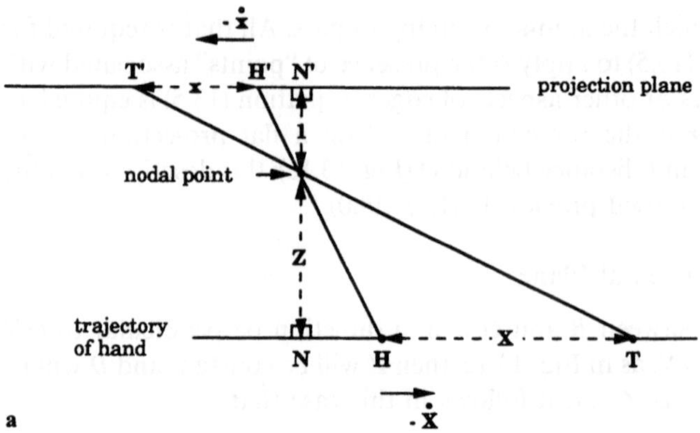

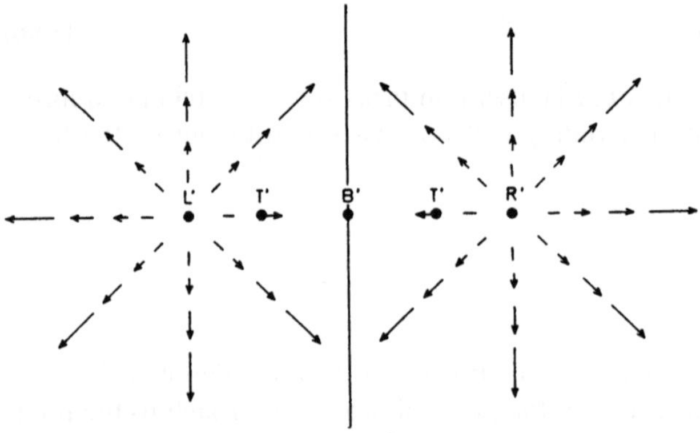

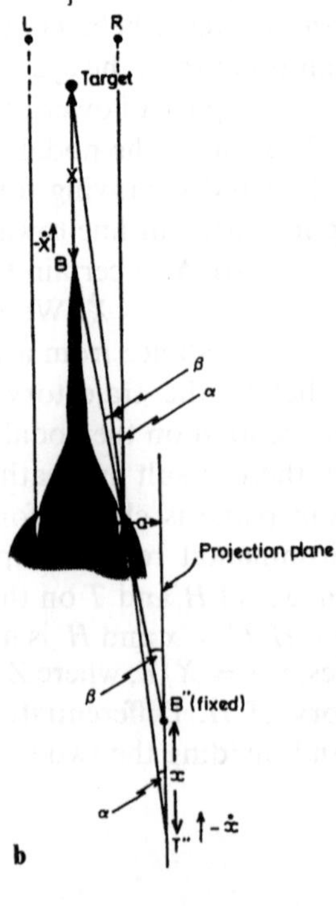

Fig. 13.2a, b. Optic specification of $\tau(X)$, time to contact under constant velocity of one object point with another. Illustrated by **a** point H on an infant's hand moving straight towards target point T on a toy, and **b** a hummingbird's billtip B moving straight towards target point on feeder. In each case, the value of $\tau(X)$ is given by the value of the optic tau function $\tau(x)$ defined on the optic projection plane. For the hummingbird, the bill point B being on course for the target is optically specified by the fact that the optic image T' of the target is in a symmetrical position with respect to centres L' and R' of optic outflow for the two eyes

Thus, the value of the tau function of the reaching distance X is given by the tau function of the optical distance x. The same applies to the control of speed of linear approach of any point to any other point. Examples are using a mouse to move the cursor to a destination on a computer screen, a bird pecking seed (cf. Chaps. 8 and 9) or spearing fish (cf. Chap. 15), and a hummingbird controlling the movement of its bill toward a food source (Fig. 13.2b; see also Tresilian 1991).

13.2.5 Timing Interceptive Acts Under Acceleration

Interceptive acts have to be timed also under conditions where approach velocity is not constant, as when a bird prepares to grasp a branch as it slows down to land. Catching something falling, landing from a jump and seizing a dodging prey are other examples. If the approach is accelerative (or decelerative) then, strictly speaking, $\tau(Z)$ for the approach surface is an overestimate (or underestimate) of the time to contact with the surface. However, as contact is approached, $\tau(Z)$ becomes a closer and closer estimate of time to contact t_c and, providing the ratio of acceleration or deceleration to velocity is not too large, the overestimate is negligibly small when the time to contact is less than about 300 ms (the formula is $\tau(Z) - t_c = 0.5\,(\ddot{Z}/\dot{Z})\,t_c^2$; see also Lee et al. 1983).

In general, $\tau(Z)$ would be simpler to use than actual time to contact since the latter could not be detected if the approach acceleration were unpredictable, which is often the case – for example, a bird being buffeted by the wind or a predator chasing a dodging prey. Also, simple robust mechanisms which are adequate are, in general, preferable to more complex ones. Therefore, it is likely that perceptual mechanisms for timing interceptive actions have evolved which use $\tau(Z)$ rather than engaging in more elaborate computations involving estimates of acceleration. Experiments indicate that $\tau(Z)$ is indeed used in timing actions during accelerative approaches. It is used by gannets plunge-diving into the sea to time wing-closure so as to achieve streamlined entry into the water (Lee and Reddish, 1981), by humans falling onto their feet to time pre-activation of leg muscles to act as shock absorbers (Sidaway et al. 1989) and by humans leaping to hit a falling ball to time leg and arm actions (Lee et al. 1983).

13.2.6 Action-Scaling Space

A bird acting in the environment needs to perceive the distances and sizes of things. However, it is not distance or size in metres or any other arbitrary unit that is required. What the bird needs is information scaled to its action system. For example, when hopping between branches it needs information that specifies the magnitude of leg thrust required to clear the gap. In general, things need to be perceived in terms of the type and magnitude of action that could be or is to be applied to them. It is the *affordances* of things that need to be perceived (Gibson 1979). With these points in mind, let us consider the spatial information available in the optic flow field. Referring back to Fig. 13.1, for surface element B we have

$$\tau(Z) = Z/\dot{Z}\,, \tag{13.8}$$

$$D = Z \tan \phi\,. \tag{13.9}$$

$\tau(Z)$ is optically specified [Eqs. (13.4) and (13.5)] and so is ϕ. Therefore we have three unknowns (Z, $\dot{Z}$ and D) and two equations. If one unknown can be specified, the equations are solvable for the other two.

Specifying an unknown in terms of some quantity X, thus making the equations soluble, is to scale the information in units of X. For example, if an animal has binocular stereopsis, distances could be scaled in units of interocular distance (see Sect. 3.2.1). Binocular stereopsis, however, has limited distance range (see Sect. 3.4.5), and animals must therefore have monocular ways of action-scaling space. The following are two possible ways in which scaling might be achieved.

13.2.6.1 Scaling in Terms of Action Cycles

Most natural locomotor activity is cyclical, whether in the air, on the ground or in water. The frequency of wingbeats, strides or undulations of the body tends to be regular, although it can be modulated to fit the demands of the environment. To see how this allows action-scaling of space, consider a bird flying towards its nest on a straight course at a constant speed with a constant wing-beat period T. Suppose that at the start of a particular wing-beat cycle – the nth cycle – the time-to-contact with a surface element on the nest is $-\tau_n(Z)$. Then

$$\text{number of wing-beat cycles to reach nest} = -\tau_n(Z)/T\,. \tag{13.10}$$

In other words, the "distance" Z to the surface element is $\tau_n(Z)/T$ wing-beat cycles. Hence, from Eq. (13.9), the D-coordinate of the surface element is given by

$$D = \tau_n(Z)\tan \phi/T \text{ wing-beat cycles}\,. \tag{13.11}$$

Thus distances in action space are, in principle, perceivable in terms of how many wing-beat cycles are required to cover the distance.

This way of action-scaling space entails the bird "knowing" the period T of its wing-beat cycle. The knowledge could be intrinsic. Alternatively, T could be measured from the optic flow field by registering the change in time-to-contact from the beginning to the end of a cycle. That is

$$\text{period of wing-beat cycle} = T = -\tau_n(Z) + \tau_{n+1}(Z)\,, \tag{13.12}$$

where $-\tau_n(Z)$ is the time-to-contact at the beginning of the nth cycle and $-\tau_{n+1}(Z)$ is the time-to-contact at the end of the cycle.

Visual regulation of gait when running over irregularly spaced stepping stones (Warren et al. 1986) is another example of the use of the information expressed in Eqs. (13.10) and (13.11). Here, "step cycle" replaces "wing-beat cycle". On the basis of optical information about time to reach a stepping stone, the runner regulates the period of the step cycle (by the vertical impulse applied to the ground) in order to reach a stone at the end of a cycle. It would be instructive to study related behaviour in birds approaching narrow apertures, where the wings must be folded in order to pass through.

13.2.6.2 Scaling in Terms of Head Acceleration

Many animals bob their heads, apparently to gain spatial information. For example, gerbils frequently bob their heads up and down before jumping (Ellard et al. 1984), as do cats and locusts (Collett 1978). Pigeons and other birds regularly bob their heads forward and back relative to the body when walking. When the head is flexing back it is being held approximately still relative to the environment. During this "hold" phase object movements would be easier to detect. However, generating a hold phase is not the sole function of head bobbing. A pigeon also bobs its head in a sinusoidal fashion when running or flying in to land on a perch, and in these cases the speed of body translation precludes a hold phase (Davies and Green 1988). It has been suggested (Frost 1978; Davies and Green 1988) that one of the functions of bobbing the head, so that it is *moving* relative to the environment, is to enhance optic flow and thereby obtain more accurate information about distances of surfaces and objects. One way in which this could be achieved is by scaling distance in units of head acceleration.

Suppose the acceleration of the head relative to the environment and in the direction of body translation is A. Then for any surface element in the environment, differentiating Eq. (13.8) (viz. $\tau(Z) = Z/\dot{Z}$) with respect to time gives $\dot{\tau}(Z) = 1 + AZ/\dot{Z}^2 = 1 + A\tau(Z)/\dot{Z}$, whence

$$\dot{Z} = A\tau(Z)/[\dot{\tau}(Z) - 1] \,. \tag{13.13}$$

Thus, from Eqs. (13.8) and (13.9), Z and D are specified by $\tau(Z)$, $\dot{\tau}(Z)$ and ϕ in units of the acceleration A of the head relative to the environment by

$$Z = [\tau(Z)^2/(\dot{\tau}(Z) - 1)]\,A \,, \tag{13.14}$$

$$D = [\tau(Z)^2 \tan\phi/(\dot{\tau}(Z) - 1)]\,A \,. \tag{13.15}$$

Acceleration A could, in principle, be registered directly by the vestibular system. Alternatively, if body acceleration were zero, or small compared to head acceleration, then head acceleration could be measured in terms of the head action itself. Either way would, in principle, allow a bird approaching its nest to scale the distance between the prospective flight path of its eye and its nest in intrinsic acceleration units which could be related to the reaching distance of its feet.

13.2.7 Theory of Control of Velocity of Approach

A bird flying to a perch has to regulate visually its velocity of approach in order to land successfully. If the bird brakes too hard it will stop short, drop and miss the perch; if it does not brake hard enough it will be unable to check its momentum when its feet hit the perch and will tip forward. In either case injury could result. We will first outline a theory of how the bird might visually control its braking and then report tests of the theory based on analyses of birds approaching destinations. The theory applies to approach along any dimension, such as curvilinear approach and rotary approach along the angular dimension, as when orienting.

13.2.7.1 Constant Deceleration Approach

Maintaining constant deceleration might appear the simplest way of controlling approach to a destination. Referring to Fig. 13.3a, suppose at a certain time a

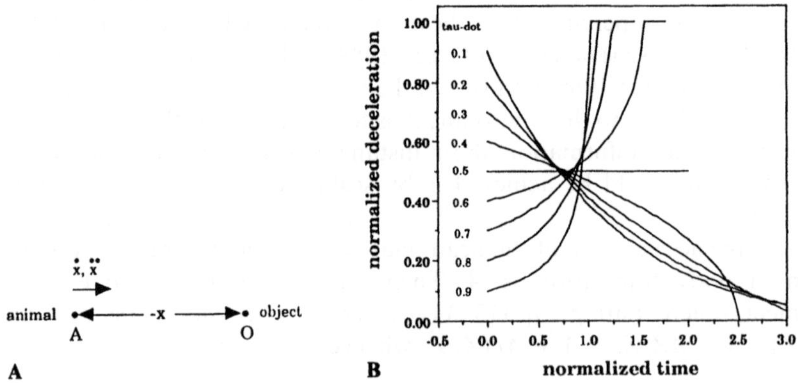

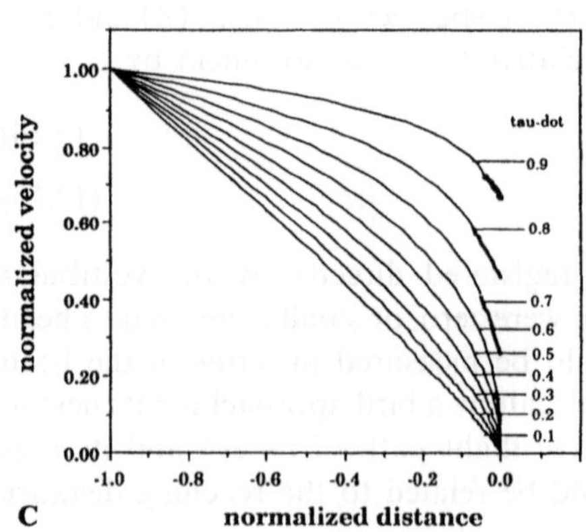

Fig. 13.3A–C. An illustration of the theory of control of velocity of approach **A** Notation for linear approach. At time t an animal has coordinate x (< 0) and is approaching a destination with speed $\dot{x}$(> 0) and acceleration $\ddot{x}$ ($\dot{x} = dx/dt$, $\ddot{x} = d^2x/dt^2$). The tau function of $x = \tau(x) = x/\dot{x}$. The rate of change of $\tau(x) = \dot{\tau}(x)$ = tau-dot. **B** Normalized curves showing how deceleration $-\ddot{x}$ would change over time if approaches were controlled by keeping *tau-dot* constant at the different values shown. **C** Corresponding normalized curves for the change in velocity $\dot{x}$ of approach with distance from the target surface

bird is a distance $-x$ from destination O and is approaching O at velocity $\dot{x}$. Suppose it now starts decelerating at constant deceleration $-\ddot{x}(>0)$. Then its stopping distance will be $-\dot{x}^2/(2\ddot{x})$ and so it will stop short of, stop at, or collide with O according to whether $-\dot{x}^2/(2\ddot{x})$ is less than, equal to, or greater than $-x$. Therefore, to *stop at O* the bird would appear to have to know its distance away $(-x)$ and its velocity of approach $(\dot{x})$ in order to set its deceleration $(-\ddot{x})$ appropriately. However, this is not necessary. A simpler solution exists in terms of $\tau(x)$.

13.2.7.2 Stopping at a Destination

The rate of change with respect to time of $\tau(x)$ $(=\dot{\tau}(x))$ is a dimensionless quantity with the interesting property of providing information for controlling braking. Figure 13.3 shows how deceleration (b) and velocity (c) change as a surface is approached if $\dot{\tau}$ is kept constant at different values between 0.1 and 0.9. The equations of these curves, and their derivations, are given in Lee et al. (1992a, b). To avoid collision it is sufficient to register the value of $\dot{\tau}(x)$, adjust braking so that $\dot{\tau}(x) \leqslant 0.5$ and then keep braking constant. This procedure would generally result in stopping short of the surface.

A general procedure to *stop at* a surface is to adjust braking so that $\dot{\tau}(x)$ stays *constant* at a value k, $0 < k \leqslant 0.5$. As Fig. 13.3c shows, if $\dot{\tau}$ lies within this range, velocity will fall to zero at the target surface, so that the animal or person stops at it. In order to keep $\dot{\tau}$ at a constant value within this range, deceleration must be reduced as the surface is approached (Fig. 13.3b); in other words, the brakes must be steadily slackened off (except for $k = 0.5$, when deceleration is constant). Analysis of braking behaviour of test drivers indicated that they followed the stop-at procedure with $k = 0.425$ (Lee 1976).

13.2.7.3 Controlling Collision

If $\dot{\tau}(x)$ is kept constant at a value k between 0.5 and 1.0, then braking has to get progressively *harder* as the object is approached (Fig. 13.3b). In fact, stopping at a destination in this way theoretically requires reaching infinite braking force. A realistic procedure – the *controlled-collision procedure* – is to keep $\dot{\tau}(x)$ constant at a value between 0.5 and 1.0 until maximum braking power is reached, and then maintain this braking force. This would result in the animal colliding with the destination but in a controlled way (Fig. 13.3c).

Figure 13.4 summarizes the theory of control of braking by presenting the effects of keeping either $\dot{\tau}$ or deceleration constant on motion towards a surface from different starting values of $\dot{\tau}$.

13.2.8 Experiments on Control of Velocity of Approach by Eye

To test whether birds control their velocity of approach to a destination by keeping $\dot{\tau}(x)$ constant, film analyses were made of a hummingbird's aerial

Value of tau-dot ($\dot{\tau}$)	Implied movement of animal	Effect of keeping acceleration/ deceleration constant	Effect of keeping tau-dot constant
$\dot{\tau} > 1$	Accelerating	Collides ($\dot{\tau}$ decreases to 1)	Collides
$\dot{\tau} = 1$	Constant velocity	Collides ($\dot{\tau}$ constant)	Collides
$0.5 < \dot{\tau} < 1$	Decelerating	Collides ($\dot{\tau}$ increases to 1)	Controlled collision (braking increases)
$\dot{\tau} = 0.5$	Decelerating	Stops at	Stops at (braking constant)
$0 < \dot{\tau} < 0.5$	Decelerating	Stops short ($\dot{\tau}$ decreases)	Stops at (braking decreases)

Fig. 13.4. Summary of implications of the theory of control of velocity of approach. If $0 < \dot{\tau} < 0.5$, deceleration monotonically decreases (*curves for* $\dot{\tau} = 0.1 - 0.4$ in Fig. 13.3B) and the animal stops just as the destination is reached (see Fig. 13.3C). If $\dot{\tau} = 0.5$, deceleration is constant (*horizontal line* in Fig. 13.3B) and again the animal stops just as the destination is reached ($\dot{\tau} = 0.5$ *line* in Fig. 13.3C). If $0.5 < \dot{\tau} < 1$, deceleration increases monotonically, as shown by the *curves for* $\dot{\tau} = 0.6 - 0.9$ in Fig. 13.3B; *thick lines* at the top of the curves correspond to reaching a deceleration ceiling. Thus, the animal reaches the destination with a certain reduced speed (see Fig. 13.3C) and makes a controlled collision with it

docking on a feeder tube (Lee et al. 1991) and of pigeons flying to a perch (Lee et al. 1993). For the hummingbird, the distance X between the bill tip and the entrance to the feeder tube was measured on each frame and from these data the value of $\tau(X)$ was calculated for each frame (cf. Fig. 13.2b). Similar measurements were made on the pigeon data, except that in this case X was the distance between the feet in landing position and the perch. If the birds had been controlling their braking for the feeder or perch by keeping $\dot{\tau}(X)$ constant then, during their deceleration, $\tau(X)$ should approach zero linearly with time to contact. Alternatively, if they had been keeping deceleration constant, $\dot{X}$ should decrease linearly over time. The data for both species supported the constant $\dot{\tau}(X)$ hypothesis (Fig. 13.5a, b) as against the constant $\dot{X}$ hypothesis. In particu-

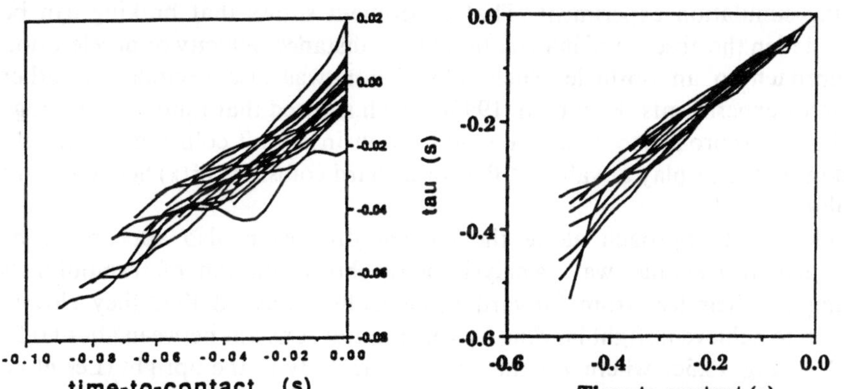

Fig. 13.5. Results of experiments testing theory of control of velocity of approach. Plots of $\tau(x)$ against time to contact for individual trials are shown. *Upper left* a hummingbird flying to dock on a feeder tube (Lee et al. 1991); *upper right* pigeons flying to land on a perch (Lee et al. 1993); *lower left* a somersaulter controlling rotation to vertical so as to land upright on her feet (Lee et al. 1992a); *lower right* a ghost bat flying to pass through a small aperture (Lee et al. 1992b). The theory predicts that, during deceleration to the destination $\dot{\tau}(x)$ is kept constant and so a plot of $\tau(x)$ against time to contact should be a straight line through the origin

lar, the results indicated that the birds were following the controlled-collision procedure. As they braked, they held $\dot{\tau}(X)$ constant at a mean value of 0.71 for the hummingbird and 0.77 for the pigeons. Thus the hummingbird passed its bill into the feeder rather than stopping at the opening, and the pigeons struck the perch with their feet while still moving forward.

Humans, too, appear to control velocity of approach by keeping $\dot{\tau}(X)$ constant. It has already been mentioned that deceleration data of test drivers stopping for a stationary obstacle indicated a constant $\dot{\tau}(X)$ of 0.425 (Lee 1976). Recently, Warren and Yilmaz (per. comm.) found a constant $\dot{\tau}(x)$ of 0.470 in a braking simulation experiment. That experiment shows that braking can be controlled in the absence of information about distance, velocity or deceleration of approach to an obstacle. Such information was also excluded in other simulator experiments (Kim et al. 1993), which showed that humans can judge whether an approach to a surface would result in a "soft collision" [$\dot{\tau}(x)$ held constant in the display at value $\leqslant 0.5$] or a "hard collision" [$\dot{\tau}(x)$ held constant at value > 0.5].

Velocity of approach along the *angular* dimension also appears to be controlled in a similar way. Analysis of the body rotation of trampolinists landing on their feet from forward somersaults indicated that they slowed rotation to achieve upright landing by keeping $\dot{\tau}(\alpha)$ constant between about 0.55 and 0.67 (Fig. 13.5c), where α is the angle of the body to the upright (Lee et al. 1992a).

13.3 Flying by Ear

Echolocating bats can chase insects through foliage and generally get around as skilfully as birds can using vision. Griffin and others (Griffin 1958; Griffin et al. 1958; Grinnell and Griffin 1958) measured the locomotor skill of bats, finding, for example, that *Myotis lucifugus* can almost perfectly avoid vertical wires of only 0.3 mm diameter when flying at 3–4.4 m s^{-1} (Griffin 1958, p. 357). *Asellia tridens* can reliably negotiate wires as thin as 0.065 mm (Gustalfson and Schnitzler 1979). Bats can also build up a precise spatial memory of their environment, remembering the location of apertures to an accuracy of 2 cm (Neuweiler and Mohres 1967).

All species of echolocating bats use broadband sonar signals which normally contain a glissando from a high to a low frequency. Frequencies range from about 10 to 100 kHz. The echoes from the broadband signals appear to be useful in obtaining information about the location, shape and texture of targets (Simmons 1989). Some species also use narrowband signals of constant frequency which appear to be helpful in perceiving target motion from echo Doppler shifts (Simmons and Kick 1983). Some bats sharpen their perception by lowering their emission frequency so as to exactly compensate for the Doppler shift due to their own flight movement (Schnitzler 1973).

The change in echo delay (the time interval between sound emission and returning echo) is also used to detect object movement with high precision. For example, Simmons et al. (1990) presented simulated jittering targets to *Eptesicus fuscus* and found that the bat had an extremely fine echo-delay acuity of 10 ns. This corresponds to a movement in depth of only 0.0015 mm. When detecting insects on the wing, such fine discrimination seems to be facilitated by the bat's directional emissions, directional hearing, middle ear muscle contractions and head aiming, all of which tend to stabilize echo-amplitudes (Kick and Simmons 1984). Suga (1988) has uncovered some of the elegant brain architecture which subserves these abilities.

For guiding flight through the cluttered environment, narrow-field abilities such as the above – on which most research has concentrated – can be only part of the story. The fact is a flying bat behaves as though it obtains a *wide-angle* (acoustic) view of the environment (Webster 1967). How is this possible? When, for example, a bat is chasing a flying insect through foliage, its cries are reflected back from different directions from the surrounding surfaces. This array of reflected sound is continuously changing because the bat is moving, and so there is an acoustic flow field incident on the bat's auditory system which is analogous to the optic flow field incident on a bird's or insect's visual system.

The acoustic flow field is structured by virtue of the differing *sound* reflecting properties of environmental surfaces, just as the optic flow field is structured by their differing light-reflecting properties. Thus the acoustic flow field can be conceived as a bundle of differentiable acoustic cones analogous to the optic cones constituting the optic flow field (Fig. 13.1a). The acoustic flow field is not generally isomorphic with its optic counterpart because acoustic texture elements of surfaces (facets and patches of different sound reflectance) do not necessarily coincide with optic texture elements. But, in basic structure, the acoustic flow field is the same as the optic flow field. Therefore the theory, described above, of the information specified by the structure of the optic flow field applies equally to the acoustic flow field.

How a bat might register the structure of the acoustic flow field is a question that has, surprisingly, not been addressed experimentally. The ability must be grounded on registering the direction of sounds. There is more known about this in humans and in owls (see Chap. 14) than in bats. Humans can detect the direction of sounds by time and intensity differences at the two ears, by changes in the spectral composition of the sound due to multiple reflections in the pinnae and head shadow effects, and by the changes in all these which result when there is movement of the head or sound source (see Moore 1989 for review).

Furthermore, sounds are not simply localized one at a time. When in front of an orchestra the array of instrumental sounds can be perceived simultaneously, just as a visual scene is taken in as a whole. Studies of sound localization in bats indicate that many species use interaural differences in intensity and arrival time (Schnitzler and Henson 1980), while some species can perform well monaurally. But how bats use the multiple reflections at the pinnae and interaural time and intensity differences to register the structure of the acoustic flow field when they are in flight is a problem in much need of study.

Whatever the mechanisms are that allow bats to register acoustic flow fields, it is clear from their natural behaviour that they do so. The sight of thousands of bats streaming out of a cavern at dusk and neatly steering around each other amply attests to this. Bats evidently have access through echolocation to the same form of information as described above for the optic flow field. Moreover, because the bat has control over the (acoustic) illumination of its environment, it has access to information additional to that available optically. In particular, information about $\tau(Z)$ is available in several different forms, as will now be shown.

13.3.1 Acoustic Taus

The optic specifications of tau described above are but particular examples of the following general theorem. If S is a sensory variable that is a power function of distance r to a destination (i.e. $S = kr^\alpha$, where k, α are constants) then

$$\tau(r) = \alpha\tau(S) . \tag{13.16}$$

The proof is straightforward. If $S = kr^\alpha$, then, differentiating with respect to time, $\dot{S} = k\alpha r^{\alpha-1}$. Hence, $\tau(S) = S/\dot{S} = (1/\alpha)r/\dot{r} = (1/\alpha)\tau(r)$; i.e. $\tau(r) = \alpha\tau(S)$.

As an example consider optical specification of tau by direction relative to the locomotor axis, as described above by Eq. (13.5). There the stimulus variable S is $\cot\phi$, and the distance r is Z. Since $\cot\phi = (1/D)Z$, then $S = (1/D)r$, where $(1/D)$ is constant. Thus, applying the theorem, $\tau(r) = \alpha\tau(S)$, where $\alpha = 1$. That is, $\tau(Z) = \tau(\cot\phi)$, which is Eq. (13.5).

It follows from the general theorem that bats could, in principle, detect $\tau(r)$ by registering the tau function of an acoustic variable that is a power function of r. Three possible tau functions will now be described. They are illustrated in Fig. 13.6.

13.3.1.1 Tau Functions of Angles Subtended at Head by Directions of Echoes

These acoustic variables are analogous to the optic variables described in Sect. 13.2. Consider the small angle θ subtended at the head by any two elements on the approach surface close to the direction of locomotion and distance Z from the head. The sensory variable θ equals aZ^{-1} for constant a. Hence, from Eq. (13.16)

$$\tau(Z) = -\tau(\theta) . \tag{13.17}$$

In general, if ϕ is the angle between the direction of locomotion and the direction of an echo from an element of the approach surface, then sensory variable $\cot\phi = Z/b$, where b is the distance shown in Fig. 13.6. Distance b is constant. Hence, from Eq. (13.16)

$$\tau(Z) = \tau(\cot\phi) . \tag{13.18}$$

It should be noted that $\tau(Z)$ is, in principle, acoustically specified *within each pulse-echo pair* as well as between pairs. This is because each sound pulse emitted by a bat is a rapid sequence of "punctuations", each corresponding to a particular aspect of the sound waveform, such as the peaks. Angles θ and ϕ are specified at each punctuation of the echo and change during the course of the echo if the bat is moving toward a surface.

13.3.1.2 Tau Function of Echo Delay

Echo delay d is the time interval between sound being emitted and the echo returning from a surface element. Each punctuation of the sound pulse emitted by a bat has its own echo delay. Referring to Fig. 13.6, for each punctuation of

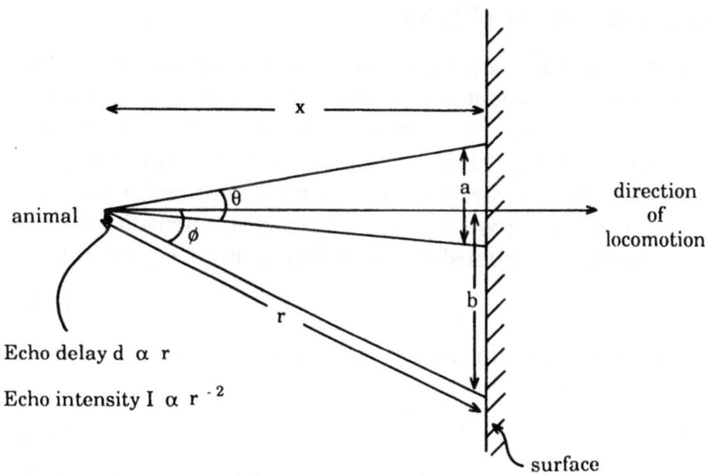

Fig. 13.6. Three acoustic variables that could be used in registering $\tau(x)$ and $\dot{\tau}(x)$: (1) Angle θ (small) defined by the directions of echoes from any two surface elements close to the direction of locomotion. More generally, angle ϕ, defined by the direction of locomotion and the direction of echo from any surface element; $\tau(x) = -\tau(\theta) = \tau(\cot\phi)$. (2) The echo delay ($d$), equal to the time between the emission of sound and the return of the echo; $\tau(x) = \tau(d)\cos^2\phi$. (3) The intensity ($I$) of the echo of sound of uniform intensity; $\tau(x) = -2\,\tau(I)\cos^2\phi$

the pulse, the sensory variable d equals $2r/c$, where r is the current distance away of the surface element and c is the velocity of sound. As a bat approaches a surface, the echo delays d from successive punctuations progressively decrease. Since the punctuations occur at high frequency the echo delays may be considered to be decreasing continuously in time. Hence, from Eq. (13.16), we have

$$\tau(r) = \tau(d)\,. \tag{13.19}$$

If the surface element lies close to the direction of locomotion, then to first-order approximation, $r = Z$. Hence,

$$\tau(Z) = \tau(d)\,. \tag{13.20}$$

In general, if ϕ is the angle between the direction of locomotion and the direction of the echo from an element of the approach surface then, referring to Fig. 13.6, $Z^2 = r^2 - b^2$. Differentiating with respect to time, $Z\dot{Z} = r\dot{r}$. Dividing the equations, $\tau(Z) = Z/\dot{Z} = Z^2/r\dot{r} = \tau(r)\cos^2\phi$. Hence, from Eq. (13.20)

$$\tau(Z) = \tau(d)\cos^2\phi\,. \tag{13.21}$$

Since $\tau(d)$ and $\cos^2\phi$ – and hence $\tau(Z)$ and $\dot{\tau}(Z)$ – are defined within each pulse-echo pair, information for timing preparatory actions during approach to a surface and for controlling velocity of approach can, in principle, be derived from a single sound pulse and its echo, as well as from successive pulses.

13.3.1.3 Tau Function of Intensity of Echo

If I is the intensity of the echo of a punctuation of a sound pulse of constant intensity reflected off a surface element, then I will follow the inverse square law, that is $I = kr^{-2}$, where r is the current distance of the surface element reflecting the sound and k is a constant (Fig. 13.6). The echo intensities I from successive punctuations will therefore progressively increase as a bat approaches a surface. The punctuations occur at high frequency and so the echo intensities may be considered to be increasing continuously in time. Hence, from Eq. (13.16)

$$\tau(r) = -2\tau(I)\,. \tag{13.22}$$

If the element lies close to the direction of locomotion, then to first-order approximation, $r = Z$, and so

$$\tau(Z) = -2\tau(I)\,. \tag{13.23}$$

In general, if ϕ is the angle between the direction of locomotion and the direction of the echo from an element of the approach surface then, as shown above, $\tau(Z) = \tau(r)\cos^2\phi$, and so from Eq. (13.23)

$$\tau(Z) = -2\tau(I)\cos^2\phi\,. \tag{13.24}$$

As with echo delay, $\tau(Z)$ and $\dot{\tau}(Z)$ are specified within single echoes.

13.3.2 Experiments on Control of Velocity of Approach by Ear

Do bats, using echolocation, control velocity of approach in the same way as birds do using vision? To test this, three ghost bats (*Macroderma gigas*) were trained to fly down a horizontal tunnel and out of the other end through an aperture (Lee et al. 1992b). The aperture was made small enough so that the bats had to slow down and regulate their flight to pass through it. To eliminate possible use of vision, the bats' eyes were covered. Their movement down the tunnel was monitored opto-electronically and $\dot{Z}$ and $\tau(Z)$ calculated at successive positions on their flight path (Z = distance between bat and aperture). If the bats had been controlling their braking for the aperture by keeping $\dot{\tau}(Z)$ constant, then the plot of $\tau(Z)$ against time to contact with the aperture should be a straight line through the origin. Alternatively, they may have been keeping deceleration $-\ddot{Z}$ constant, in which case $\dot{Z}$ should decrease linearly with time to contact.

Figure 13.5d shows the plots of $\tau(Z)$ against time to contact for ten flights by one of the bats. The mean value of r^2, 0.97 (SD 0.017), was very close to unity, the index of perfect linearity. The mean value of r^2 for the corresponding plots of $\dot{Z}$ against time to contact was 0.689 (SD 0.179) and was significantly less ($p < 0.005$). This was also the case for the other bats. The experiment thus supports the hypothesis that bats control velocity of approach by keeping $\dot{\tau}(Z)$ constant using echolocation just as birds do using vision. Which acoustic tau they were using could not be determined from the experiment and is a question which remains to be answered.

Whether humans use echolocation to control velocity of approach in a similar way to bats is not known. However, a recent study by Schiff and Oldak (1990) casts some light on the question. Films of approaching trucks, cars and speaking people were presented to subjects using the sound only, the picture only and both together. During the simulated approach the film was turned off and subjects had subsequently to indicate when the vehicle or person would have contacted them. For time-to-contacts up to 4 s, judgments with sound only were as accurate as with picture only or with both picture and sound. Since the sound was recorded and played back through a single channel, the information the subjects were using must have been available monaurally. A likely candidate is the tau function of intensity, since Rosenblum et al. (1987) found that intensity change is the most effective information for locating moving sound sources.

13.4 Concluding Remarks

This chapter has outlined a theory of prospective control of flight based on information available in optic and acoustic flow fields. The analysis has concentrated on the control of approach to a destination because that problem is fundamental to many different actions. Though the discussion has centred on control of linear approach, as when a bird is flying straight to a perch, the theory applies equally to control of approach along any dimension, such as control of rotary approach when orienting eyes, head or body.

Three aspects of control of approach have been addressed; timing preparatory actions, action-scaling space, and controlling velocity of approach. It has been shown how information for each aspect of control is provided by the tau function of an optic or acoustic variable. The tau function of a variable has been defined formally as the ratio of the variable to its rate of change. However, this should not be taken to imply that registering a tau function necessarily entails separately registering the variable and its rate of change and then computing their ratio. The tau function might be registered without these intervening steps – just as acceleration, formally defined as rate of change of velocity, is registered by the vestibular system without first registering velocity.

Indeed, if it is the tau function of an optic or acoustic variable rather than the variable itself that provides information for control, then it is likely that mechanisms will have evolved for registering the tau function directly. The existence of such a visual mechanism is suggested by the discovery of neurons in the nucleus rotundus of pigeon which, during a simulated approach of an object, fire optimally when the tau function of distance reaches a particular value (Wang and Frost 1992; see Sect. 12.2.3). It is to be hoped that further work assessing the effect on action control of the systematic manipulation of sensory input variables will help uncover the mechanisms underlying the wonders of animal flight.

Acknowledgements. Reported research by the author was supported by grants from the Medical Research Council and the US Air Force European Office of Aerospace Research and Development.

References

Berthoz A, Pozzo T (1988) Intermittent head stabilization during postural and locomotory tasks in humans. In: Amblard B, Berthoz A, Clarac F (eds) Posture and gait: development, adaptation and modulation. Elsevier, Amsterdam, pp 189–198

Collett TS (1978) Peering – a locust behaviour pattern for obtaining motion parallax information. J Exp Biol 76:237–241

Collett TS (1980) Some operating rules for the optomotor system of a hoverfly during voluntary flight. J Comp Physiol 138:271–282

Daniel BM, Lee DN (1990) Development of looking with head and eyes. J Exp Child Psychol 50:200–216

Davies MNO, Green PR (1988) Head bobbing during walking, running and flying: relative motion perception in the pigeon. J Exp Biol 138:71–91

Davies MNO, Green PR (1990) Optic flow-field variables trigger landing in hawk but not in pigeons. Naturwissenschaften 77:142–144

Ellard CG, Goodale MA, Timney B (1984) Distance estimation in the Mongolian gerbil: the role of dynamic depth cues. Behav Brain Res 14:29–39

Friedman MB (1975) Visual control of head movements during avian locomotion. Nature 255:67–69

Frost BJ (1978) The optokinetic basis of head-bobbing in the pigeon. J Exp Biol 74:187–195

Gibson JJ (1961) Ecological optics. Vision Res 1:253–262

Gibson JJ (1979) The ecological approach to visual perception. Houghton Mifflin, Boston

Griffin DR (1958) Listening in the dark. Yale University Press, New Haven

Griffin DR, Novick A, Kornfield M (1958) The sensitivity of echolocation in the fruit bat *Rousettus*. Biol Bull 115:107–113

Grinnell AD, Griffin DR (1958) The sensitivity of echolocation in bats. Biol Bull 114:10–22

Gustalfson Y, Schnitzler HU (1979) Echolocation and obstacle avoidance in the hipposiderid bat *Asellia tridens*. J Comp Physiol 131:161–167

Junger W, Dahmen HJ (1991) Response to self-motion in waterstriders: visual discrimination between rotation and translation. J Comp Physiol 169:641–646

Kelber A, Zeil J (1990) A robust procedure for visual stabilisation of hovering flight position in guard bees of *Trigona* (*Tetragonisca*) *angustula* (Apidae, Meliponinae). J Comp Physiol 167:569–577

Kick SA, Simmons JA (1984) Automatic gain control in the bat's sonar receiver and the neuroethology of echolocation. J Neurosci 4:2725–2737

Kim N-G, Turvey MT, Carello C (1993) Optical information about the severity of upcoming contacts. J Exp Psychol Hum Percept Perform 19:179–193

Koenderink JJ (1986) Optic flow. Vision Res 26:161–180

Lee DN (1974) Visual information during locomotion In: McLeod RB, Pick HL (eds) Perception: essays in honour of James J. Gibson. Cornell University Press, New York, pp 250–267

Lee DN (1976) A theory of visual control of braking based on information about time-to-collision. Perception 5:437–459

Lee DN (1980) The optic flow field: the foundation of vision. Philos Trans R Soc Lond B 290: 169–179

Lee DN, Lishman JR (1975) Visual proprioceptive control of stance. J Hum Mov Stud 1:87–95

Lee DN, Reddish PE (1981) Plummeting gannets: a paradigm of ecological optics. Nature 293:293–294

Lee DN, Young DS (1985) Visual timing of interceptive action. In: Ingle DJ, Jeannerod M, Lee DN (eds) Brain mechanisms and spatial vision. Nijhoff, Dordrecht, pp 1–30

Lee DN, Young DS, Reddish PE, Lough S, Clayton TMH (1983) Visual timing in hitting an accelerating ball. Q J Exp Psychol 35A:333–346

Lee DN, Reddish PE, Rand DT (1991) Aerial docking by hummingbirds. Naturwissenschaften 78:526–527

Lee DN, Young DS, Rewt D (1992a) How do somersaulters land on their feet? J Exp Psychol Hum Percept Perform 18:1195–1202

Lee DN, van der Weel FR, Hitchcock T, Matejowsky E, Pettigrew JD (1992b) Common principle of guidance by echolocation and vision. J Comp Physiol 171:563–571

Lee DN, Davies MNO, Green PR, van der Weel FR (1993) Visual control of velocity of approach by pigeons when landing. J Exp Biol 180:85–104

Moore BCJ (1989) An introduction to the psychology of hearing. Academic Press, London

Neuweiler G, Mohres FP (1967) Die Rolle des Ortsgedächtnisses bei der Orientierung der Großblatt-Fledermaus *Megaderma lyra.* Z Vergl Physiol 57:147–171

Pfaff M, Varju D (1991) Mechanisms of visual distance perception in the hawk moth *Macroglossum stellatarum.* Zool Jahrb Physiol 95:315–321

Rosenblum LD, Carello C, Pastore RE (1987) Relative effectiveness of three stimulus variables for locating a moving sound source. Perception 16:175–186

Schiff W (1965) Perception of impending collision: a study of visually directed avoidant behaviour. Psychol Monogr 79 (604)

Schiff W, Detwiler ML (1979) Information used in judging impending collision. Perception 8:647–658

Schiff W, Oldak R (1990) Accuracy of judging time-to-arrival: effects of modality, trajectory and gender. J Exp Psychol Hum Percept Perform 16:303–316

Schnitzler HU (1973) Control of Doppler shift compensation in the greater horseshoe bat *Rhinolophus ferrumequinum.* J Comp Physiol 82:79–92

Schnitzler HU, Henson OW (1980) Performance of airborne animal sonar systems: I Microchiroptera. In: Busnel RG, Fish JF (eds) Animal sonar systems. Plenum Press, New York, pp 109–181

Sidaway B, McNitt-Gray J, Davis G (1989) Visual timing of muscle preactivation in preparation for landing. Ecol Psychol 1:253–264

Simmons JA (1989) A view of the world through the bat's ear: the formation of acoustic images in echolocation. Cognition 33:155–199

Simmons JA, Kick JA (1983) Interception of flying insects by bats. In: Huber F, Mark H (eds) Behavioral physiology and neuroethology: roots and growing points. Springer, Berlin Heidelberg New York, pp 267–279

Simmons JA, Ferragamo M, Moss CF, Stevenson SB, Altes RA (1990) Discrimination of jittered sonar echoes by the echolocating bat *Eptesicus fuscus*: the shape of the target images in echolocation. J Comp Physiol A 167:589–616

Suga N (1988) Parallel-hierarchical processing of biosonar information in the mustached bat. In: Nachtigal PE, Moore PWB (eds) Animal sonar: processes and performance. Plenum Press, New York, pp 149–159

Todd JT (1981) Visual information about moving objects. J Exp Psychol Hum Percept Perform 7:795–810

Tresilian JR (1991) Empirical and theoretical issues in the perception of time to contact. J Exp Psychol Hum Percept Perform 17:865–876

Von Hofsten C (1980) Predictive reaching for moving objects by human infants. J Exp Child Psychol 30:369–382

Wang Y, Frost BJ (1992) Time to collision is signalled by neurons in the nucleus rotundus of pigeons. Nature 365:236–238

Warren WH, Hannon DJ (1990) Eye movements and optical flow. J Opt Soc Am A 7:160–169

Warren WH, Young DS, Lee DN (1986) Visual control of step length during running over irregular terrain. J Exp Psychol Hum Percept Perform 12:259–266

Warren WH, Mestre DR, Blackwell AW, Morris MW (1991) Perception of circular heading from optical flow. J Exp Psychol Hum Percept Perform 17:28–43

Webster FA (1967) Some acoustical differences between bats and men. In: Dufton R (ed) International conference on sensory devices for the blind. St Dunstan's London, pp 63–88

14 Directional Hearing in Owls: Neurobiology, Behaviour and Evolution

S.F. VOLMAN

14.1 Introduction

Birds are primarily active in the daytime, and most birds rarely venture forth from their roosts at night. Many species of owls, however, are nocturnal predators, and they exhibit a number of sensory adaptations for this existence. Most owls have large, frontally directed eyes (see Sect. 1.6.2) that are tubular in shape, with retinas that are densely packed with rod photoreceptors. These features help maximize image brightness and visual acuity at low light levels. (cf. Martin 1986; Chap. 1). Owls' auditory systems are also specialized for detecting and localizing prey. Owls have large heads, and many species have a complete or partial facial ruff of stiff, sound-reflective feathers that acts as an effective sound collector (Fig. 14.1). In addition, the ears of some owls are oriented asymmetrically in the vertical plane, which gives them the ability to perceive both the azimuth and elevation of a sound source simultaneously and with high precision. In contrast, owls with symmetrical ears most likely must listen from at least two different head positions to locate a sound in both axes.

14.2 Bilateral Ear Asymmetry and Sound Localization in Owls

Bilateral ear asymmetry is achieved by a variety of morphological adaptations, and it has evolved independently five to seven times among the owls (Norberg 1977). In some species, asymmetry is apparent in the skull, whereas in others the position or orientation of various soft tissues of the outer ear differs between the left and right sides (Figs. 14.2 and 14.3). It is clearly advantageous for an owl hunting for concealed, terrestrial prey from the air or from a perch to be able to localize sounds quickly and deftly in two dimensions. Relative to an owl's head, the vertical axis corresponds to the prey's position on the ground in front of the owl, and the horizontal axis to the prey's lateral position. An owl with asymmetrical ears, which can orient towards a brief sound in a single movement, should be better able to find quarry that makes only intermittent sounds and also to avoid detection by the prey. In addition, simultaneous horizontal and

Department of Zoology, The Ohio State University, 1735 Neil Avenue, Columbus, OH 43210, USA

M.N.O. Davies and P.R. Green (Eds.)
Perception and Motor Control in Birds

Fig. 14.1. Some owls with symmetrical and asymmetrical ears. The heads are drawn approximately in scale with their actual sizes. *Upper row* The barn owl *Tyto alba* (*left*) and the long-eared owl *Asio otus* (*right*) have asymmetrical ears. Both of these owls have prominent facial ruffs and similar sized heads with average skull width of 4.1 cm. *Lower row* The burrowing owl *Athene cunicularia* (*left*) and the great horned owl *Bubo virginianus* (*right*) have symmetrical ears. Burrowing owls have poorly developed facial ruffs and an average skull width of 3.1 cm, whereas the ruffs of great horned owls are well developed, and their mean skull width is 5.7 cm (Data on skull widths from Volman and Konishi 1990)

Fig. 14.2. A barn owl with its facial disk feathers removed to show the pre-aural flaps (*arrows*) and the sound-reflective facial ruff. The left flap is located higher on the head than the right one. High frequency sounds from below the horizon are louder in the left ear, and those from above, louder on the right. The flaps conceal the small ear openings (Drawing is after a photograph provided by M. Konishi)

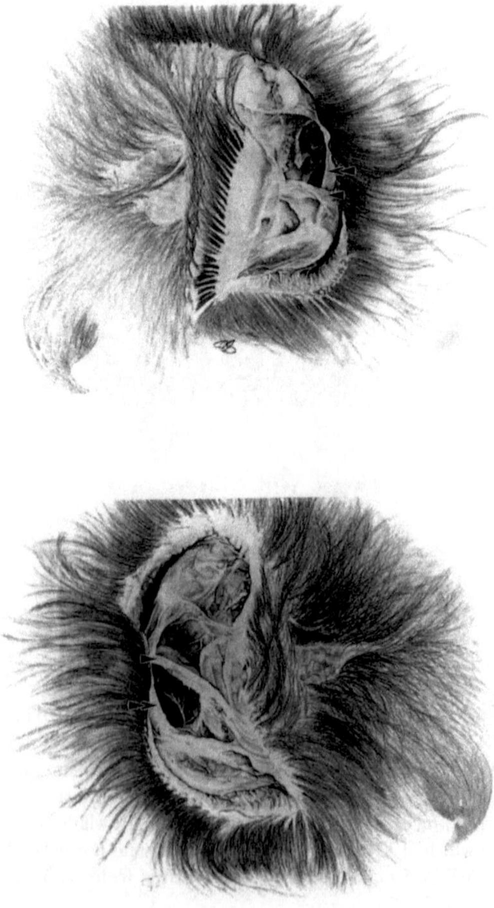

Fig. 14.3. Lateral views of the left and right ears of a long-eared owl. The pre-aural flap is folded forward to reveal the long ear slits and asymmetrical structures. On the left side, the opening to the auditory meatus (*double arrow*) is higher than on the right, and a nearly horizontal, dermal septum (*arrow*) can be seen below the opening. On the right, the septum is above the ear opening. The drawings are from photographs of a preserved head, and the apparent difference in the lengths of the ear slits is an artifact of fixation

vertical localization is probably necessary for an owl to track a moving target (Norberg 1987).

In general, species of owls with asymmetrical ears are more likely to be extremely nocturnal and/or to specialize on small rodent prey (Norberg 1987; Voous 1989). A variety of behavioural studies support the conclusions that owls use sound to find their prey and that asymmetrical owls rely more on their hearing than do symmetrical species. The asymmetrical barn owl (*Tyto alba*) can localize prey solely by ear (Payne and Drury 1958; Payne 1962, 1971). In complete darkness, barn owls reliably and accurately fly down from a perch to strike directly at a hidden speaker or a mouse moving in dry leaves, and they can even orient their talons to the longitudinal axis of the mouse. Barn owls will also strike at a hidden speaker playing short bursts of broadband sound (Konishi 1973b). The owl turns its head towards the speaker, waits for a second sound burst, and then begins its strike. If the sound ceases before the owl begins to fly, its strike is less accurate, but it can re-orient during flight to a sound presented from the same, or a different, sound source.

Payne (1971) was able to train another asymmetrical species, the saw-whet owl (*Aegolius acadicus*), to strike auditory targets in the dark, but neither of two symmetrical species, the great horned owl (*Bubo virginianus*), or the screech owl (*Otus asio*), would even attempt to fly when the lights were turned out. On the other hand, Marti (1974), using a less demanding task in which a hungry owl, live mice, and dry leaves were placed in a dark room, found that owls with asymmetrical ears (a barn owl and a long-eared owl, *Asio otus*) and those with symmetrical ears (a great horned owl and a burrowing owl, *Athene cunicularis*) were all eventually able to find their prey. Norberg (1987) cites several field observations of the asymmetrical great grey owl (*Strix nebulosa*) plunging through 20 cm or more of snow to capture rodents in snow tunnels, which presumably could be located only by their sounds.

Owls orient toward an auditory target even if the sound ceases before the owl begins to turn (Konishi 1973b; Knudsen et al. 1979; Frost et al. 1989; Beitel 1991). Their orienting responses are thus "open loop" and do not depend on auditory feedback received during the head turn. Orientation towards a sound source is accomplished entirely, or almost entirely, by head movements. Barn owls do not move their eyes more than 3–4°, and gaze shifts measured from both head and eye movements do not differ appreciably from measurements of head position alone (du Lac and Knudsen 1990). The great horned owl's eye movements are limited to less than about 2° (Steinbach and Money 1973). Thus orientation accuracy can be measured in the dark with magnetic search coils mounted on the owl's head, or from films of head turns towards hidden speakers.

Barn owls make errors of less than 2° in azimuth and elevation when orienting towards broadband sounds located within a central cone of 10° radius; their errors are less than 4° for angles between 10° and 30°; and they fail to localize targets in the frontal hemifield originating beyond 30° with an average error of about 6° (Knudsen et al. 1979). Barn owls seem able to tell front from back, although this ability has not been systematically tested. When the target is behind their heads, they almost always orient to the correct quadrant (H. Wagner pers. comm.). Saw-whet owls, which also have asymmetrical ears, localize the azimuth of targets in the frontal hemisphere with an average accuracy of 1.75°, and they orient to within 1° of targets that are less than 15° from the midline (Frost et al. 1989). Their elevational accuracy has not been measured.

The great horned owl is the only symmetrical species whose sound localization skill has been studied, and only for azimuthal orientation in the horizontal plane (Beitel 1991). For locations more than 30° from the vertical midline, the open-loop responses of these owls are as accurate as those of barn owls. Great horned owls also orient to the correct quadrant for sounds in the posterior sound field, but for these positions the initial turns are usually short of the target by an average of about 40°. In all these species, the head turns are ballistic, and their precision does not improve if the sound remains on throughout the turn, although a second, corrective turn may be executed if the sound continues after the initial turn is completed.

Bilateral ear asymmetry in owls was first mentioned in the scientific literature by Streets (1870) and Collett (1871). But according to Norberg (1987), Stresemann (1934) was the first to suggest that asymmetry played a role in sound localization. He inferred this function by observing that owls often tilt their heads about the vertical plane during intent listening, temporarily placing one ear opening above the other. Pumphrey (1948) noted that the two ears of an asymmetrical owl would have different directions of maximum sensitivity in the vertical plane, and thus binaural comparisons might be used for two-dimensional sound localization. Norberg (1968, 1978) refined Pumphrey's ideas and tested their applicability on model and specimen heads of Tengmalm's owls (*Aegolius funereus*) which have a marked asymmetry of the skull. Norberg placed small microphones in the ear canals and broadcast sounds of various frequencies from different spatial positions to determine whether there were significant and systematic differences between the two ears in the level or timing of the sounds. From these measurements he concluded that owls with asymmetrical ears could determine the azimuthal position of a sound source from the interaural time and sound-level differences present in the low frequency components of a sound, whereas the interaural level differences of higher frequency components could be used for vertical localization. With the exception that owls can in fact use the interaural time disparities of relatively high frequency sounds, Norberg's deductions have been essentially borne out, particularly in the extensive studies of barn owls.

The horizontal separation of the ears, and any other factor that contributes to differences in the length of the path by which a sound reaches the two ears, give rise to interaural time differences (ITDs). In the several owls for which ITD has been measured as a function of position, it changes almost linearly along a line connecting the two ears (Norberg 1968, 1978; Moiseff and Konishi 1981; Moiseff 1989a; Olsen et al. 1989; Volman and Konishi 1989). In symmetrical species, this line is the horizontal meridian, whereas in asymmetrical owls, it is tilted about 12° from the horizon (Fig. 14.4). The changes in ITD with target position are relatively insensitive to frequency within the hearing range of owls, but the rate of change with position increases with the distance between the two ears. Thus owls with large heads, such as the great horned, do not require as much temporal resolution to achieve the same degree of spatial resolution as the barn and saw-whet owls.

Time differences are present in both the initial onset of a sound and in its ongoing waveform. Ongoing time disparity is often the more reliable measure of ITD for biological sounds, which do not always have clear onsets. Although owls may use information present in onset disparities, behavioural and physiological evidence suggests they primarily rely on ongoing time differences (Moiseff and Konishi 1981; Wagner 1990, 1991a, 1992). But the use of ongoing time differences limits the types of sounds that can be localized and also puts some exceptional demands on owls' auditory systems. First there is the problem of phase ambiguity, because ongoing time differences are measured by the phase delay between the two ears (Fig. 14.5). If an owl used only low frequencies to localize sounds, the maximum time delay, which occurs when the sound source

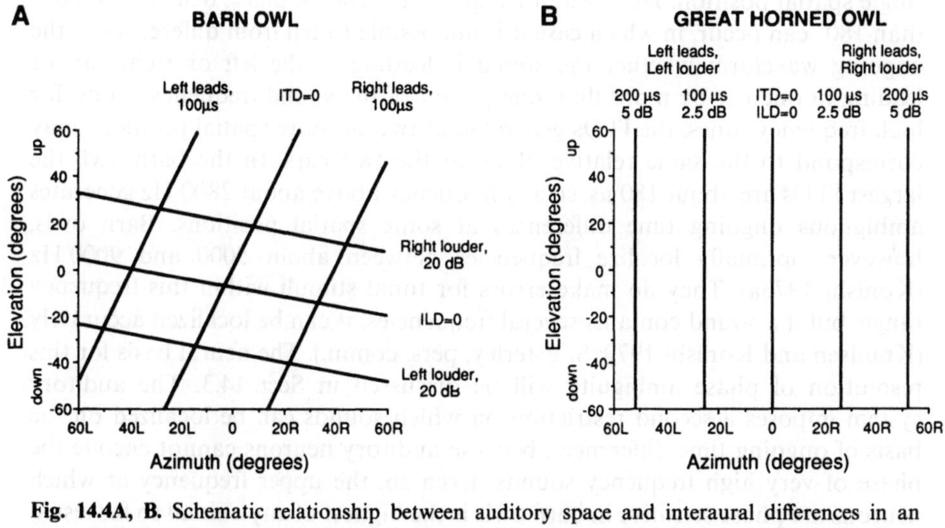

Fig. 14.4A, B. Schematic relationship between auditory space and interaural differences in an asymmetrical and a symmetrical owl. **A** Barn owl iso-ITD and iso-ILD contours drawn from equations for average orienting turns produced by broadband dichotic stimuli (Moiseff 1989b). The actual contours are more curved than these, but show the same general trend: ITD and ILD form a non-orthogonal bi-coordinate map of auditory space. When only ITD varies, the phantom targets shift mostly in their horizontal position, but when ILD alone is varied, the responses shift more nearly vertically. For direct measures of ITD and ILD contours produced by sounds of different frequencies see Moiseff (1989a), Olsen et al. (1989), and Brainard et al. (1992). **B** Great horned owl iso-ITD and iso-ILD contours. The ITD map was synthesized from neurophysiological data and the ILD map from cochlear microphonic responses to a 4 kHz tone (Volman Konishi 1989). Both ITD and ILD change systematically with azimuth. The magnitudes of ITD shifts are larger than in the barn owl, but the ILD shifts are smaller

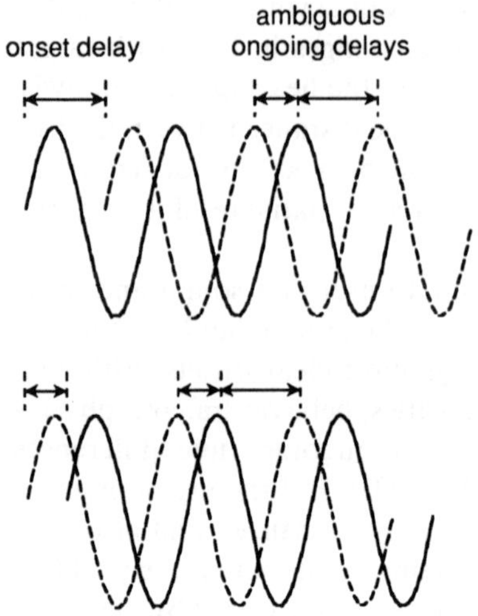

Fig. 14.5. Graphic representation of interaural time differences. Ongoing time differences are equivalent to *phase* differences between the two waveforms. Two different onset delays (*upper and lower*) produce the same sets of ongoing delays

is directly opposite one ear, would always produce a phase delay of less than 180°. Every phase delay would then correspond to a unique ITD and therefore a single spatial position. However, for higher frequencies, phase delays of greater than 180° can occur, in which case it is impossible to tell from differences in the ongoing waveform whether the sound is leading in the left or right ear, or leading in one ear by more than one period of the sound frequency. Thus, for high frequency tones, the ITDs generated at two or more spatial positions may correspond to the same relative phase at the two ears. In the barn owl, the largest ITDs are about 180 μs, so any frequency above about 2800 Hz generates ambiguous ongoing time differences at some spatial positions. Barn owls, however, optimally localize frequencies between about 5000 and 9000 Hz (Konishi 1973a). They do make errors for tonal stimuli within this frequency range, but if a sound contains several frequencies, it can be localized accurately (Knudsen and Konishi 1979; S. Esterley, pers. comm.). The neural basis for this resolution of phase ambiguity will be discussed in Sect. 14.3. The auditory system imposes a second restriction on which sounds can be localized on the basis of ongoing time differences, because auditory neurons cannot encode the phase of very high frequency sounds. Even so, the upper frequency at which neurons are phase sensitive in barn owls is the highest of any animal so far tested (Sullivan and Konishi 1984).

Interaural intensity or sound-level differences (ILDs) are produced when the head or external ear structures differentially obstruct and attenuate the sound paths to the two ears. As with ITD, the frequencies that generate useful ILDs are constrained, but in the opposite direction. A substantial sound shadow exists only for frequencies whose wavelengths are small compared to the shadowing obstruction, thus higher frequencies produce larger ILDs. Interaural sound-level differences have been measured in barn owls with cochlear microphonic recordings and with small microphones inserted in the ear canals (Payne 1971; Knudsen 1980; Coles and Guppy 1988; Moiseff 1989a; Olsen et al. 1989; Brainard et al. 1992). The pattern of ILDs is quite complex because, for each frequency, there is a different pattern of peaks and nulls in the sound field. In general, however, at frequencies below 4–5 kHz, sounds are shadowed chiefly by the head, and ILDs of less than about 10 dB are generated when the sound source is displaced mainly in the *horizontal* plane. At higher frequencies, ILDs of at least 20 dB can occur at some positions. Owing to the additional sound shadowing from the asymmetrical ear flaps, high-frequency sounds from above the horizontal meridian are generally louder at the right ear, whereas those from below are generally louder at the left ear. Thus, as frequency increases, ILDs not only become larger, but they also increasingly reflect the *vertical* position of the sound source. Therefore, a sound emanating from any point in the two-dimensional plane and containing a combination of several frequencies in the range of 5 to 9 kHz, will generate a unique combination of a single ITD and a frequency-dependent set of ILDs.

Barn owls will orient towards a "phantom target" created by dichotic stimuli presented through earphones (Moiseff and Konishi 1981; Moiseff 1989b). If the ITD of a dichotic, broadband stimulus is varied while ILD is held constant, the owl behaves as if it were perceiving targets at positions that differed mainly in the horizontal plane. Conversely, if ILD alone is varied, the owl orients to phantom targets along a more vertical axis (Fig. 14.4A). The actual binaural signals a barn owl would receive from an external sound source are more complex than the single ITDs and ILDs presented in these dichotic stimuli, but these simple combinations are nevertheless sufficient to produce orienting responses to positions throughout a two dimensional plane (Moiseff 1989b). In accordance with the results from dichotic stimulation, if one ear of a barn owl is plugged so as to attenuate sound, errors in the owl's orienting responses are made mainly in the vertical plane (Knudsen and Konishi 1979).

14.3 Neural Mechanisms for Sound Localization in Barn Owls

The neural processing of interaural difference cues has been most extensively studied in the barn owl, and this work has been the subject of two recent comprehensive reviews (Konishi et al. 1988; Takahashi 1989). The reader is referred to these, and to the newer studies cited here, for detailed descriptions of experimental results. Here I will summarize what is known about the barn owl's sound localization system and then use this information as the basis for comparison with the less well-studied systems of other asymmetrical and symmetrical owls.

The study of how the barn owl's auditory system computes sound location from interaural cues began with the discovery of an auditory space map in the owl's midbrain (Knudsen and Konishi 1978a, b). Each neuron in the external subdivision of the inferior colliculus, ICx[1], responds to a free-field sound stimulus from only a very limited region of space. There "space-specific" neurons are arranged into a two dimensional, topographic map. On each side of the brain, sites in the most medial, anterior part of ICx respond to stimuli from a small portion of ipsilateral space. Those at the anterior pole of the nucleus have receptive fields directly in front of the owl, and at progressively more posterior positions, increasingly more contralateral spatial positions are represented (Fig. 14.6). In the dorsal-ventral dimension, the receptive fields of ICx neurons progress in an orderly manner from above to below the horizontal meridian.

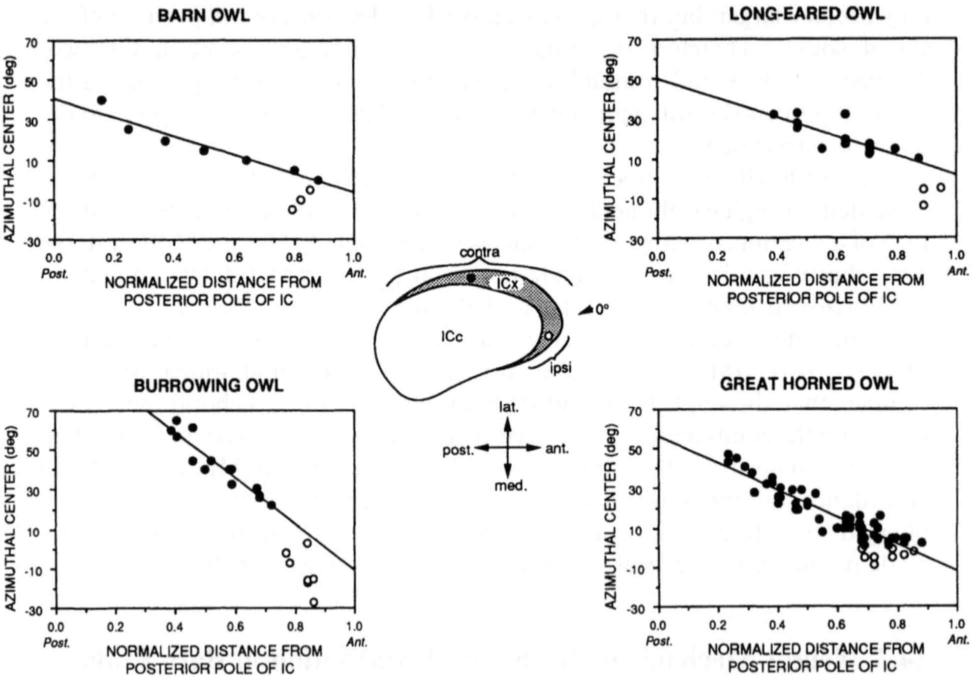

Fig. 14.6. Maps of auditory space in the ICx of four species of owls. The best azimuths of ICx units are plotted relative to their normalized distance from the posterior pole of the inferior colliculus. *Filled symbols* and *positive numbers* represent contralateral azimuths; *open symbols* and *negative numbers*, ipsilateral azimuths. Regression lines were calculated for the contralateral azimuths only. The *inset* shows a dorsal view of the inferior colliculus, with the central and exterior subdivisions and the locations of the contra- and ipsilateral receptive fields indicated (Redrawn from Volman and Konishi 1990)

[1] The bird nucleus mesencephalicus lateralis, pars dorsalis (MLd) is a field homologue of the mammalian inferior colliculus (Karten 1967; Carr 1992). The subdivisions of the barn owl's MLd were described by Knudsen (1983); although these subdivisions may not be strictly homologous to subdivisions of the mammalian IC, collicular nomenclature has been used in most studies of the barn owl and will be followed here.

Recently, it has been found that ICx neurons are also sensitive to the direction of apparent motion of acoustic stimuli (Wagner and Takahashi 1990). Compared to neurons in the central subdivision of IC (ICc), and in all lower levels of the barn owl's auditory system, neurons in the ICx are broadly tuned for frequency and often respond best to broadband sounds. Their preferred frequencies tend to be between 6 and 7 kHz, but responses to frequencies anywhere from about 3 to 9 kHz are common. Thus, ICx neurons favour those combinations of frequencies that barn owls localize most easily (Konishi 1973a; Knudsen and Konishi 1979).

Space-specific auditory neurons in the ICx, and in the optic tectum to which the ICx projects, respond to dichotic stimuli presented through earphones (Moiseff and Konishi 1981; Olsen et al. 1989; Brainard et al. 1992). Each space-specific neuron responds only to a small range of time and sound-level disparities, with its best ITD and ILD closely correlated with the ITDs and ILDs generated by a free-field stimulus in the centre of its spatial receptive field. Like the orienting responses owls make to dichotic stimuli, the physiological data show that the azimuth of a neuron's receptive field depends most strongly on its ITD selectivity to broadband stimuli, whereas its elevation tuning depends more on its ILD selectivity.

Neural processing at several levels of the owl's auditory system (Figs. 14.7 and 14.8) is necessary to extract the correct interaural differences. Beginning with the two cochlear nuclei, and up to the level of the IC, time and intensity information are processed in separate pathways (Moiseff and Konishi 1983; Takahashi et al. 1984; Takahashi and Konishi 1988a, b). Functional segregation of the two cochlear nuclei, the nucleus magnocellularis in the time pathway and the nucleus angularis in the intensity pathway, has been demonstrated by selective inactivation of each nucleus (Takahashi et al. 1984). When one cochlear nucleus is partially inactivated by injection of the local anaesthetic lidocaine, predictable changes are recorded in the responses of space-specific neurons in ICx. For example, if nucleus angularis on one side of the brain is injected, ILD selectivity shifts towards the injected side; louder sounds are now required on this side to compensate for the inactivation. Such injections have no effect on ITD tuning. Conversely, if nucleus magnocellularis is injected unilaterally, the ITD tuning of space-specific neurons broadens and shifts while ILD selectivity remains unchanged.

The distinct specializations of the two cochlear nuclei are also indicated by differences in their physiology (Sullivan and Konishi 1984; Sullivan 1985) and anatomy (Takahashi and Konishi 1988a; Carr and Boudreau 1991). Primary auditory fibres encode timing information by phase locking to a tonal stimulus. They fire action potentials significantly more often during a particular phase of the stimulus cycle, even if they do not produce an action potential on every cycle. Primary auditory fibres terminate on the somata of nucleus magnocellularis neurons with large calycine endings, which are suited for the preservation of precise timing information across this synapse. Phase locking to frequencies as high as 8–9 kHz has been found in nucleus magnocellularis (Sullivan and

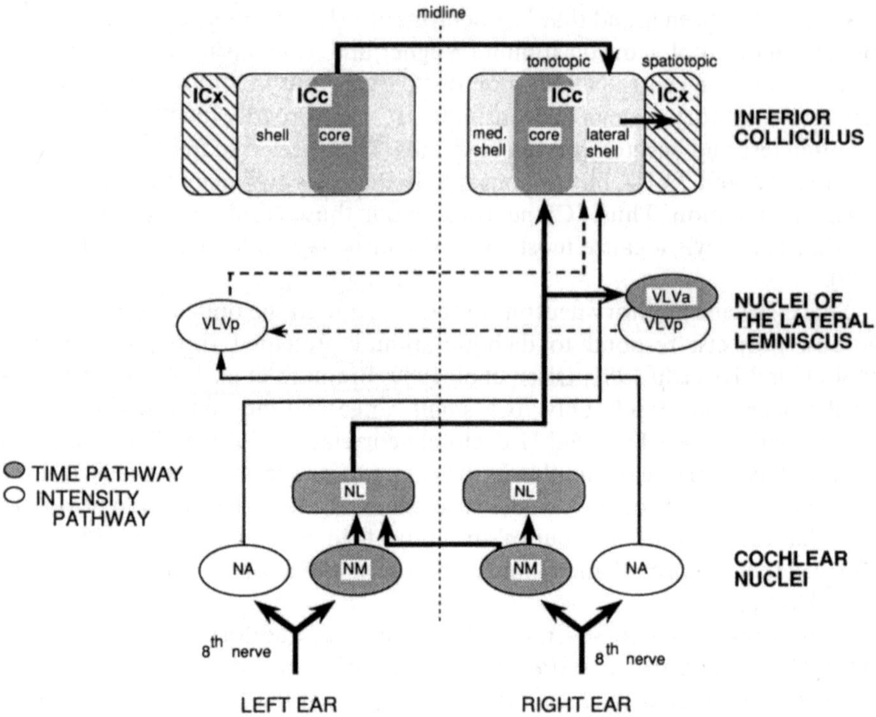

Fig. 14.7. Pathways for the computation of interaural cues. Not all known connections are shown, nor are all areas discussed in the text. *Dark stippling* designates the time pathway, and *open symbols* the intensity pathway. In the inferior colliculus, the medial shell of the ICc apparently contains neurons sensitive only to ILD, but the lateral shell and the ICx encode both time and intensity information; *NA* nucleus angularis; *NM* nucleus nagnocellularis; *NL* nucleus laminaris; *VLVp* nucleus ventralis lemnisci lateralis pars posterior; *VLVa* nucleus ventralis lemnisci lateralis pars anterior

Konishi 1984). By contrast, in nucleus angularis the primary auditory fibres branch repeatedly, and their synaptic endings are distributed over the somata and dendrites of the postsynaptic cells. Nucleus angularis neurons phase lock poorly to high frequency tones, but they are better able than those in nucleus magnocellularis to signal changes in the sound level by changes in their firing rates.

Nucleus laminaris is the first *binaural* nucleus in the time pathway. It is a large nucleus, which for much of its extent appears as a multilayered slab of very large cells. The tonotopic axis of nucleus laminaris runs roughly from caudal to rostral, with each frequency represented in a wide, medio-lateral slice (Takahashi and Konishi 1988a). Fibres from the ipsilateral and contralateral magnocellular nuclei, which display the same frequency preferences, cross the nucleus laminaris in opposite directions from its dorsal and ventral borders respectively, and they converge on the soma or very short dendrites of the laminaris neurons (Carr and Konishi 1990). As the magnocellularis afferents cross laminaris, action potentials from one side become progressively delayed relative to those from the other side (Sullivan and Konishi 1986; Carr and Konishi 1988, 1990). Thus, laminaris neurons are in a position to encode interaural delays by responding maximally when action potentials from the two magnocellular nuclei coincide. Individual laminaris neurons do respond preferentially to stimuli with particular interaural delays (Moiseff and Konishi 1983; Carr and Konishi 1990), but it has so far proved impossible to record enough laminaris neurons in sequence to demonstrate conclusively a map of interaural delay in the post-synaptic cells. Nevertheless, the presence of such a map can be inferred from the orderly changes in time delay recorded from magnocellular afferents within nucleus laminaris (Sullivan and Konishi 1986; Carr and Konishi 1988, 1990; Konishi et al. 1988).

Nucleus laminaris neurons respond preferentially to certain interaural delays, but they cannot signal a unique ITD. Rather, they will respond maximally to all interaural delays that contain a particular interaural *phase* difference (IPD), equivalent to a set of ITDs separated by integral multiples of the period of the neuron's best frequency (Fig. 14.8). On the cellular level, laminaris neurons face the same phase-ambiguity problem as an owl localizing a tonal stimulus, and one would expect these neurons to respond to a sound at more than one location. Such experiments have not been carried out in laminaris, but neurons in the central subdivision of the inferior colliculus, which also have multiply peaked ITD tuning curves, often have more than one distinct receptive field (Knudsen and Konishi 1978b).

Between the central and external subdivisions of the inferior colliculus the "true" ITD of a complex sound is extracted from the multiply peaked ITD responses of neurons with different best frequencies (Fig. 14.8). Afferents from nucleus laminaris project to the "core" of ICc, and here unique ITDs are encoded in a dorsal-ventral array, across the tonotopic axis of the colliculus (Wagner et al. 1987). Core neurons with multiple ITD peaks are aligned so that across all frequencies only one peak is common to all cells (Fig. 14.8). In addition, this common peak changes in an orderly manner from approximately anterior to posterior, forming an "array map" of ITD. The ITDs represented in the ICc core correspond to *ipsilateral* space, but core neurons project to the ICc lateral shell on the other side of the brain, which results in a representation of *contralateral* space within the lateral shell, similar to that in the ICx (Takahashi et al. 1989). Inputs from each column of the array in the lateral shell are integrated by single ICx neurons, which consequently respond to multiple frequencies, but have only a single large ITD peak when stimulated with broadband noise (Takahashi and Konishi 1986; Wagner et al. 1987). In many of these steps, inhibitory processes serve to sharpen the ITD tuning (Wagner 1990; Fujita and Konishi 1991).

ICx neurons are also tuned to a small range of ILDs, but the steps involved in the derivation of this selectivity are not as well understood as the computation of ITD. As discussed at the end of Sect. 14.2, a broadband sound from a

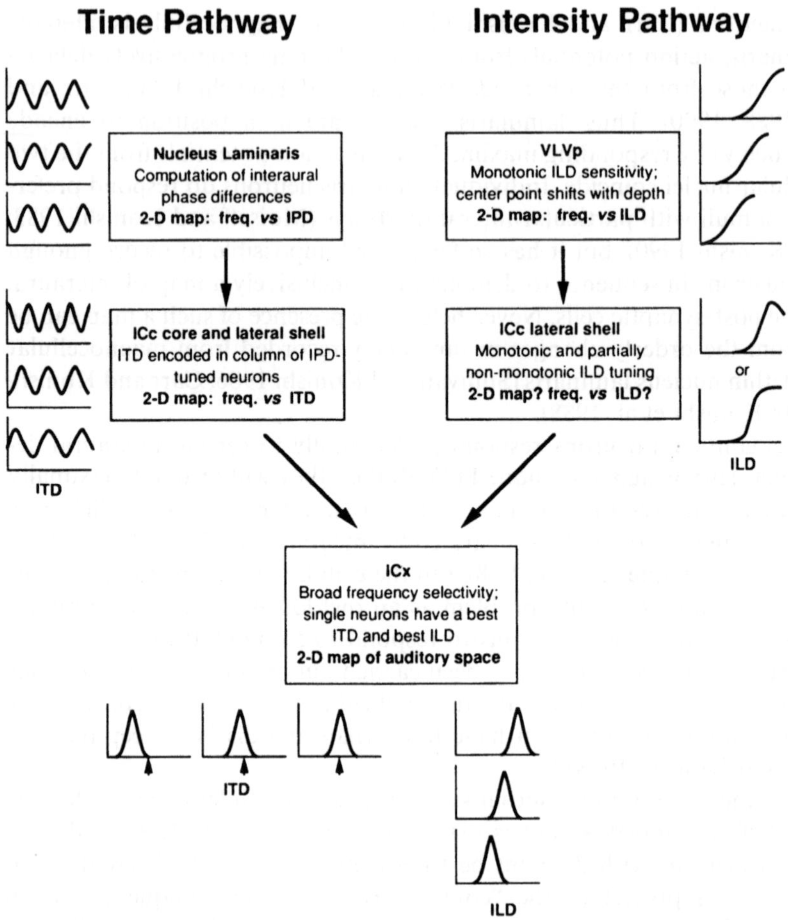

Fig. 14.8. Schematic representation of neuronal response properties in the time and intensity pathways. *Time pathway*: idealized representative tuning curves of responses to different ITDs. Neurons in nucleus laminaris are sharply tuned for frequency, and they respond to every ITD that contains an integral multiple of a particular interaural phase difference (IPD). The IPD to which a neuron is most sensitive shifts systematically across the nucleus. Therefore these responses are phase ambiguous. In the core and lateral shell of ICc, the tonotopic axis runs from dorsal to ventral. The ITD tuning curves in a column across different frequencies layers share a single common peak. Neurons from separate frequency laminae in the lateral shell converge onto space-specific neurons in the ICx, which respond best to a single ITD for broadband stimuli. ITD selectivity in ICx forms an anterio-posterior map of azimuth. *Intensity pathway*: Idealized representative tuning curves of responses to ILDs. In the VLVp, neurons are excited via the contralateral ear and inhibited via the ipsilateral ear (ipsilateral ear louder is to the left on the abscissa, and contralateral ear louder to the right). The ILD tuning curves are monotonic and do not have a single ILD peak. The half-maximal point of these tuning curves shifts systematically from dorsal to ventral in the nucleus, and therefore different populations of VLVp neurons would be active for different ILDs. In the lateral shell, ILD tuning curves have a variety of shapes, but it is not yet known if they form a map. ICx neurons have non-monotonic ILD tuning curves, the peaks of which vary, depending on frequency, with both elevation and azimuth (See text for further details)

single spatial location generates ILDs that are different for each frequency, and a single frequency may produce the same ILD at different spatial locations. These spectral disparities contribute to the difficulty of understanding how ILD is processed. In addition, there must be a neural asymmetry or "inversion" somewhere in the intensity pathway before the level of the ICx. When a sound source containing higher frequencies is located above the horizon it is generally louder in the right ear and sounds from below the horizon are louder in the left ear. Therefore, on the left side of the brain dorsal sites in the ICx respond preferentially to sounds that are louder in the contralateral ear, whereas on the right side dorsal sites prefer ILDs favouring the ipsilateral ear. These differences in contralateral-ipsilateral interactions are also likely to increase the challenge of deciphering the ILD pathway. Both of these complexities render the correct correspondence between ILD and spatial position highly sensitive to the size and shape of an individual owl's head and ears. Barn owls require visual information to calibrate their auditory map in the optic tectum (Knudsen 1988a, b), and inappropriate ILD tuning appears to account for most of the topographic distortion, including inversions, found in blind-reared owls (Knudsen et al. 1991).

The nucleus ventralis lemnisci lateralis pars posterior (VLVp) is the first binaural nucleus in the intensity pathway (Fig. 14.7; Takahashi and Konishi 1988b). VLVp neurons are excited by sounds in the contralateral ear and inhibited or unaffected by sounds in the ipsilateral ear (Moiseff and Konishi 1983; Manley et al. 1988). The excitation is by way of a direct projection from the contralateral nucleus angularis, whereas the inhibition arrives indirectly via the VLVp on the opposite side of the brain (Takahashi and Keller 1992). VLVp neurons have sigmoidally shaped, monotonic ILD tuning curves as shown in Fig. 14.8. Neurons with this sort of tuning are highly sensitive to changes in ILD over the steeply sloped part of the curve, but no further changes in ILD can be signalled once the cell is firing maximally. However, not all cells in VLVp are inhibited equally and there is an apparent ventral to dorsal gradient of ipsilateral inhibition within the nucleus. Thus the particular contralateral-louder ILD at which a VLVp neuron begins to respond, and the ILD to which it responds maximally, both gradually decrease from dorsal to ventral in the nucleus (Fig. 14.8; Manley et al. 1988). Although VLVp neurons have not been studied with free-field stimuli, one would expect their receptive fields to have only an upper or lower border, but not both. In contrast, ICx neurons have receptive fields which are restricted to a small range of elevations bounded from both above and below.

At the level of the lateral shell, where the time and intensity pathways first converge, a variety of types of ILD tuning curves are encountered (Fujita and Konishi 1989; Takahashi 1989). Some are monotonic, whereas others have distinct or partial peaks. The inputs that contribute to these responses arise from several lower levels including the medial shell of ICc, the VLVp, and directly from the nucleus angularis. The role of these various inputs in shaping the ILD response is not yet clear (Adolphs and Takahashi 1989; Takahashi 1989). Further convergence from the lateral shell to the ICx results in neurons with peaked, non-monotonic ILD tuning curves, which respond to only a small range of binaural stimuli (Fig. 14.8; Moiseff and Konishi 1983; Takahashi et al. 1984). As in the ITD pathway, frequency convergence is also important for resolving spatial ambiguity in the intensity pathway, because a single tone may produce the same ILD when broadcast from more than one spatial location (Brainard et al. 1992).

The ICx projects topographically to the optic tectum (Knudsen and Knudsen 1983) where individual neurons respond to both auditory and visual stimuli from the same region of space (Knudsen 1982; Olsen et al. 1989). Much of the recent detailed analysis of auditory space-tuned neurons has been carried out in the optic tectum (Knudsen 1984; Olsen et al. 1989; Brainard et al. 1992). Auditory responses in the tectum have not been systematically compared with those in the ICx, so it is not clear whether further neural processing occurs between these two areas, although there is some indication that receptive fields are slightly smaller in the optic tectum. Even in the tectum, the receptive fields are large compared to the barn owl's behavioural accuracy. The average tectal receptive field is 30° in azimuth and 50° in elevation (Knudsen 1982), with a central "best area" of 16°–18° by 30°–35° (Olsen et al. 1989; Knudsen et al. 1991). Therefore, more precise spatial resolution must depend on comparing or averaging the activity across a population of space-tuned neurons (cf. Brainard et al. 1992).

The space-specific receptive fields of ICx and tectal neurons strongly suggest that they are involved in producing orienting responses to sound, and there is evidence that this is the case. Microstimulation of a site in the tectum produces a head turn similar to that evoked by an auditory or visual stimulus that activates the same site (du Lac and Knudsen 1990). The pathways by which the tectum produces motor responses are currently being explored (e.g. Masino and Knudsen 1990). Additional evidence for the role of the ICx and tectum in orientation to sounds comes from lesion studies. Owls with partial lesions of the ICx or tectum either misorient or fail to turn towards auditory targets presented in locations corresponding to the lesioned part of the spatial map (Wagner 1991b). There may however be other parallel routes for the production of orienting responses to auditory stimuli, because the deficits from these lesions are not permanent.

14.4 Comparative Physiology of Sound Localization Among the Owls

The fact that ear asymmetry has arisen independently several times among the owls raises two questions. Do other, unrelated asymmetrical owls use strategies for sound localization similar to the barn owl's? And are the mechanisms underlying directional hearing in symmetrical species pre-adaptive for the evolution of asymmetry? Some insight into these questions has been gained by recent comparative neurophysiological studies of two symmetrical and an additional two asymmetrical species (Fig. 14.1; Wise et al. 1988; Volman and Konishi 1989, 1990; Volman 1990).

The two symmetrical species are the great horned and the burrowing owls. Great horned owls, which are one of the largest owls, are nocturnal or occasionally crepuscular. They have a moderately well developed facial ruff and large ear openings. The two ear openings seem to be the same size, although in the closely related Eurasian eagle owl (*Bubo bubo*) the left ear opening may be about 10% smaller in height than the right ear (Norberg 1977). Burrowing owls are one of the smaller owls. They are crepuscular or diurnal, have a poorly developed facial ruff and small ear openings. Phylogenetically, burrowing owls belong to a group of genera that appear to be the stem from which other clades have arisen (S. Coats pers. comm.). No owls in this group have asymmetrical ears, which suggests that symmetry in the burrowing owl represents the primitive condition. The two asymmetrical species in these comparative studies, the long-eared and saw-whet owls, are both considered strictly nocturnal. Their ear asymmetry seems to have evolved independently from each other and from that in the barn owl (Norberg 1977). The long-eared owl's elaborate ears are shown in Fig. 14.3. Not only are they asymmetrical, they are also more mobile than the ears of most other birds. Both the shape of the ear slit and the pre-aural flap change position when long-eared owls assume an alert posture (Ilyichev 1975). Unfortunately, the effects of these movements on sound localization have not been well studied. The long-eared owl has a well developed facial ruff that extends above the eyes, like that in the barn owl. Saw-whet owls have only a moderate facial ruff, but they have long, slot-shaped ear openings and a marked asymmetry in the temporal part of the skull, which directs the right ear opening upward and the left downward (Norberg 1977, 1978; L. Wise and B. Frost pers. comm.). In the following discussion, the response properties of ICx neurons are compared among these four species and the barn owl, although only partial data are available, especially for the long-eared and saw-whet owls.

In all five species, ICx neurons have spatial receptive fields that are restricted in azimuth and arrayed topographically (Fig. 14.6; Wise et al. 1988; Volman and Konishi 1989, 1990). The great horned and barn owls have the narrowest receptive fields, with mean "best areas" (where response is at least 50% of the maximum) of 16°–18°. In the saw-whet and long-eared owl respectively, the fields are approximately 22° and 29°. The receptive fields in the burrowing owl are much wider, averaging about 44°. These receptive field widths scale with head size except in the barn owl, and possibly in the saw-whet owl, where they are narrower than expected (Volman and Konishi 1990).

In the auditory maps of most of these owls, there are no receptive fields centred at more than about 15° ipsilateral, and the central 40–50° of contralateral space appears to be over emphasized (Fig. 14.6). The burrowing owl's map appears to be the least "focused". It extends further ipsilaterally, and represents contralateral space more uniformly, than the maps in the other owls. In the symmetrical owls, the receptive fields are restricted only in azimuth and not in elevation, whereas in the two additional asymmetrical owls. they are also somewhat limited in elevation, but less so than in the barn owl. In some recording sequences through the ICx of the saw-whet owl there is a clear progression of preferred elevation from upper to lower fields, which suggests the presence of a two-dimensional map similar to that in the barn owl (L. Wise and B. Frost pers. comm.) The long-eared owl has the least distinct elevation tuning among the asymmetrical owls. This might however be an artifact of recording from anaesthetized birds, because these owls may use their mobile outer ears to intensify interaural sound-level differences when they are awake.

The responses of ICx neurons suggest that all owls use ITD as a cue for horizontal sound position. Preferred ITD is topographically mapped (Wise et al. 1988; Voman and Konishi 1989, 1990), and it is correlated with receptive field azimuth (Volman and Konishi 1989). Furthermore, the neural mechanisms for computing ITD appear to be similar in all owls. In the seven species of owls that have been looked at, including the five being discussed here, nucleus laminaris is hypertrophied compared to the presumed ancestral, single-layered form found in birds such as chickens and pigeons (Carr 1989, 1992; S. Volman unpubl. observ.). The owls' multilayered nucleus laminaris provides an anatomical substrate for fine resolution of interaural time differences. The structure of this nucleus appears to be optimized in asymmetrical species, in which the proportional thickness is greatest relative to the distance between the ears (Carr 1989, 1992). Nevertheless, the great horned owl with a relatively smaller nucleus laminaris can localize sounds in azimuth as accurately as does the barn owl, at least at the more peripheral locations (Beitel 1991), and its receptive fields are equally narrow (Volman and Konishi 1990). Comparable mechanisms for processing ITD are also suggested by the similarity of response properties of neurons in the inferior colliculus. The ICx neurons of all species respond to a broader range of frequencies than those in the ICc, and they have single receptive fields and single large ITD peaks with noise stimuli, but multiple receptive fields and multiply peaked ITD tuning curves when stimulated with tones. It is noteworthy that these response patterns are present even in the burrowing owl, which has a small head and hears only low frequencies. This suggests that the neural solution to phase ambiguity arose in the ancestors of modern owls.

There is a clear difference in the range of frequencies audible to symmetrical versus asymmetrical owls, evidenced by data from 15 species (Trainer 1946; Schwartzkopff 1962; van Dijk 1973; Calford 1988; Fay 1988; Wise et al. 1988). Acute hearing in symmetrical species appears limited to frequencies below 6 or 7 kHz, whereas asymmetrical owls have low threshold hearing to at least 8 or 9 kHz. These audibility differences are also reflected in the frequency response of space-specific neurons in the ICx. The highest best frequencies of ICx neurons recorded in asymmetrical owls are about 8.5 kHz in the barn owl and 6.5 kHz in the long-eared and saw-whet owls; in the symmetrical great horned and burrowing owls, they are about 5.5 and 4.5 kHz respectively. These differences in high-frequency hearing have consequences for the usefulness of ILD cues in directional hearing, because the magnitude of this binaural cue is frequency dependent.

Do symmetrical owls use the ILD cue? In the great horned owl, ILD changes smoothly from left ear louder to right ear louder as a speaker is moved from left to right. However, maximum ILDs produced by a 4-kHz tone are less than about 10 dB (Volman and Konishi 1989). ILDs have not been measured in the burrowing owl, but one would expect them to be even smaller than in the great horned owl, because their head is half the size. The ILD tuning curves recorded for ICx neurons in both these symmetrical species are very similar in shape and width to those in the barn owl, and the best ILD is correlated with the best ITD, suggesting that ILD may be an additional cue for azimuth. But ILD cannot provide very accurate azimuthal information because the tuning curves are wide *relative* to the ILDs generated by sounds in these owls hearing ranges. Regardless of how symmetrical owls may use ILD information, the similarity of their ILD tuning to that in the barn owl implies that major new pathways for processing ILD were probably not required for ear asymmetry to evolve (Volman 1990).

14.5 Evolution of Bilateral Ear Asymmetry

During his phylogenetic study of bilateral ear asymmetry, Norberg (1977) came across an individual Tengmalm's owl with a small septum across its left ear opening, reminiscent of the septa in long-eared owls (Fig. 14.3). A chance abnormality of this sort might begin the process of selection for ear asymmetry in a population of owls for which it is functionally useful. For this to occur, however, the auditory system of such an owl would have to be able to exploit the anomalous asymmetry, presumably to improve directional hearing in the vertical plane. There seem to be only two conditions necessary for such an achievement. First, the owl would have to hear frequencies that are high enough to generate vertical asymmetries in the sound fields of the two ears. In addition, for each azimuthal position, the auditory system would have to encode and correctly interpret a *set* of ILDs representing different vertical positions.

Must high frequency hearing precede ear asymmetry? The upper frequencies heard by symmetrical owls may be sufficient to exploit ear asymmetry, in which case the extended hearing range of asymmetrical owls might have arisen secondarily to improve two-dimensional hearing further. This possibility is buttressed by the fact that high frequency hearing has been found only in owls with asymmetrical ears, although the hearing ranges of many symmetrical owls remain to be tested. Alternatively, asymmetry may have arisen only in owls that were already able to hear high frequencies, in which case there should be some advantage to high-frequency hearing even in a symmetrical owl. Higher frequencies would, of course, improve the use of ILD cues for azimuthal localization in a symmetrical owl. Another possible advantage is that *temporal* resolution may be better at high frequencies. For example in the barn owl the average ITD tuning is nearly 12 μs sharper at 7 kHz that at 3.5 kHz, which is equivalent to narrowing a receptive field by about 4° in azimuth (Volman and Konishi 1989). Nevertheless, this sharpening might not result from the use of high frequencies per se, but rather the neural interactions that produce sharper ITD tuning may simply be concentrated among high frequency neurons.

How might an owl that fortuitously develops ear asymmetry come to associate the correct binaural cues with each location in space? In the barn owl, these relationships are known to be plastic and to depend on sensory experience during post-natal development (see Knudsen 1988b). As an owl grows, spatial binaural cues are modified as the size and shape of its head and external ears change (Knudsen et al. 1984; Haresign and Moiseff 1989). Immature barn owls are able to adjust their orienting responses when binaural cues are altered by unilateral ear occlusion (Knudsen et al. 1984), and the auditory space map in the optic tectum shifts accordingly (Knudsen 1985). Thus, while a barn owl is growing, its auditory system becomes calibrated to its head and ears. It seems reasonable to assume that similar developmental adjustments occur in symmetrical owls. It might even be possible to investigate experimentally whether a normally symmetrical owl with an induced ear asymmetry could develop improved, two-dimensional sound localization and a two-dimensional space map in its ICx.

14.6 Future Directions

The comparative studies of directional hearing in owls indicate that selective pressure for accurate sound localization has produced a variety of morphological specializations in the auditory periphery that nevertheless depend on similar binaural cues and neural mechanisms. However, there are a number of questions remaining about sound localization in owls.

The physiological evidence and sound-field measurements of symmetrical species suggest that they should not be able to determine the elevation of a sound source without rotating their heads. But no behavioural studies have yet addressed this prediction. Nor do we know the physical basis or neural processes owls use to determine when a sound originates from behind their heads, although the behavioural evidence shows they can do so. We also have no information about sound localization in owls with only minor ear asymmetries. For example, it is unclear what advantage is conferred by having one ear opening larger than the other, yet differences of this sort are found in the Eurasian eagle owl and all or most members of the genera *Strix* and *Ciccaba* (Norberg 1977). Long-eared and some other owls can move their ears, but we do not know to what extent these shape changes enchance binaural difference cues.

Asymmetrical owls probably would not have evolved their two-dimensional system if they were unable to encode both ITD and ILD over the same relatively high frequency range, in which both types of information are useful. Although it is theoretically possible to determine the azimuth and elevation of a broadband sound from different frequency components, this could easily lead to confusion if there were more than one noise source. I have also argued elsewhere (Volman 1990; Volman and Konishi 1990) that the evolution of asymmetry probably depends on the seemingly minor use of ILD by symmetrical owls. If these owls determined azimuth exclusively from sound-level differences over part of their frequency range, as is the case in mammals, then it is difficult to see how ILD could come to encode mainly elevation in asymmetrical owls. Both these considerations lead to the conclusion that a complete explanation of the owl's remarkable directional hearing depends on understanding when and how their ancestors first developed the ability to encode the phase of high frequency sounds and to solve the phase ambiguity problem.

Acknowledgements. I am grateful to Susan Mauersberg for drawing Figs. 14.1, 14.2 and 14.3, and to Hermann Wagner, Mark Konishi and the editors for their comments on earlier drafts of the chapter.

References

Adolphs R, Takahashi T (1989) Inhibition shapes responses to interaural level difference in the inferior colliculus of the barn owl. Soc Neurosci Abstr 15:1116

Beitel RE (1991) Localization of azimuthal sound direction by the great horned owl. J Acoust Soc Am 90:2843–2846

Brainard MS, Knudsen EI, Esterly SD (1992) Neural derivation of sound source location: Resolution of spatial ambiguities in binaural cues. J Acoust Soc Am 91:1015–1027

Calford MB (1988) Constraints on the coding of sound frequency imposed by the avian interaural canal. J Comp Physiol 162:491–502

Carr CE (1989) Comparative anatomy of the brainstem auditory pathways in owls. In: Erber J, Menzel R, Pfluger H-J, Todt D (eds) Neural mechanisms of behavior. Proc 2nd Int Congr Neuroethology Part II. Thieme, Stuttgart, p 116

Carr CE (1992) The evolution of the central auditory system in reptiles and birds. In: Popper A, Webster D, Fay R (eds) The evolutionary biology of hearing. Springer, Berlin Heidelberg, New York, pp 511–543

Carr CE, Boudreau RE (1991) Central projections of auditory nerve fibers in the barn owl. J Comp Neurol 314:306–318

Carr CE, Konishi M (1988) Axonal delay lines for time measurement in the olw's brainstem. Proc Natl Acad Sci USA 85: 8311–8315

Carr CE, Konishi M (1990) A circuit for the detection of interaural time differences in the brainstem of the barn owl. J Neurosci 10:3227–3246

Coles RB, Guppy A (1988) Directional hearing in the barn owl (*Tyto alba*). J Comp Physiol 157:149–160

Collett R (1871) On the asymmetry of the skull in *Strix tengmalmi*. Proc Zool Soc Lond 1871: 739–743

du Lac S, Knudsen EI (1990) Neural maps of head movement vector and speed in the optic tectum of the barn owl. J Neurophysiol 63:131–146

Esterly SD, Knudsen EI (1987) Tuning for interaural difference cues varies with frequency for space-specific neurons in the owl's optic tectum. Soc Neurosci Abstr 13:1468

Fay RR (1988) Hearing in vertebrates. Hill-Fay Associates, Winnetka, Ill, pp 197–216

Frost BJ, Baldwin PJ, Csizy ML (1989) Auditory localization in the northern saw-whet owl, *Aegolius acadicus*. Can J Zool 67:1955–1959

Fujita I, Konishi M (1989) Transition from single to multiple frequency channels in the processing of binaural disparity. Soc Neurosci Abstr 15:114
Fujita I, Konishi M (1991) The role of GABAergic inhibition in processing of interaural time difference in the owl's auditory system. J Neurosci 11:722–739
Haresign T, Moiseff A (1989) Early growth and development of the common barn-owl's facial ruff. Auk 105:699–705
Ilyichev (1975) Localization in birds: adaptive mechanisms for passive localization in owls. Knowledgebooks, Moscow (in Russian)
Karten HJ (1967) The organization of the ascending auditory pathway in the pigeon (*Columbia livia*) I. Diencephalic projections of the inferior colliculus (nucleus mesencephali lateralis, pars dorsalis). Brain Res 11:134–153
Knudsen EI (1980) Sound localization in birds. In: Popper AN, Fay RR (eds) Comparative studies of hearing in vertebrates. Springer, Berlin Heidelberg New York, pp 289–322
Knudsen EI (1982) Auditory and visual maps of space in the optic tectum of the owl. J Neurosci 2:1177–1194
Knudsen EI (1983) Subdivisions of the inferior colliculus in the barn owl (*Tyto alba*). J Comp Neurol 218:174–186
Knudsen EI (1984) Auditory properties of space-tuned units in the optic tectum of the owl. J Neurophysiol 52:709–723
Knudsen EI (1985) Experience alters the spatial tuning of auditory units in the optic tectum during a sensitive period in the barn owl. J Neurosci 5:3094–3109
Knudsen EI (1988a) Early blindness results in a degraded auditory map of space in the owl's optic tectum. Proc Natl Acad Sci USA 85:6211–6214
Knudsen EI (1988b) Experience shapes sound localization and auditory unit properties during development in the barn owl. In: Edelman GM, Gall WE, Cowan WM (eds) Auditory function: neurobiological bases of hearing. Wiley, New York, pp 137–149
Knudsen EI, Knudsen PF (1983) Space-mapped auditory projections from the inferior colliculus to the optic tectum in the barn owl (*Tyto alba*). J Comp Neurol 218:187–196
Knudsen EI, Konishi M (1978a) A neural map of auditory space in the owl. Science 200:795–797
Knudsen EI, Konishi M (1978b) Space and frequency are represented separately in the auditory midbrain of the owl. J Neurophysiol 41:870–884
Knudsen EI, Konishi M (1979) Mechanisms of sound localization in the barn owl (*Tyto alba*). J Comp Physiol 133:13–21
Knudsen EI, Blasdel GG, Konishi M (1979) Sound localization by the barn owl (*Tyto alba*) measured with the search coil technique. J Comp Physiol 133:1–11
Knudsen EI, Esterly SD, Knudsen PF (1984) Monaural occlusion alters sound localization during a sensitive period in the barn owl. J Neurosci 4:1001–1011
Knudsen EI, Esterly SD, du Lac S (1991) Stretched and upside-down maps of auditory space in the optic tectum of blind-reared owls: acoustic basis and behavioural correlates. J Neurosci 11:1727–1747
Konishi M (1973a) Locatable and nonlocatable acoustic signals for barn owls. Am Nat 107:775–785
Konishi M (1973b) How the owl tracks its prey. Am Sci 61:414–424
Konishi M, Takahashi TT, Wagner H, Sullivan WE, Carr CE (1988) Neurophysiological and anatomical substrates of sound localization in the owl. In: Edelman GM, Gall WE, Cowan WM (eds) Auditory function: neurobiological bases of hearing. Wiley, New York, pp 721–745
Manley GA, Köppl C, Konishi M (1988) A neural map of interaural intensity difference in the brain stem of the owl. J Neurosci 8:2665–2676
Marti CD (1974) Feeding ecology of four sympatric owls. Condor 76:45–61
Martin GR (1986) Sensory capacities and the nocturnal habit of owls (Strigiformes). Ibis 128:266–277
Masino T, Knudsen EI (1990) Horizontal and vertical components of head movement are controlled by distinct neural circuits in the barn owl. Nature 345:434–437
Moiseff A (1989a) Binaural disparity cues available to the barn owl for sound localization. J Comp Physiol 164:629–636

Moiseff A (1989b) Bi-coordinate sound localization by the barn owl. J Comp Physiol 164:637–644
Moiseff A, Konishi M (1981) Neuronal and behavioral sensitivity to binaural time differences in the barn owl. J Neurosci 1:40–48
Moiseff A, Konishi M (1983) Binaural characteristics of units in the owl's brainstem auditory pathway: precursors of restricted spatial receptive fields. J Neurosci 3:2553–2562
Norberg RÅ (1968) Physical factors in directional hearing in *Aegolius funereus* (Linne.) (Strigiformes), with special reference to the significance of the asymmetry of the external ears. Ark Zool 20:181–204
Norberg RÅ (1977) Occurrence and independent evolution of bilateral ear asymmetry in owls and implications on owl taxonomy. Philos Trans R Soc Lond Biol 280:375–408
Norberg RÅ (1978) Skull asymmetry, ear structure and function, and auditory localization in Tengmalm's owl, *Aegolius funereus* (Linne). Philos Trans R Soc Lond Biol 282:325–410
Norberg RÅ (1987) Evolution, structure and ecology of northern forest owls. In: Nero RW (ed) Biology and conservation of northern forest owls. Symp Proc. USDA Forest Service Gen Tech Rep RM-142 US Dept Agric, Fort Collins, CO, pp 9–43
Olsen JF, Knudsen EI, Esterly SD (1989) Neural maps of interaural time and intensity differences in the optic tectum of the barn owl. J Neurosci 9:2591–2605
Payne RS (1962) How the barn owl locates prey by hearing. Living Bird 1:151–159
Payne RS (1971) Acoustic location of prey by barn owls (*Tyto alba*). J Exp Biol 54:535–573
Payne RS, Drury WH (1958) *Tyto alba*, marksman of the darkness. Nat Hist NY 67:316–323
Pumphrey RJ (1948) The sense organs of birds. Ibis 90:171–199 (Reprinted with some additions 1949, Smithson Inst, Annu Rep, pp 305–330)
Schwartzkopff J (1962) Zur Frage des Richtungshörens von Eulen (Striges). Z Vergl Physiol 45:570–580
Steinbach MJ, Money KE (1973) Eye movements of the owl. Vision Res 13:889–891
Streets TH (1870) Remarks of the cranium of an owl. Proc Acad Nat Sci Phila 1870:73
Stresemann E (1934) Sauropsida, Aves. In: Kükenthal WG, Krumbach (eds) Handbuch der Zoologie, vol 7, part 2. de Gruyter, Berlin, pp 769–899
Sullivan WE (1985) Classification of response patterns in cochlear nucleus of barn owl: correlation with functional response properties. J Neurophysiol 53:201–216
Sullivan WE, Konishi M (1984) Segregation of stimulus phase and intensity coding in the cochlear nucleus of the barn owl. J Neurosci 4:1787–1799
Sullivan WE, Konishi M (1986) Neural map of interaural phase difference in the owl's brainstem. Proc Natl Acad Sci USA 83:8400–8404
Takahashi T (1989) The neural coding of auditory space. J Exp Biol 146:307–322
Takahashi T, Keller CH (1992) Commissural connections mediate inhibition for the computation of interaural level differences in the barn owl. J Corp Physiol, 70:161–169
Takahashi T, Konishi M (1986) Selectivity for interaural time difference in the owl's midbrain. J Neurosci 6:3413–3422
Takahashi TT, Konishi M (1988a) Projections of the cochlear nuclei and nucleus laminaris to the inferior colliculus of the barn owl. J Comp Neurol 274:190–211
Takahashi TT, Konishi M (1988b) Projections of nucleus angularis and nucleus laminaris to the lateral lemniscal nuclear complex of the barn owl. J Comp Neurol 274:212–238
Takahashi T, Moiseff A, Konishi M (1984) Time and intensity cues are processed independently in the auditory system of the owl. J Neurosci 4:1781–1786
Takahashi TT, Wagner H, Konishi M (1989) Role of commissural projections in the representation of bilateral auditory space in the barn owl's inferior colliculus. J Comp Neural 281: 545–554
Trainer JE (1946) The auditory acuity of certain birds. PhD Thesis, Cornell University, Ithaca, New York
van Dijk T (1973) A comparative study of hearing in owls of the family Strigidae. Neth J Zool 23:131–167
Volman SF (1990) Neuroethological approaches to the evolution of neural systems. Brain Behav Evol 36:154–165

Volman SF, Konishi M (1989) Spatial selectivity and binaural responses in the inferior colliculus of the great horned owl. J Neurosci 9:3083–3096
Volman SF, Konishi M (1990) Comparative physiology of sound localization in four species of owls. Brian Behav Evol 36:196–215
Voous KH (1989) Owls of the northern hemisphere. MIT Press, Cambridge
Wagner H (1990) Receptive fields of neurons in the owl's auditory brainstem change dynamically. Eur J Neurosci 2:949–959
Wagner H (1991a) A temporal window for lateralization of interaural time difference by barn owls. J Comp Physiol 169:281–289
Wagner H (1991b) Sound-localization deficits induced by lesions in the owl's auditory space map. Soc Neurosci Abstr 17:1483
Wagner H (1992) On the ability of neurons in the barn owl's inferior colliculus to sense brief appearances of interaural time difference. J Comp Physiol 170:3–11
Wagner H, Takahashi T (1990) Neurons in the midbrain of the barn owl are sensitive to the direction of apparent acoustic motion. Naturwissenschaften 77:439–442
Wagner H, Takahashi T, Konishi M (1987) Representation of interaural time difference in the central nucleus of the barn owl's inferior colliculus. J Neurosci 7:3105–3116
Wise LZ, Frost BJ, Shaver SW (1988) The representation of sound frequency and space in the midbrain of the saw-whet owl. Soc Neurosci Abstr 14:1095

15 Tuning of Visuomotor Coordination During Prey Capture in Water Birds

G. KATZIR

15.1 Introduction

In order to perform the necessary motor activities for prey capture, a predator must obtain spatio-temporal information on the orientation and distance of the prey, and on the relative motion between it and the prey (cf. Chap. 13). Predatory acts are useful in the study of visuomotor coordination mechanisms and the means by which they are tuned to achieve optimal direction and timing of movements. "Sit and wait" predatory patterns are especially useful, as they involve relatively distinct motor patterns, are of short duration and are relatively easy to record. A "sit and wait" predator stalks prey slowly, or even keeps motionless until the prey is close enough, and then performs a fast capturing movement (O'Brian et al. 1990). This is often a "point of no return", after which prey either is captured or escapes. For example, after slowly approaching and aiming at an insect, a chameleon (*Chameleo* spp.) rapidly "shoots" its long tongue at it. To avoid over- or undershooting of the tongue, distance must be accurately estimated prior to "shooting", as no motor corrections are performed later (Harkness 1977; Flanders 1985). Predators such as toads, preying mantids and herons, which capture prey using a rapid movement, must perform in a similar manner (Mittelstaedt 1957; Curio 1976; Ewert 1980).

Alternatively, predators such as raptors (hawks, falcons) actively follow their prey (Howland 1974; Weihs and Webb 1984), often at high speeds. Here, monitoring of relative speeds and paths is needed to enable interception and estimation of the moment of contact. This is especially important if the prey is on or close to the ground. The visuomotor mechanisms involved in approaching prey rapidly may not differ from those of alighting on a bough, yet the velocities and risks involved may be greater (Davies and Green 1990, 1991; see Chaps. 13 and 16). The prey, unlike the bough, will attempt to detect the predator's approach and to escape.

Avian species which prey on fish rely on mechanisms of visuomotor coordination which must deal with the particular visual problems created at the interface between two optical media, air and water. Avian predation on fish ("piscivoury") occurred as early as the Cretaceous (Swinton 1975) and is

Department of Biology, the University of Haifa at Oranim, Tivon 36910, Israel

M.N.O. Davies and P.R. Green (Eds.)
Perception and Motor Control in Birds

presently found mainly in Sphenisciformes (penguins), Pelecaniformes (e.g. gannets, cormorants), Procellariiformes (e.g. petrels), Ciconiiformes (e.g. herons), Charadriiformes (e.g. terns), Coraciiformes (e.g. kingfishers) and Falconiformes (osprey, sea eagles) (Welty and Baptista 1988). Piscivorous birds are mostly grouped on the basis of their feeding and fishing techniques (Ashmole 1971; Alerstam 1990, Fig. 1). For the present discussion it seems more appropriate to divide birds into two groups on the basis of the visual/optical environment within which they operate (Fig. 15.1; Lythgoe 1979). The first group, "underwater feeders", consists of birds which dive from the surface into the water where they pursue fish. The second group, "surface plungers and strikers", includes either birds which plunge dive from the air and may continue their pursuit actively (e.g. gannets), or birds which wade in shallow water and strike with their bill at submerged prey (e.g. herons). Although active pursuit of prey is found predominately among underwater feeders, it is also exhibited in certain herons and other species of the second group. "Sit and wait" predation is found in some surface plungers and strikers, such as herons, egrets and kingfishers.

Underwater feeders such as penguins, cormorants and diving ducks, operate visually in both air and water, and must therefore shift from aerial to aquatic vision (Walls 1967). In particular, as discussed in Sect. 1.4.2.2, these birds suffer from loss of the refractive power of the cornea when the eyes are submerged (Walls 1967; Howland and Sivak 1984; Martin and Young 1984; Sivak et al. 1985). Flattening of the cornea, found in penguins (see Sect. 1.4.2.2) provides one

Fig. 15.1. A selection of avian species which feed on fish or other aquatic organisms. Surface plunge divers and strikers: *1* gull; *2* storm petrel; *3* heron; *4* kingfisher; *5* tern; *6* pelican; *7* gannet. Underwater feeders: *8* puffin; *9* penguin (After Ashmole 1971; Alerstam 1990)

solution to this problem, by minimizing the loss of accommodative power accompanying submergence. An alternative solution is to achieve a large range of accommodation, which can be up to 50 D or more in cormorants and diving ducks (Levy and Sivak 1980). In these birds, contraction of the iris sphincter results in the formation of a rigid disc with a central pupil, while contraction of the ciliary muscle pushes the malleable lens against this disc. The central lens bulges through the pupil, allowing the birds to accommodate the 70–80 D needed to focus light on the retina when the eye is submerged (Sivak et al. 1985).

In contrast, surface plungers and strikers locate prey and commence their capturing movements when their eyes are above the water. They must therefore cope with visual problems at the interface between the two optical media. It is on these problems that this chapter will focus.

15.2 Surface Plungers and Strikers

Birds which plunge dive or strike at their submerged prey must estimate the prey's position, movement and distance to the substrate, as well as their own dive or strike speed and the moment of contact with the prey. Unique to this situation is the need to estimate prey depth underwater, as well as the moment of their own contact with the water surface (Lee and Reddish 1981). Moreover, these have to be performed against the obstructing effects of light reflection and refraction at the water surface (Lythgoe 1979).

Daylight incident on the water surface is either reflected back into the air, or penetrates the surface and is refracted (Jenkins and White 1976). Light reflection and refraction have important consequences for birds that need to look down into the water. Reflection from the sky at the water surface imposes a disabling glare that reduces visual contrasts. Refraction displaces and distorts the image of a submerged object. Moreover, water surface movement creates "dynamic lenses", affecting both reflected and refracted light rays (Schenck 1957; McFarland and Loew 1983; Loew and McFarland 1990). Due to these factors, looking from the air into the water is most difficult, as light becomes an important *obstacle* to vision. These visual problems may thus affect visually guided behaviour in birds more than in any other group of terrestrial vertebrates.

15.2.1 Light Reflection

The radiance from the water surface comprises light reflected from the sun and sky and from the inherent radiance reflected back from the water volume itself. On a calm surface the brightest feature is the reflection of the sun. Reflection is greater at oblique angles of incidence, causing the reflected skylight to be brighter near the horizon than when looking directly downwards (Lythgoe 1979).

Spectral reflectance of the surface glare is similar to that of the sky. Light upwelling from the water has a spectral radiance dominated by the absorption and scatter of the water itself. Water begins to absorb light strongly at wavelengths greater than 575 nm, where reflections from the sky are therefore often brighter than the upwelling radiance of the water. To see well through the surface a visual mechanism should be most sensitive to wavelengths in which the sky reflections are relatively poor and the upwelling light is rich. Visibility through the surface would be best at 425–525 nm for blue water, at 500–550 nm for blue-green water and at 520–570 nm for green water. A visual mechanism most sensitive to wavelengths greater than 575 nm will allow clear viewing of very shallow, but not deep, objects (Lythgoe 1979).

15.2.2 Light Refraction

Light rays are refracted at the air-water interface in accordance with Snell's law:

$$\mathrm{Sin}\, i/\mathrm{Sin}\, r = N, \tag{15.1}$$

where i is the angle of the incident ray, r is the angle of the refracted ray relative to the normal, and N is a constant (1.33 for pure water). Due to refraction, an underwater prey will appear to a bird in the air somewhere along the line of refraction, and so somewhat higher than it actually is (Fig. 15.2a; Jenkins and White 1976; Horvath and Varju 1990). Therefore, to hit an underwater prey accurately, corrections for the disparity are required.

As well as displacing and distorting the aerial image, light refraction at the water surface has a magnifying effect, so reducing the apparent brightness of a submerged object. Furthermore, it affects the directional distribution of underwater light through water surface movement (Dill 1977; Lythgoe 1979).

15.2.3 Surface Movement

Surface movement affects the aerial image in several ways. First, when the surface is moved by wind causing ripples or waves, reflected light is fragmented into a multitude of glitter points. Second, surface movement produces continuous changes in light penetrating the water. The wavelets act as cylindrical lenses of constantly changing curvature and project a network of bright light into the water (Schenck 1957; Fig. 15.2b). Thus in clear shallow water, periodic bright bands of light produced by the refraction of sunlight at the wavy surface sweep across the bottom at surface wave velocity (McFarland and Loew 1983; Loew and McFarland 1990). Third, the constantly changing "surface lenses" cause the image of a stationary subsurface object to be continuously distorted and displaced (in both the vertical and horizontal planes).

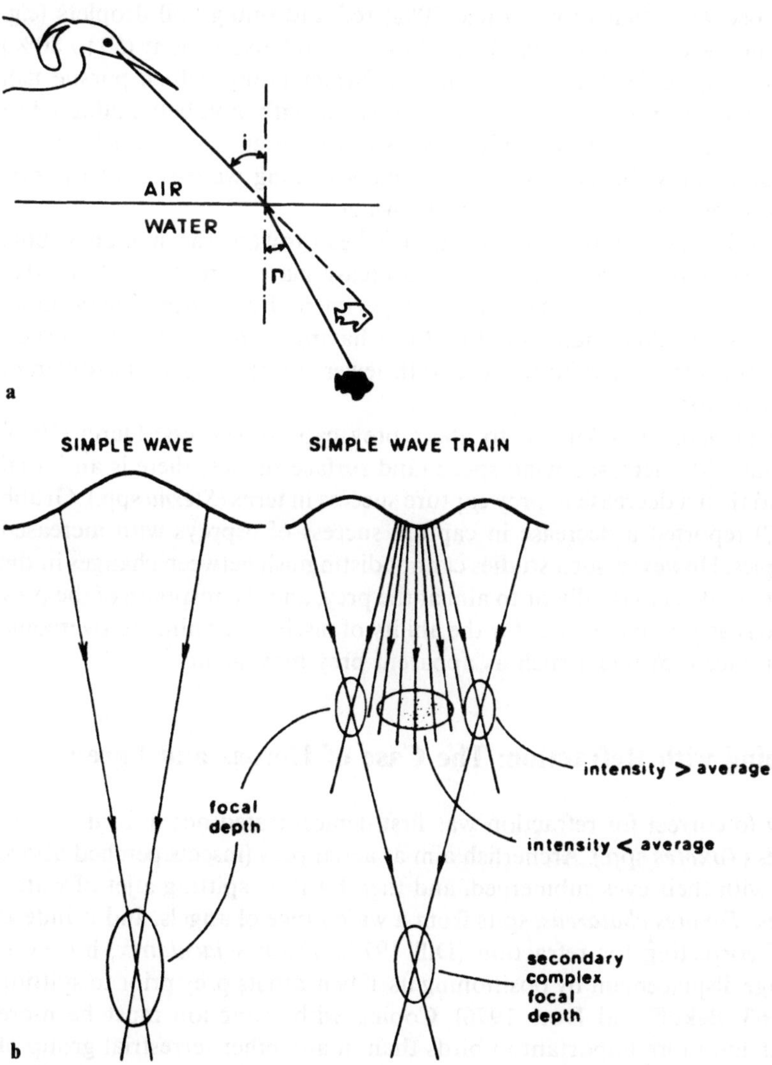

Fig. 15.2. **a** Disparity of the prey's image due to light refraction at the air/water interface. Light is refracted in accordance with Snell's Law. As $i > r$ the apparent image (*white fish*) will be seen along the line of refraction, above the real fish (*black*) (Katzir and Intrator 1987). **b** The formation of "dynamic lenses" by surface waves, acting to focus and defocus light penetrating the surface (After Loew and McFarland 1990)

15.2.4 Coping with Light Reflection and Surface Movement

It has been suggested that the oil droplets contained in avian retinal cones are used for overcoming reflected light. These droplets contain dissolved carotenoid pigments, and may act as chromatic filters (cf. Martin 1985). Sea birds were divided by Muntz (1972) into two groups, based on the distribution of retinal oil droplets; those with relatively few (ca. 20%) red and orange oil droplets (e.g. razorbills, shearwaters, cormorants) and those with relatively many (50 to 80%) such droplets (e.g. gulls, terns). Birds of the former group, which pursue fish underwater, may be more sensitive to shorter wavelengths and better adapted to seeing objects well beneath the surface. Birds of the latter group, which plunge dive to capture prey, benefit from the richness in long wave-sensitive cones, which help them to see better through the water.

There are indications that, by tilting their head toward the sun, great blue herons (*Ardea herodias*) behave so as to decrease glare (Krebs and Partridge 1973). However, at present, there is no experimental evidence (behavioural, psychophysical or other) indicating that birds having more oil droplets in their retinas perform differently from those with fewer oil droplets under different chromatic conditions.

Surface movement is known to affect prey capture success. Dunn (1973) reported that with increased wind speed (and surface ripples) there is an initial increase and then a decrease in prey capture success in terns (*Sterna* spp.). Grubb (1976, 1977) reported a decrease in capture success of ospreys with increased surface ripples. However, such studies cannot distinguish between changes in the birds' ability to detect visually or to aim at the prey, and the response of the prey itself. There is as yet no evidence for the ability of piscivorous birds to overcome effects of surface movement such as apparent prey movement.

15.3 Coping with Refraction: The Case of Herons and Egrets

The ability to correct for refraction was first demonstrated not in birds but in archerfishes (*Toxotes* spp.). Archerfish aim at aerial prey (insects perched above the water) with their eyes submerged, and then hit it by spitting a jet of water. One species, *Toxotes chatareus*, spits from a wide range of angels, and is indeed capable of correcting for refraction (Dill 1977). *Toxotes jaculatrix*, however, avoids image displacement by positioning itself beneath its prey prior to spitting (Luling 1963; Bekoff and Dorr 1976). Coping with refraction must be more widespread and more important in birds than in any other terrestrial group of vertebrates. It has been demonstrated to date only in the reef heron, the little egret and the squacco heron (Katzir and Intrator 1987; Katzir et al. 1989; Lotem et al. 1991; Katzir et al. in prep.) as well as in the pied kingfisher, *Ceryle rudis* (Labinger et al. 1991; Katzir and Shechtman in prep.).

Egrets and herons commonly stalk underwater prey while they wade in shallow water. To capture prey an egret performs a rapid strike, which commences while its head is held above the water (Hancock and Kushlan 1984). If the egret's eyes are not directly above the prey, the prey's apparent and real positions will be disparate. The magnitude of the disparity will be determined by the egret's eye position relative to the prey at the moment of strike. For example, an egret aiming at a small prey that is 20 cm underwater, while its eyes are 20 cm above the water and at a horizontal distance of 20 cm, faces a vertical disparity of approximately 8 cm between real and apparent prey positions. Can egrets compensate for disparity caused by refraction, or do they avoid the problem by, for example, striking only straight downwards?

15.3.1 Prey Capture by Little Egrets in the Field

In little egrets (*Egretta garzetta garzetta*) the proportion of successful strikes observed in the field varies between 40 and 90% (Hafner et al. 1982). Observations on strike angles of little egrets foraging in shallow fish ponds indicate that nearly one-half of the strikes (43%; n = 2244; 131 focal birds) were successful (Lotem et al. 1991). The strike angle was vertical in only 25% of the cases, while the majority (64%) were intermediate, and the remainder (11%) were acute (see Fig. 15.3). If there is no correction for refraction, a lower capture success would be expected at more acute angles and at deeper strikes, where the disparity is greater. However, the results of Lotem et al. (1991) suggested the opposite; capture success increased with increasing acuteness of strike angle, while strike depth had no apparent effect on success. Light refraction therefore appears to have little effect on little egrets' capture success and the birds are probably able to correct for the disparity between real and apparent prey positions. The increase in capture success at more acute strike angles may be because fish are less able to detect the bird's strike; an aerial predator close to the horizon will

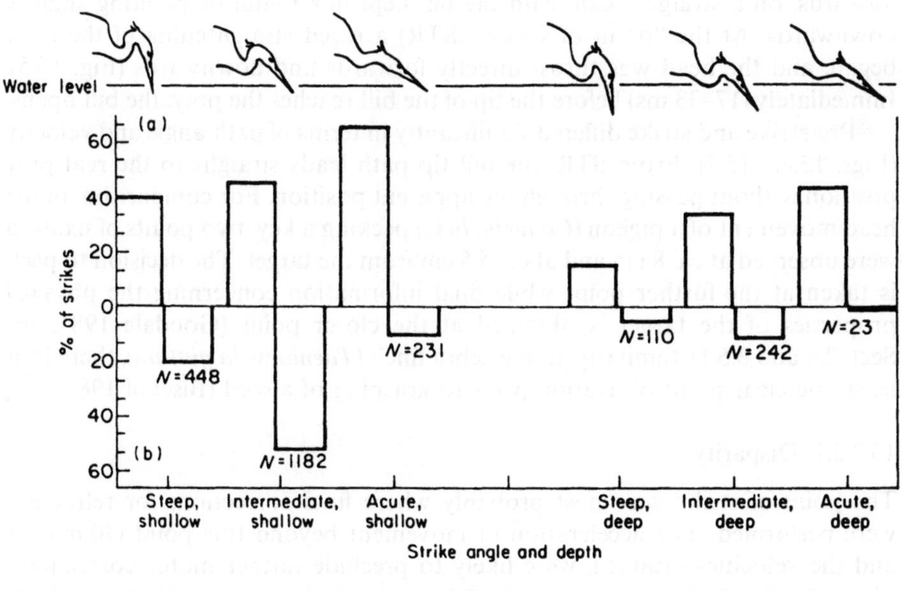

Fig. 15.3. a Capture success by little egrets at different combinations of depth and strike angle (percentage of strikes on which prey was captured). **b** Frequency of occurrence of the different strike angle/depth combinations (percentage of all strikes observed); N number of strikes observed (Lotem et al. 1991)

appear dimmer and smaller relative to a similar object overhead, being at the edge of the "Snell's window" (Walls 1967).

Field studies are handicapped by problems in measuring important parameters. The magnitude of the refraction problem in the above example cannot be computed, as the relative spatial positions of the bird and the prey, and therefore the visual angles involved, were unknown. Moreover, physical factors (e.g. water turbidity) and prey factors (e.g. species, size, colour and movement) are important in determining prey capture success, and in the field these factors obscure the effect of disparity itself. For these reasons, laboratory experiments must be performed.

15.3.2 Prey Capture by Reef Herons in Captivity

Visually guided behaviour during prey capture of reef herons (*Egretta gularis schistacea*)[1] was tested in captivity. The herons were tested in a setup in which small, stationary, underwater "prey" (fish) could be presented at different sighting angles (Katzir and Intrator 1987; Katzir et al. 1989; Fig. 15.4). When striking at stationary, submerged prey, reef herons literally did not miss over a wide range of sighting and striking angles. This clearly demonstrates their ability to cope with the disparity between real and apparent prey images.

In the movement of the heron's head, two phases were distinguished; a "pre-strike" and a "strike" (Figs. 15.5, 15.6a). During pre-strike, the head moved forwards, on a straight path, with the bill kept horizontal or pointing slightly downwards. At the "point of strike" (STR) a rapid straightening of the neck began, and the head was thrust directly forwards and downwards (Fig. 15.5). Immediately (17–33 ms) before the tip of the bill reaches the prey, the bill opens.

Pre-strike and strike differed significantly in terms of path angle and velocity (Figs. 15.6a, 15.7). From STR, the bill tip path leads straight to the real prey position without passing through the apparent position. For comparison, in the head movement of a pigeon (*Columba livia*) pecking a key, two points of fixation were observed at ca. 8 cm and at ca. 5.5 cm from the target. The decision to peck is taken at the further point while final information concerning the physical properties of the target is obtained at the closer point (Goodale 1983; see Sect. 2.4 and 3.5.1). Similarly, in the zebra finch (*Taeniopygia guttata*), there is at least one clear point of fixation prior to grasping of a seed (Bischof 1988).

15.3.2.1 Disparity

The point of strike was most probably where final corrections for refraction were performed. The acceleration of movement beyond this point (70 m s^{-2}), and the velocities attained, were likely to preclude further motor corrections.

---

[1] The reef heron is considered by some as a subspecies of the little egret (cf. Hancock and Kushlan 1984)

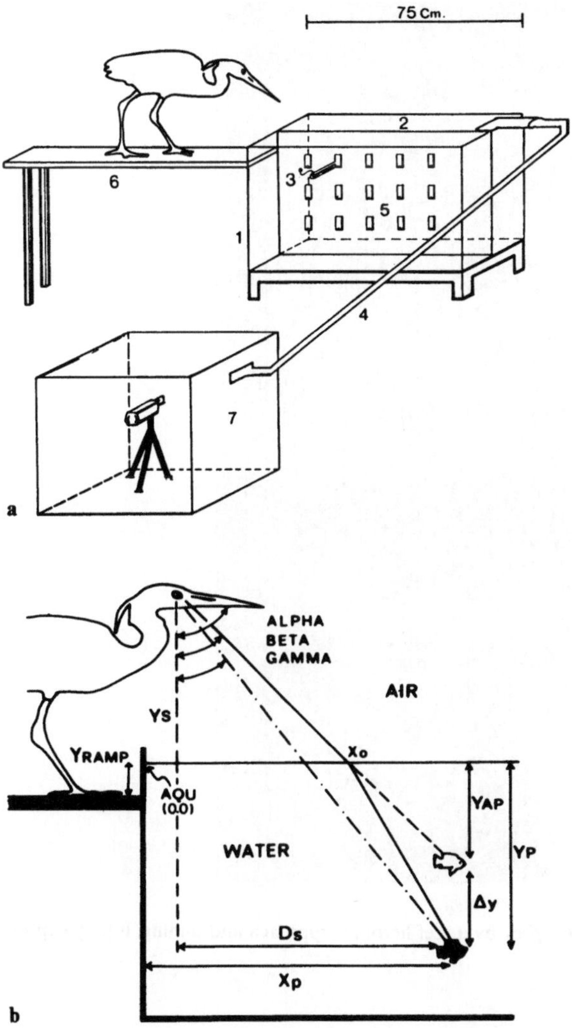

Fig. 15.4. **a** Reef herons' experimental setup: *1* aquarium; *2* feeder; *3* perspex rods for holding prey; *5* openings for prey presentation; *6* ramp; *7* hide. **b** Parameters measured from films, with X-axis determined by the water surface: Y_p prey depth; *Ds* horizontal distance between eye and prey at the moment of strike; *Alpha* angle between eye-bill line and the vertical; *Beta* angle of sighting of apparent prey relative to the vertical; X_0 the horizontal distance between the real prey position, and where it will appear on the surface; Y_{ap} apparent prey position ($Y_{ap} = X/\tan i$). Apparent prey position was regarded here has being directly above the real prey. (Katzir and Intrator 1987)

Also, underwater vision must have been impaired by the nictitating membranes over the eyes, and by air bubbles drawn in by the penetration of the bill into the water (see Fig. 15.5b). At STR the eye – bill angle (Alpha) and the eye to apparent prey angle (Beta) were correlated with prey depth ($r = 0.804$ and $r = 0.827$ respectively, $p < 0.01$), and with prey distance ($r = 0.889$ and $r = 0.852$

a

b

Fig. 15.5a, b. Striking of underwater prey by a reef heron; **a** approach and aiming; **b** prey capture (Katzir and Intrator 1987)

respectively, $p < 0.01$), as well as with each other ($r = 0.977$, $p < 0.01$; Fig. 15.8a). The line of sighting to the prey was approximately 7.5° below the bill. In other words, if the eyes were immobile in their orbits, the image of the prey would fall on a given area of the retina during consecutive STRs.

In most birds, the visual fields are largely monocular and only a limited portion in front of and below the bill is binocular (cf. Martinoya et al. 1981; Martin 1986; Sect. 1.6). In the pigeon, an image from the binocular portion falls on the "red field" or "area dorsalis" of the retina (upper temporal quadrant). This area was suggested by different authors as the area of acute, close range binocular vision (cf. Martinoya et al. 1981; Goodale 1983; Martin 1986; Bischof 1988; Sect. 3.4.2). While pecking at a seed, or probing for a worm, the eyes of starlings converge simultaneously and sighting of the prey is performed beneath

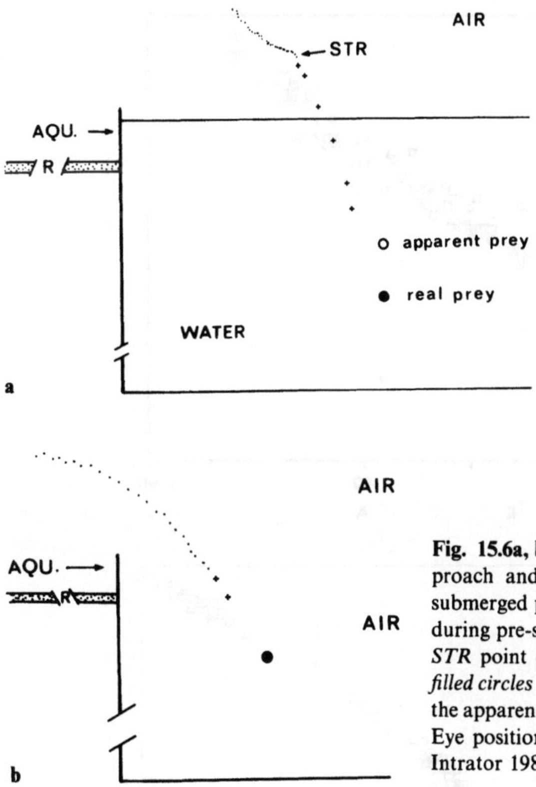

Fig. 15.6a, b. Consecutive eye positions during approach and striking at **a** submerged and **b** unsubmerged prey by a reef heron. *Dots* eye positions during pre-strike; + + eye positions during strike; *STR* point of strike; *R* ramp; *Aqu* aquarium wall; *filled circles* real prey position. In **a**, the *open circle* is the apparent prey position, calculated at point *STR*. Eye positions are at 18.2 ms intervals (Katzir and Intrator 1987)

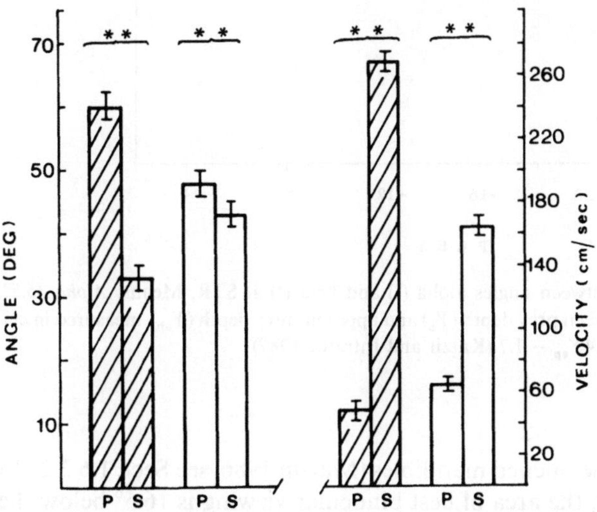

Fig. 15.7. Mean path angles (*left*) and velocities (*right*) during pre-strike (*P*) and strike (*S*). *Cross-hatched bars* submerged prey; *open bars* unsubmerged prey. *Vertical lines* are standard errors. (Katzir and Intrator 1987)

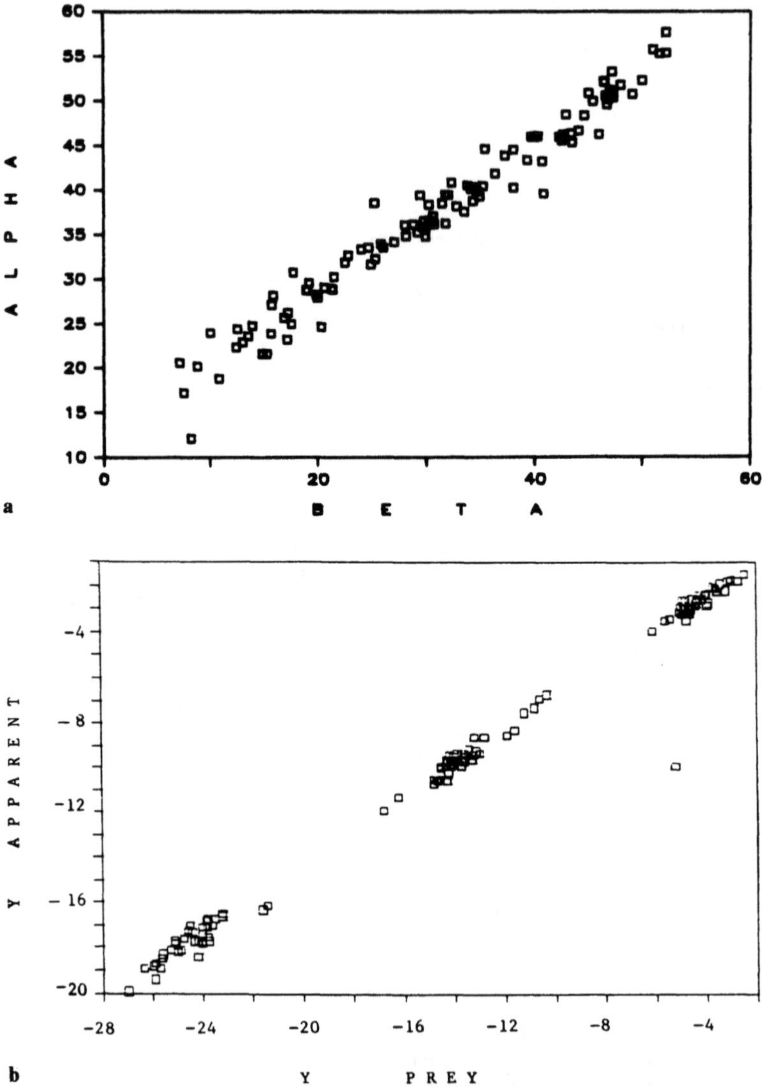

Fig. 15.8. **a** The relationship between angles alpha (°) and beta (°) at STR. Means: *Alpha* 35.5°, *Beta* 28°. **b** The relationship between prey depth (Y_p) and apparent prey depth (Y_{ap}), measured in cm, at STR. Regression line: $Y_p = 1.4Y_{ap} - 1.7$ (Katzir and Intrator 1987)

the bill, or even between the opened mandibles (Martin 1986; see Sect. 1.6.2.2). In zebra finches, for example, the area of best binocular viewing is 16.5° below the bill yet, unlike the pigeon, they tend to fixate their food monocularly (Bischof 1988). Scanning of distant objects in most avian species studied is performed by each eye independently, and the central fovea is most probably the site of focusing distant objects in the monocular field (cf. Moroney and Pettigrew 1987).

At present it is not clear if herons, upon reaching STR, fixate monocularly or binocularly. Also, the retinas of these species have not been studied in detail, and the observed "line of sighting" cannot yet be related to specific retinal areas. Preliminary measurements of the visual fields and the degree of eye movements in four heron species were performed by Martin and Katzir (in prep.). These were the reef heron, the squacco heron (*Ardeola ralloides*), the cattle egret (*Bubulcus ibis*) and the night heron (*Nycticorax nycticorax*). The largest degree of binocular overlap occurred in these species below the bill. Eye movements showed marked interspecific difference, being greatest in the little egret, and least in the night heron.

The finding that seems of major importance is that at STR apparent prey depth and real prey depth were linearly correlated (Fig. 15.8b). This unique feature of STR was supported by analysis of arbitrary points along the herons' head path (e.g. 10, 20 or 30 frames prior to STR) which revealed no other point with these relationships. As gape size was not correlated with prey parameters, the herons most probably did not compensate for greater disparities by striking with a wider gape.

15.3.2.2 Striking at Prey Not in Water

The visuomotor patterns observed during strikes at submerged prey raise a question; do reef herons perform differently when prey is not in water? Observations indicate that strikes at unsubmerged prey did indeed differ markedly from those at submerged prey. The path of the bill tip here was directed at the prey from relatively far away. Although a point of strike was observed, the difference between pre-strike and strike was much smaller. Eye path during strikes followed the straight line between STR and prey position (Fig. 15.6b). Moreover, at STR, the angle of sighting of unsubmerged prey was merely 2.5° below the eye – bill line (Katzir and Intrator 1987).

15.3.3 A Model for Coping with Light Refraction and Its Verification

Reef herons determine the prey's real position during their approach and strike at it directly. They do not merely avoid the problem of refraction by striking vertically downwards. How do they correct? An attempt to answer the question was made by Katzir and Intrator (1987). They formulated a model which incorporates the empirical finding that real and apparent prey depths are linearly correlated ($Y_p = 1.4Y_{app} - 1.7$) and the index of refraction ($n = 1.33$) (Fig. 15.9). At STR real prey depth is 1.4 times greater than apparent prey depth. If one chooses an arbitrary prey position, then the respective apparent position can be calculated. For each pair of real and apparent prey positions, there is a point on the water surface (X_0) which also fulfils Snell's law [Eq. (15.1)], and all

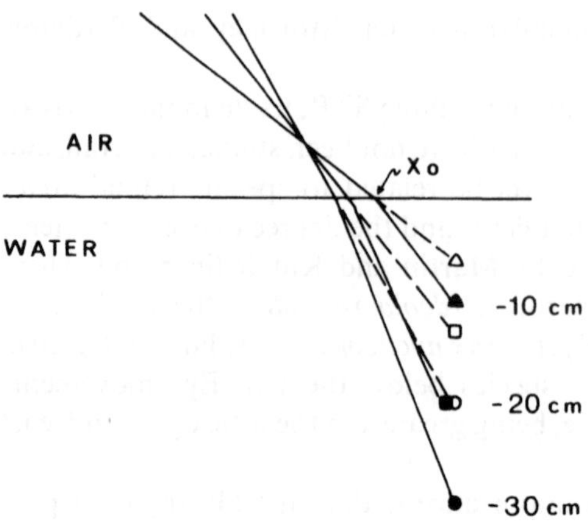

Fig. 15.9. The lines predicted by the model for which $Y_p = 1.4Y_{ap} - 1.7$ and $\sin i/\sin r = 1.33$ (see text). *Broken lines* trajectories of the refracted rays; *black figures* real prey positions; *open figures* corresponding apparent positions (Katzir and Intrator 1987)

the points on the line extrapolated from Y_{app} through X_0 conform to both the index of refraction and the empirical finding (Fig. 15.9). The lines connecting different Y_{app} positions and X_0 tend to converge. If a heron sights prey along any one line, it may apply its "rule of thumb for correction" and strike appropriately at the real prey position. It is not apparent, however, how a heron determines that it is on the right line of sighting. In captivity it is able to determine it correctly, achieving a 100% capture success. In the field it may be hindered by prey movement or surface movement, so that attempting to keep the correct Y_p/Y_{app} ratio leads to a lower success. Interestingly, Dill (1977) demonstrated that at the point at which the archerfish "decides" to spit its jet of water, a linear relationship exists between the apparent elevation of the prey relative to the fish's eyes, and the true elevation of the prey relative to the nose.

The model described above predicts that reef herons will be more limited in their ability to correct for refraction when viewing prey at very acute angles. This does not contradict the results presented in Sect. 15.3.1 (Fig. 15.3) of a higher capture success in the field for acute strikes. The herons must obviously find the balance between accuracy of striking and the probability of being detected by the fish. Moreover, it is expected that the frequency of misses should be related to the differences between the expected and the calculated ratios of prey depth to apparent prey depth. Experiments in which reef herons could view and strike at stationary submerged prey at acute angles only (Fig. 15.10a) did indeed demonstrate that the proportion of misses was markedly greater than in the unrestricted situation (Katzir et al. 1989). As sighting angle became more acute, the frequency of missing the prey increased, caused by the bill tips sliding just above or just below the prey. Prey depth (Y_p) and apparent prey depth (Y_{app}) were correlated within successful strikes ($r = 0.899$, $p < 0.05$) as well as within unsuccessful strikes ($r = 0.969$, $p < 0.01$). The lines' equations are $Y_p = 1.35\,Y_{app} - 3.77$ for successful strikes (compared with $Y_p = 1.4Y_{app} - 1.7$ for the unrestricted situation) and $Y_p = 1.13\,Y_{app} - 7.5$ for unsuccessful strikes (Fig. 15.10b). As predicted, for any given prey depth, the larger the difference between the observed and the predicted apparent prey depth, the higher the probability of missing a prey.

15.3.4 Prey Capture in Cattle Egrets and Squacco Herons in Captivity

Species, such as little egrets and reef herons, which feed predominantly on fish, frequently need to cope with refraction. At the other end of the scale, cattle egrets feed on terrestrial vertebrates and invertebrates and rarely if ever on fish, and are most probably the only terrestrial forager heron species. Are cattle egrets capable of coping with refraction? Do they have "rules for correction" as do the reef herons? In an attempt to answer these questions Katzir et al. (in prep.) compared visually guided capture of prey in cattle egrets and in squacco herons. Squacco herons are roughly the size of cattle egrets, yet are strictly aquatic in habit.

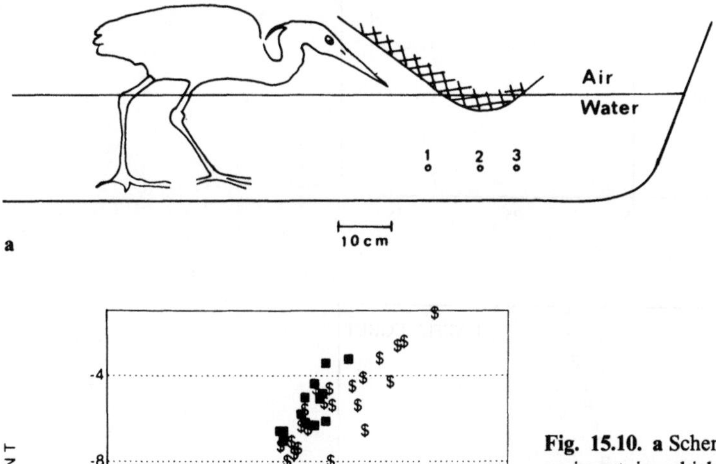

a

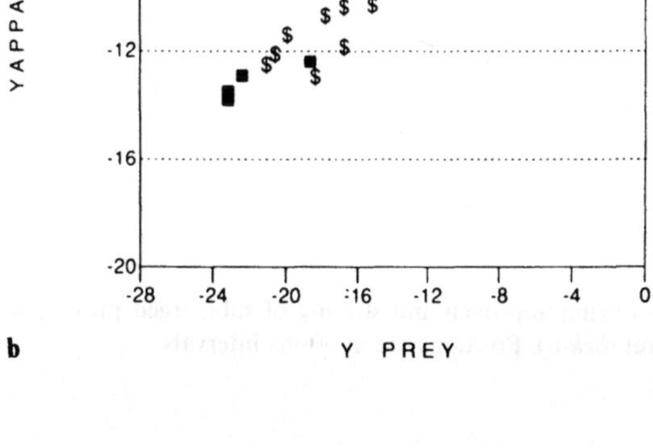

b

Fig. 15.10. a Schematic view of the experiment in which reef herons could view and strike at submerged prey at acute angles only. *1*, *2* and *3* are prey presentation points of increased acuteness. **b** The relationship between prey depth (Y_p) and apparent prey depth (Y_{ap}) at STR for reef herons restricted to strikes at acute angles. $ Successful attempts; *solid squares* unsuccessful attempts (Katzir et al. 1989)

The experiments, similar to those with reef herons (Sect. 15.3.2), indicate that the visuomotor patterns during prey capture in cattle egrets and in squacco herons differ in several respects. First, while inexperienced cattle egrets miss submerged prey frequently, inexperienced squacco herons strike accurately from their very first trials. With repeated prey presentations there is a clear improvement in the cattle egrets' capture success, implying a process of learning. This is especially interesting, as the cattle egrets do not miss prey presented unsubmerged. Second, cattle egrets show neither a clear change in head path angle, nor a change in acceleration, to define a "point of strike" (Fig. 15.11a). In contrast, the head path of squacco herons showed a distinct "point of strike" (STR), very similar to that of reef herons (Fig. 15.11b). Moreover, there are indications that in squacco herons apparent prey depth and real prey depth at STR are linearly correlated (Fig. 15.12). It is therefore possible that squacco

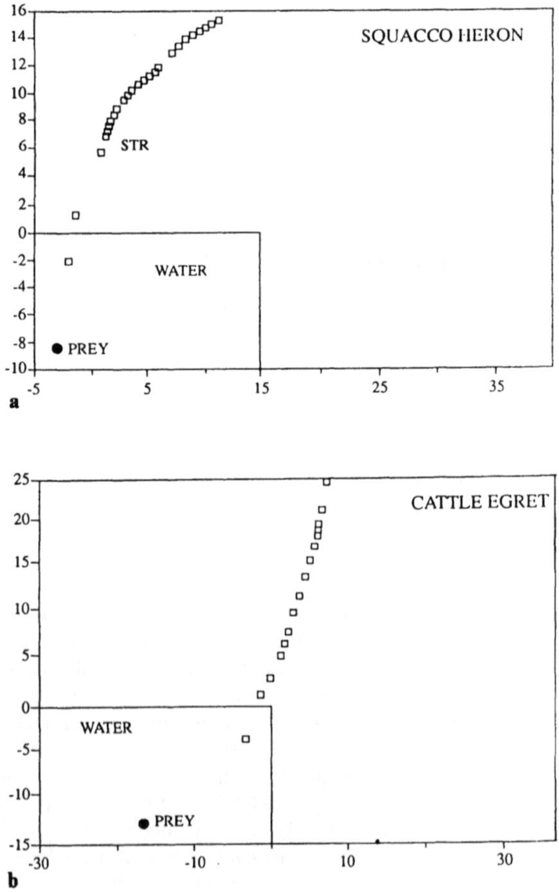

Fig. 15.11a, b. Consecutive eye positions during approach and striking of submerged prey by a squacco heron (*above*), and by a cattle egret (*below*). Positions are at 40-ms intervals

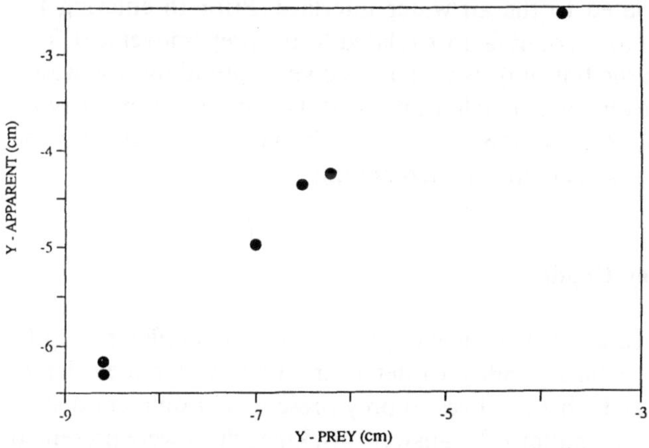

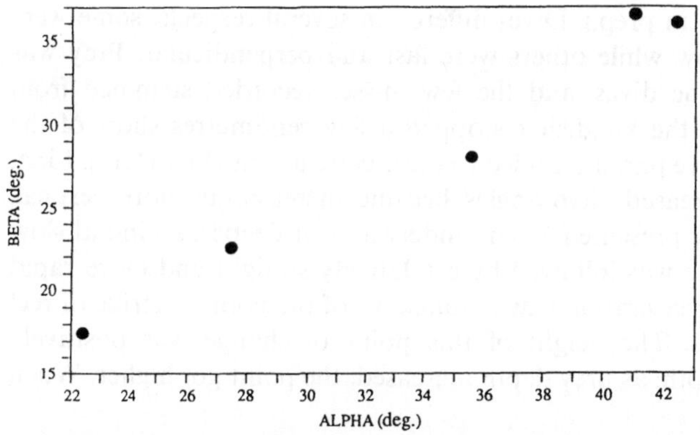

Fig. 15.12. *Above* The relationship between prey depth (Y_p) and apparent prey depth (Y_{ap}) at the point of strike (STR) for a squacco heron. *Below* The relationship between angles alpha (°) and beta (°) at STR, for the same species

herons and reef herons are employing a similar technique to overcome refraction. At STR, the squacco herons kept their angle Beta (eye to apparent prey) somewhat steeper (10° below the bill) than the reef herons approximately (6° – 7°).

15.4 Visually Guided Prey Capture in Pied Kingfishers

Pied kingfishers (*Ceryle rudis*) capture fish by plunge diving into the water. This is mostly done from hovering flight several metres above the water, or from a perch (Douthwaite 1976; Reyer et al. 1988). As it hovers, the kingfisher keeps its head stable relative to movements of its body and wings (Lee 1980; Lee and Young 1986). During prey search, detection, hover and dive, it is confronted with visual problems related to the air/water interface. Prior to entering the water it must also take into account factors related to the prey (movement, size, depth and distance from the bottom), as well as its own height above the water, its dive speed (to avoid over- or undershooting) and its wing position (to avoid physical damage; Lee and Reddish 1981). Much of this must be performed in the hovering position, which is energetically demanding.

15.4.1 Estimation of Prey Depth

Experiments in which pied kingfishers dive to prey presented at different depths provided evidence for the birds' ability to determine underwater prey depth. One such experiment asked whether dives to prey presented at different depths are different in their physical pattern. To answer this, kingfishers were presented with small, stationary prey at depths ranging from 0 to 60 cm underwater (Katzir and Shechtman in prep.). Dives differed in several respects: some were relatively acute and slow while others were fast and perpendicular. Prey was taken in virtually all the dives, and the few misses recorded stemmed from "undershooting", when the kingfisher stopped a few centimetres short of the prey. Examination of dive parameters led to three conclusions (Fig. 15.13). First, as prey depth was increased, dive angles became increasingly more vertical. Second, in dives to prey presented 15 cm underwater or deeper, an initial slow and rather curved phase was followed by a relatively straight and more rapid descent. This point of acceleration was reminiscent of the point of strike of reef and of squacco herons. The height of this point of change was positively correlated with prey depth: as prey depth increased, the point got higher. When

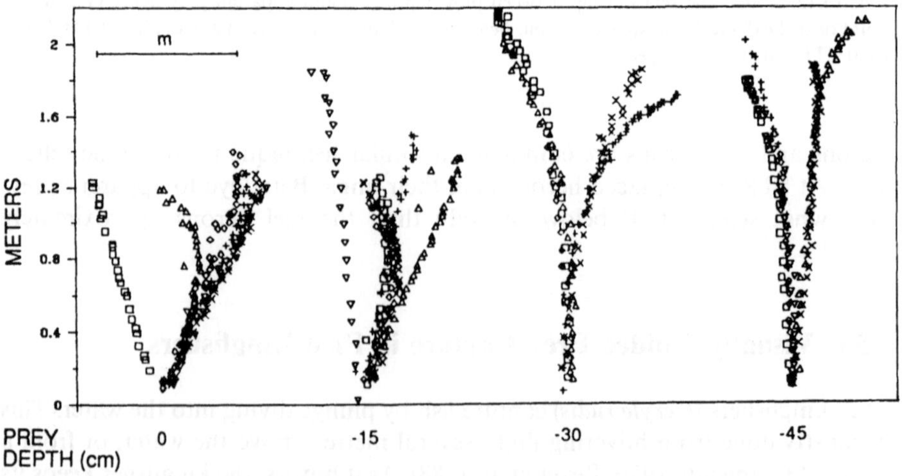

Fig. 15.13. Dive patterns of pied kingfishers to prey at different depths. Mean dive angles for prey depths 0, 15, 30 and 45 cm were 28°, 17°, 6° and 5° respectively

prey was presented at the water surface or 15 cm underwater, no such changes were observed, and dives were graded with frequent decelerations ("stops"). Third, as prey depth was increased, the speed of the final descent increased. When prey was at the surface, diving speed was approximately 2 $m\,s^{-1}$, while for prey 30 or 45 cm underwater, diving speeds showed a two-fold increase to over 4.5 $m\,s^{-1}$.

The observation that pied kingfishers do not refrain from diving at acute angles demonstrates that they are able to correct for refraction. The changes observed in dive angle and dive speed demonstrate their ability to determine prey depth, and to attain the necessary momentum needed to reach this depth. Determination of prey depth from a hovering position, several metres above the water, was further demonstrated by Labinger et al. (1991, in prep.).

15.4.2 Effect of Prey Movement on Capture Success

The experiments above indicate that herons and kingfishers are able to correct for the disparity between the real and the apparent positions of prey caused by light refraction. Moreover, in captivity, stationary prey was never missed in many hundreds of dives observed (Labinger et al. 1991; Katzir and Camhi 1993). Under natural conditions, however, pied kingfishers capture prey on only 10 to 50% of dives (Douthwaite 1976; Migongo 1978; Jackson 1984; Reyer et al. 1988).

Failure to capture fish in the wild may be due to different factors. Some are directly related to the predator's visuomotor performance, such as the abilities to cope with the light problems mentioned above or to intercept prey, while others may be related to the prey (i.e. size, colour, contrast, movement or ability to detect and avoid an approaching predator) or to the environment (i.e. amount of shelter, water turbidity). To investigate the effect of prey movement on capture success of a plunge diver, Katzir and Camhi (1993) presented captive pied kingfishers with single black mollies (*Poecilia sphenops*). The fish were presented in an aquarium with clear water and with minimal surface movement.

Kingfishers' success rate is capturing live mollies was approximately 50%, similar to the rate observed in the field. There was no difference in the depth and size between trials in which the fish was caught and those in which the fish escaped. The first observable prey movement occurred significantly earlier in escapes (when the kingfisher was a mean 0.025 s away) than in captures (0.01 s away). In some cases, the fish responded before or at the instant the bird hit water, indicating that it was using visual cues to detect the approaching bird (Fig. 15.14). Captures and escapes differed in the distance travelled by the escaping fish. In captures the distance travelled by the fish between its first observable movement and capture was 0.4 cm. In escapes the fish travelled 2 cm between its first observable movement and the moment the bird's bill reached its level. In captures, the initial fish swimming speed was approximately 24 $cm\,s^{-1}$ compared with 60 $cm\,s^{-1}$ in escapes. The two parameters of the kingfishers' dives measured (dive angle and dive speed) indicated no apparent difference.

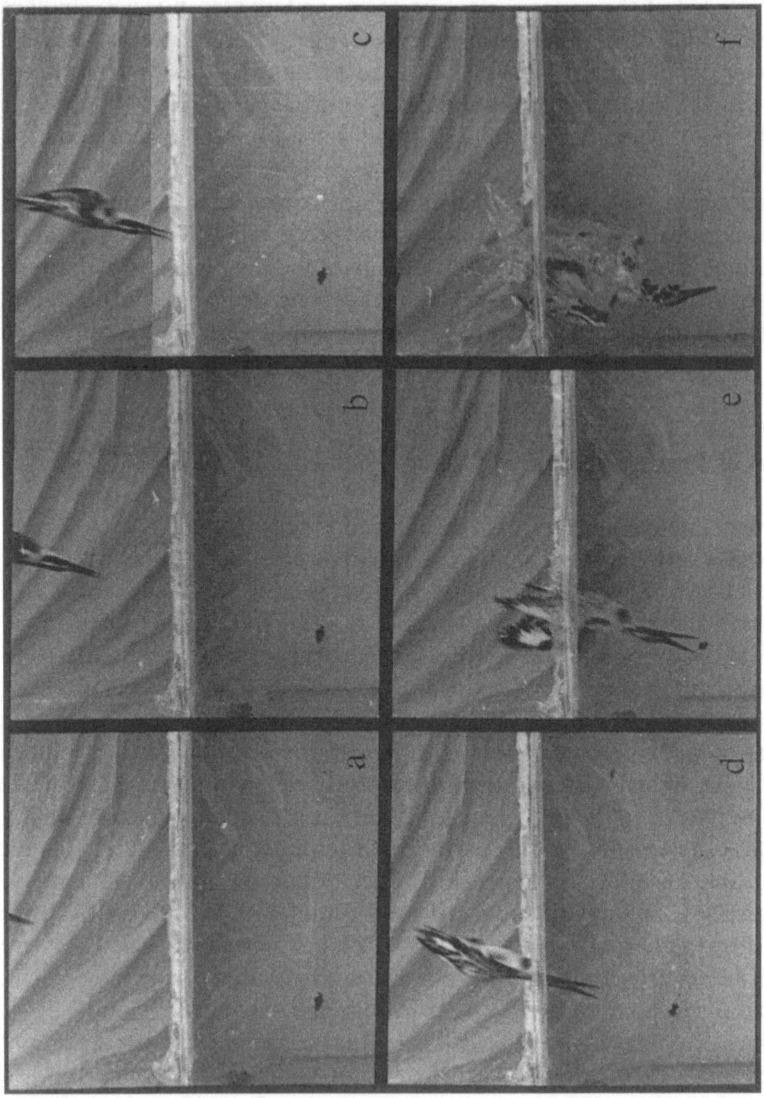
a
b
c
d
e
f

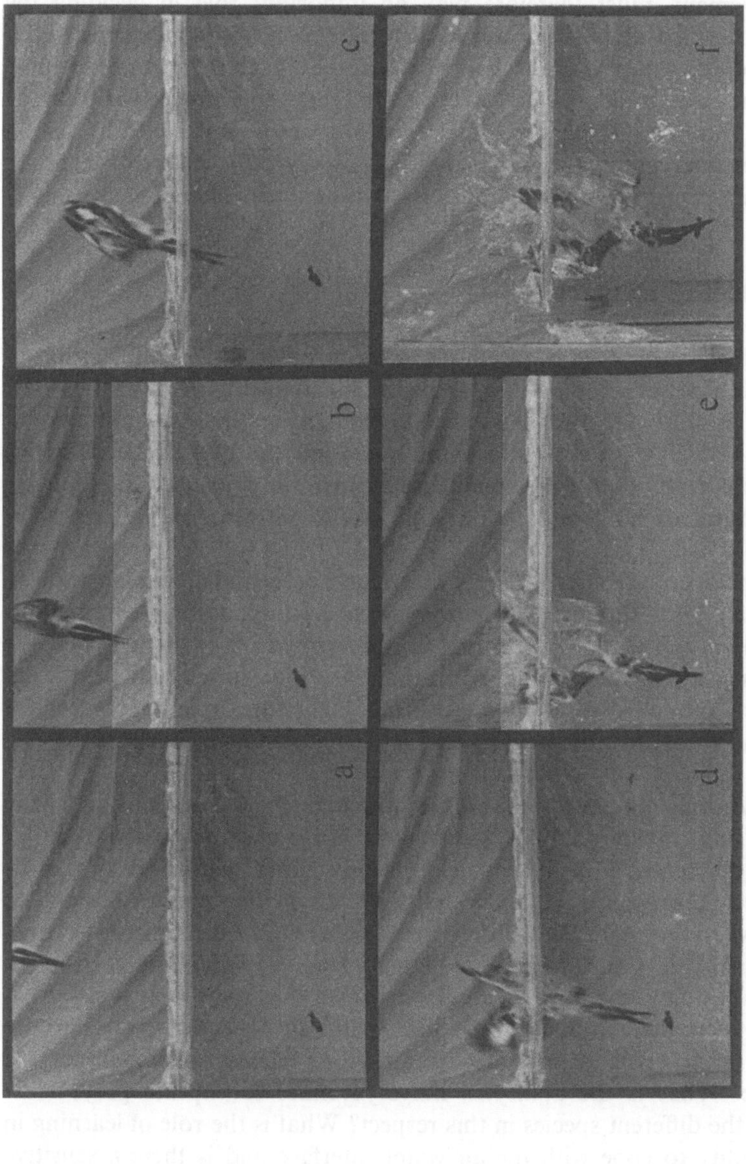

Fig. 15.14. *Above* Consecutive fish and pied kingfisher positions at termination of an unsuccessful dive, where the fish escaped. From *a* to *c* the fish was parallel to the front glass of aquarium. At *d* its body turned through 90°. At *f* the fish is near the base of the kingfisher's bill. *Below* Consecutive fish and kingfisher positions in a successful dive, where the fish was captured. The fish shows no apparent movement. Single frames are from a 16-mm film taken at 64 frames s^{-1}, and are 15.6 ms apart (Katzir and Camhi 1993)

Prey movement must therefore play an important role in deciding the outcome of the pied kingfishers' dives. Fish that responded only a few milliseconds earlier, and moved faster, had significantly higher chances of escaping. Displacing the body by somewhat less than a full register is a successful strategy for avoiding capture. There are no indications at present that the diving kingfishers are intercepting the prey and the question whether they or herons are capable of intercepting, and if so within what range, remains open.

15.5 Concluding Remarks

Piscivorous birds must cope with unique visual problems, ranging from problems of amphibious accommodation to problems created by light at the water surface. Studies on diving birds have shed much light on the structural adaptations of the optical apparatus, which can overcome the loss of the refractive power of the cornea when submerged. In contrast, only now is a preliminary picture emerging of how birds cope with visual problems at the air-water interface.

Several species of herons and egrets which have been studied demonstrate an ability to correct for the disparity between real and apparent prey positions imposed by light refraction. Corrections are preformed as the birds' eyes fixate at given points along their paths. Also, there are indications that pied kingfishers, which plunge dive from a hovering flight, are able to correct for refraction and to control their dive speed and angle so as to reach the correct prey depth.

Many questions remain open for future research. These questions may be of interest not only in the study of piscivorous birds but also in the study of interspecific differences in a finely tuned visually guided behaviour and in its ontogeny. Do herons cope with reflected light, or do they merely attempt to avoid it? How do they cope with surface movement? Do their retinas show areas of higher visual acuity and/or areas of higher carotene containing oil droplets, and if so do they correspond to the angles at which they sight their prey? What are the roles of monocular and of binocular vision during sighting and capture of the prey? Of no lesser importance are questions related to the ontogeny of prey capture. What is the ability of naive juveniles to capture prey? How different are the different species in this respect? What is the role of learning in the birds' ability to cope with the air-water interface and is there a sensitive period for this learning?

References

Alerstam T (1990) Bird migration. Cambridge University Press, Cambridge

Ashmole NP (1971) Sea bird ecology and the marine environment. In: Farner DS, King JR (eds) Avian biology, vol I. Academic Press, New York, pp 224–286

Bekoff M, Dorr R (1976) Predation by "shooting" in archerfish, *Ioxotes jaculatrix*: accuracy and sequences. Bull Psychon Soc 7:167–168

Bischof H-J (1988) The visual fields and visually guided behaviour in the zebra finch (*Taeniopygia guttata*). J Comp Physiol 163:329–337

Curio E (1976) the ethology of predation. Springer, Berlin Heidelberg New York

Davies MNO, Green PR (1990) Flow field variables trigger landing in hawk but not in pigeons. Naturwissenschaften 77:142–144

Davies MNO, Green PR (1991) The adaptability of visuomotor control in the pigeon during landing flight. Zool Jahrb Physiol 95:331–338

Dill LM (1977) Refraction and the spitting behavior of the archerfish (*Toxotes chatareus*). Behav Ecol Sociobiol 2:169–184

Douthwaite RJ (1976) Fishing techniques and foods of the pied kingfisher on Lake Victoria. Ostrich 47:153–160

Dunn E (1973) Changes in fishing abilities of terns associated with wind speed and sea surface conditions. Nature 244:520–521

Ewert JP (1980) Neuroethology. Springer, Berlin Heidelberg New York

Flanders M (1985) Visually guided head movement in the African chameleon. Vision Res 25:935–942

Goodale MA (1983) Visually guided pecking in the pigeon (*Columba livia*). Brain Behav Evol 22: 22–41

Grubb TC (1976) Why ospreys hover. Wilson Bull 89:149–150

Grubb TC (1977) Weather dependent foraging in ospreys. Auk 94:146–149

Hafner H, Boy V, Gory G (1982) Feeding methods, flock size and feeding success in little egrets, *Egretta garzetta* and squacco heron *Ardeola ralloides* in Camargue, southern France. Ardea 70:45–54

Hancock J, Kushlan J (1984) The herons handbook. Croom Helm, London

Harkness L (1977) Chameleons use accomodation to judge distance. Nature 267:346–349

Horvath G, Varju D (1990) Geometric optical investigation of the underwater visual field of aerial animals. Math Biosci 102:1–19

Howland HC (1974) Optimal strategies for predator avoidance: the relative importance of speed and manoeuvrability. J Theor Biol 47:333–350

Howland HC, Sivak JG (1984) Penguin vision in air and water. Vision Res 24:1905–1909

Jackson S (1984) Predation by pied kingfishers and whitebreasted cormorants on fishes in the Kosi estuary system. Ostrich 55:113–132

Jenkins FA, White HE (1976) Fundamentals of optics, 2nd edn. McGraw Hill, New York

Katzir G, Camhi J (1993) Escape response of black mollies (*Poecilia sphenops*) from predatory dives of a pied kingfisher (*Ceryle rudis*). Copeia (in press)

Katzir G, Intrator N (1987) Striking of underwater prey by reef herons, *Egretta gularis schistacea*. J Comp Physiol 160:517–523

Katzir G, Shechtman E, Dive patterns of pied kingfishers (*Ceryle rudis*): coping with light refraction in determining prey depth. (in prep.)

Katzir G, Lotem A, Intrator N (1989) Stationary underwater prey missed by reef herons, *Egretta gularis*: heat position and light refraction at the moment of strike. J Comp Physiol 165:573–576

Katzir G, Arad Z, Strod T, Coping with light refraction in four heron species. (in prep)

Krebs JR, Partridge B (1973) The significance of heat tilting in the great blue heron. Nature 245:533–535

Labinger Z, Katzir G, Benjamini Y (1991) Prey size choice by captive pied kingfishers, *Ceryle rudis* L. Anim Behav 42:969–975

Labinger Z, Benjamini Y, Katzir G, Prey choice in the pied kingfisher, *Ceryle rudis* L: the relationship between prey size and depth. (in prep)

Lee DN (1980) The optic flow field: the foundation of vision. Philos Trans R Soc Lond B 290: 169–179

Lee DN, Reddish PE (1981) Plummeting gannets: a paradigm of ecological optics. Nature 293:293–294

Lee DN, Young DS (1986) Gearing action to the environment. Experimental brain research, Series 15. Springer, Berlin Heidelberg New York, pp 217–230
Levy B, Sivak JG (1980) Mechanisms of accomodation in the bird eye. J Comp Physiol 137: 267–272
Loew ER, McFarland WN (1990) The underwater visual environment. In: Douglas RH, Djamgoz (eds) The visual system of fish. Chapman and Hall, London, pp 1–40
Lotem A, Katzir G, Schechtman E (1991) Capture of submerged prey by little egrets, *Egretta garzetta garzetta*: strike depth, strike angle and the problem of light refraction. Anim Behav 42:341–346
Luling KH (1963) The archerfish. Sci Am 209:100–110
Lythgoe JN (1979) The ecology of vision. Clarendon Press, Oxford
Martin GR (1985) Eye. In: King AS, McLelland J, (eds) Form and function in birds, vol 3. Academic Press, London, pp 311–373
Martin GR (1986) The eye of a passeriform bird, the European starling (*Sturnus vulgaris*): eye movement amplitude visual fields and schematic optics. J Comp Physiol 159:545–557
Martin GR, Katzir G, Visual fields in herons (Ardoidae) – panoramic vision beneath the bill. Brain Behav Evol (in press)
Martin GR, Young SR (1984) The eye of the Humboldt penguin, *Spheniscus humboldti*: visual fields and schematic optics. Proc R Soc Land B 223:197–222
Martinoya C, Rey J, Bloch S (1981) Limits of the pigeon's binocular fields and direction of best binocular viewing. Vision Res 21:1197–1200
McFarland WN, Loew ER (1983) Wave produced changes in underwater light and their relation to vision. Environ Biol Fishes 8:173–184
Migongo EWK (1978) Environmental factors affecting the distribution of malachite and pied kingfishers in Lake Nakuru national park. Thesis, University of Nairobi, Kenya
Mittelstaedt H (1957) Prey capture in mantids. In: Scheer BT (ed) Recent advances in invertebrate physiology. University of Oregon Publications, Oregon, pp 51–71
Moroney MK, Pettigrew JD (1987) Some observations on the visual optics of kingfishers (Aves, Coraciformes, Alcedinidae). J Comp Physiol 160:137–149
Muntz WRA (1972) Inert absorbing and reflecting pigments. In: Dartnall HJA, (ed) Handbook of sensory physiology, vol VII/1. Springer, Berlin Heidelberg New York, pp 529–565
O'Brian WJ, Brownman HI, Evans BI (1990) Search strategies of foraging animals. Am Sci 78:152–160
Reyer H-U, Mogongo-Bake W, Schmidt L (1988) Field studies and experiments on the distribution and foraging of pied and malachite kingfishers at lake Nakuru (Kenya). J Anim Ecol 57:595–610
Schenck H (1957) On the focusing of sunlight by ocean waves. J Opt Soc Am 47:653–657
Sivak JG, Hildebrand T, Lebert C (1985) Magnitude and rate of accomodation in diving and nondiving birds. Vison Res 25:925–933
Swinton WE (1975) Fossil birds. British Museum of Natural History, London
Walls GL (1967) The vertebrate eye and its adaptive radiation. (facsimile of 1942 edition). Hafner, New York
Weihs D, Webb PW (1984) Optimal avoidance tactics in predator prey interactions. J Theor Biol 106:189–206
Welty CJ, Baptista L (1988) The life of birds, 4th edn. Saunders College Publishing, New York

16 Multiple Sources of Depth Information: An Ecological Approach

M.N.O. Davies[1] and P.R. Green[2]

16.1 Depth Perception and the Control of Behaviour

The purpose of perceiving depth is to control behaviour within a three-dimensional world. This may involve gross locomotion of the animal within the environment, monitoring the movement of others, or movement of limbs in relation to the animal's body and the environment. The general aims in controlling gross movement of the body are to avoid harmful collisions with objects while achieving desirable goals, as when an ostrich traverses difficult terrain while running, a guillemot alights on a cliff face without crash landing, or competing male jungle fowl time their foot strikes in order to win contests. Whether the aim is to avoid or to achieve contact with an object, successful coordination of behaviour will require relatively fast processing of information.

Depth perception is also essential for the control of the movement of a part of an animal with respect to an intended target. Examples are the extension and contact of the feet in relation to a perch in the slow flight landing manoeuvres of pigeons, the fast strike of the goshawk's talons against a partridge in flight, or a chicken's peck to grasp and consume a morsel of food. In some cases, such as pecking, fast processing is not necessarily paramount.

Any textbook of visual perception identifies the generally accepted set of depth cues (e.g. Coren and Ward 1989; Goldstein 1989; Bruce and Green 1990). Members of the set include pictorial cues of linear perspective and retinal image size, the dynamic cues of the optic flow field such as motion parallax and tau (see Chap. 13), and the physiological cues of accommodation (see Chap. 2), convergence and binocular disparity (see Chap. 3). This last cue has been the focus of a large proportion of the research into visual depth perception.

Some depth cues such as accommodation, convergence or tau, provide information about *absolute* distance or time to contact, while others may enable the perceiver to judge the *relative* distance between two or more objects. Cues to relative depth include accretion/deletion, height in the image plane, visual texture and linear perspective. Binocular disparity also provides relative depth information unless combined with a range finder such as convergence (for further discussion, see Chap. 3).

[1] Department of Psychology, University College London, Gower St, London WC1E 6BT, UK
[2] Department of Psychology, University of Nottingham, Nottingham NG7 2RD, UK

M.N.O. Davies and P.R. Green (Eds.)
Perception and Motor Control in Birds

The perception of absolute depth involves two closely related goals. The first is to maximize the *accuracy* of the activity being performed, by keeping as small as possible the difference between the mean distance over which an action is performed and the target distance. The second goal is to maximize *precision*, by reducing the scatter around the mean performance. For example, granivorous birds will increase their fitness by maximizing their energy intake and/or minimizing the time to feed. To achieve either, pecking should be controlled so as to reduce placement error of the beak in relation to the food item, and this involves maximizing both accuracy and precision.

16.2 Models of Visual Depth Perception

The mechanisms of depth perception involved in timing actions will be expected to achieve optimal solutions to the problems of maximizing accuracy, precision and speed. It is well known that there is a variety of sources of depth information in the light reaching an animal's eyes, but how do mechanisms of depth perception exploit this variety in order to achieve optimal solutions to the problem? What is the relationship between the different potential cues to depth? We will approach this often neglected problem in the field of visuomotor control by considering a number of possible models.

16.2.1 The Hierarchical Model

In the history of the investigation of depth perception, an implicit hierarchy has often been assumed, in which binocular disparity is seen as the "alpha" cue to depth, with other cues taking subsidiary positions in the pecking order. It has further been assumed that the "subsidiary cues" are used only when binocular disparity information is no longer available. This view seems to have been based on anthropomorphic assumptions. Primates in general and humans in particular are sensitive to binocular disparity, and this has led to the false deduction that when binocular disparity is available it must be the dominant cue to depth (see Chap. 3).

There are important limits to the usefulness of stereopsis for the purposes of guiding locomotion, however (cf. Sect. 13.2.6). The accuracy of stereoscopic depth perception is time dependent (McKee et al. 1990). In man, White and Odom (1985) have demonstrated that global stereoscopic information is integrated over a time period of 500 ms (critical duration) to reach a minimum disparity threshold of 20–30″. The assumption that global stereopsis processing is slow because of the complexity of the neural processing involved is supported by Mayhew and Frisby (1980) and Marr and Poggio (1979). This time lag imposes a constraint on the usefulness of stereopsis where rapid change occurs in the optic array, as during the flight manoeuvres of birds. On the other hand, in circumstances where the perceiver is functionally stationary, time will be available for optimal depth estimation using stereopsis.

McKee et al. (1990) have also investigated the accuracy of stereoscopic judgements for a range of standing disparities. Again the results demonstrate that binocular disparity has its weak points as a means of timing behaviour. McKee et al. (1990) show that the smallest Weber fraction associated with stereoscopic judgements is 4–6%. This is compared to size judgement, which is rapid and gives accuracy of 2–3%, and motion parallax which allows judgements of less than 5% at speeds greater than $3° s^{-1}$. McKee et al. (1990) conclude that stereopsis is not a parsimonious solution to the problem of guiding gross movements such as walking or running, since too much time is required to reach a reasonable level of accuracy in estimating depth with a steady view. Instead they propose that stereopsis, which produces the most accurate estimates for features very near the fixation plane, is primarily used for fine control of movements (cf. Sect. 3.5.1).

If the loss of binocular disparity meant that behavioural coordination in a three-dimensional world was lost, this would be conclusive proof of its essential role. However, it has been regularly demonstrated that creatures sensitive to binocular disparity can perform adequately without this information (e.g. Collett and Harkness 1982; Davies and Green 1991a). As Goldstein (1989) concludes:

"Although no single cue is crucial for our perception of depth (we can eliminate any monocular depth cue or close one eye to eliminate binocular disparity and still see depth), the more cues we have, the better our chances of accurately deducing the three dimensions of the world from the two-dimensional information on our retinas". (Goldstein 1989, p. 245).

The assumed pre-eminence of binocular disparity has also influenced comparative investigation into visual depth perception in vertebrates, leading to the assumption that binocular overlap in the visual field automatically implies the use of binocular disparity as the main depth cue (cf. Sects. 1.6.2.3 and 3.4.1). In the absence of binocular overlap, the assumption has been that another depth cue becomes the critical cue. In either case, the emphasis has been on the identification of a single cue, whether binocular disparity or not, which has the key role in providing a particular species with depth information. We term this general approach the hierarchical model.

16.2.2 The Heterarchical Model

The hierarchical model assumes that the ability to use subsidiary cues has evolved because of the risk of losing the ability to use the dominant cue; for example, of losing stereopsis through injury to an eye. This argument is reasonable as far as it goes, but overlooks the fact that different cues will allow greater accuracy and precision of depth estimation in some circumstances than in others. We have already mentioned one example; the limitations of binocular disparity when depth changes rapidly as an observer moves about. Also, disparity is effective only at relatively short distances, within limits determined by the separation of the eyes (see Sect. 3.4.5). The same is true of accommodation (see Chap. 2), where the limit is set by depth of field (see Sect. 1.5). It is easy to identify constraints on other cues. Texture gradients will not be useful for airborne targets against an empty background; motion parallax will not be reliable where observer speed cannot be estimated; retinal size of a target requires knowledge of its actual size to be effective as a distance cue; and so on.

The likelihood of a particular distance cue achieving optimal control of behaviour will vary from one problem in visuomotor control to another. It is unlikely that animals will have evolved to select sources of depth information according to a fixed hierarchy. Rather, it seems more likely that the ordering of importance of cues will be flexible and will change according to context. For example, whether optic flow information is a more effective means of timing an action than binocular disparity may vary from one occasion to another, according to the speed of computation required, whether the target can be fixated in the binocular visual field, and so on. We would expect these and other sources of depth information to be organized in a *heterarchy* rather than a hierarchy.

The heterarchical model allows more flexibility than the hierarchical model, although the two are equivalent as long as the demands on the animal from the environment remain constant. The relationship between the two models closely parallels the development of recent models of bird navigation, in which the contributions of multiple redundant cues to the control of orientation do not follow a fixed hierarchy but instead vary with context and with individual birds' life histories (see Chap. 5). We will next discuss results from our recent work on the landing flight of birds, to illustrate the inadequacy of the Hierarchical model for a relatively straightforward visuomotor behaviour. The original aim of this series of investigations was the traditional one of identifying a single depth cue used to time landing manoeuvres, but the results obtained proved difficult to interpret in such simple terms.

16.2.2.1 Landing Flight in Birds

Our investigations into the visual depth information used by pigeons (*Columba livia*) to guide their landing flight towards a perch (e.g. Davies and Green 1988, 1990, 1991a) began with the hypothesis that the birds would use *tau*, the optic flow variable which directly specifies time-to-contact with an object or surface (Lee 1980; see Chap. 13). The method used was based on Wagner's (1982) work on landing flight in houseflies, where the coefficients of variation of a number of parameters specifying distance were measured at intervals before the landing manoeuvre began. It was assumed that the parameter which triggers landing would show least variation at an estimated reaction time of 60 ms before foot extension (see Davies and Green 1990). The parameters measured were the distance between eye and perch, the angular velocity of the perch through the visual field and the ratio tau of eye-perch distance to the bird's velocity.

The results (Davies and Green 1990) indicated that the pigeons did not rely on tau to control their landing flight, as perch distance varied less than the other two contenders (see Fig. 16.1, left). It was concluded that one of the depth cues specifying perch distance and independent of approach speed (e.g. binocular disparity, accommodation, image size) was the critical source of depth information. It is not possible to identify from these data the distance cue or cues involved, although other evidence allows us to exclude some possibilities.

First, the state of accommodation, when the perch is in focus, of the lens and/or cornea (Schaeffel and Howland 1987; see Sect. 2.3) is unlikely to control the initiation of landing flight in pigeons because of the gradient of myopia in the lower visual field (Fitzke et al. 1985; see Sect. 2.5). This implies that the far point of the eye, beyond which an object cannot be brought into focus, falls closer to the eye for objects lower in the visual field. Green et al. (1992) measured the positions of the perch in the visual field of pigeons at intervals during landing flights, and compared the calculated far point of the eye for these positions with the actual distances of the perch. For a bird with average lower field myopia, during the last 150 ms before foot extension the perch lies within the estimated depth of field around the far point of the eye. This implies that accommodation of lens or cornea would not be required to bring the perch into focus and so could not provide a signal to trigger foot extension. This could be possible in birds with a lower than average degree of myopia, but it seems that accommodation can be excluded as a *general* cue to control foot extension.

Second, if pigeons were able to compare the known dimensions of a perch with the size of its retinal image, they could obtain information about its distance. If they relied solely on this distance cue, we would expect pigeons' first flights to a novel perch to be timed by some other means, as at least one landing

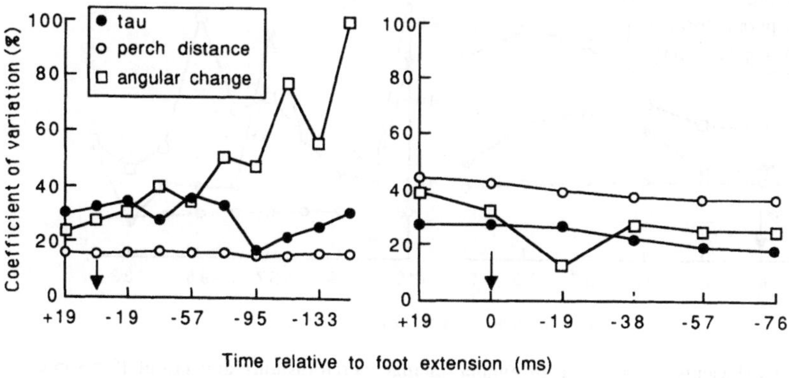

Fig. 16.1. Coefficients of variation for angular change, perch distance and tau at 19 ms intervals during landing flight. Times are before (–) and after (+) the point of foot extension, marked by a *vertical arrow*. *Left* Pigeon; *right* Harris hawk (After Davies and Green 1990)

would be required to learn the size of the perch. Some evidence against this prediction was obtained by Davies (1989) in two experiments which measured the coefficients of variation of depth parameters during landing flights of adult and newly fledged pigeons to a novel perch. In both cases, distance varied less than tau or motion parallax, just as for flights to a familiar perch (see Fig. 16.2). Because of difficulties in persuading birds to fly to the perch in these experiments, the results are based on fewer landings than those with a familiar perch, but nonetheless they fail to support the hypothesis that the *only* distance cue triggering foot extension is the retinal size of the perch.

Whatever the distance cue used to time landing, the probable reason why pigeons do not use tau (specifically, the tau function of the optical size of the perch) is the rhythmic variation in head velocity, or "head-bobbing" behaviour, which occurs during landing flight. Head-bobbing is not simply a motoric response to strenuous slow landing flight (Davies and Green 1988; Davies et al. submitted), but is visually driven. Exactly what the visual functions of head bobbing are remains an unanswered question, although the behaviour does potentially enhance motion parallax information during the thrust phase, and object motion detection during the flexion phase (Frost 1978; Davies and Green 1988). However, it is clear that the regular oscillation in the velocity of the bird's head means that it cannot rely on the tau function of the optical size of the perch to specify a unique time-to-contact, as this parameter depends upon the velocity of the eye. Figure 16.3 illustrates the problems of using tau as the controlling depth cue while head-bobbing.

When the analysis of landing flight was repeated with the Harris hawk (*Parabuteo unicinctus*), tau showed less variation just before foot extension than either perch distance or angular change (Davies and Green 1990; see Fig. 16.1, right). The Harris hawk does not head-bob in landing flight, and this result

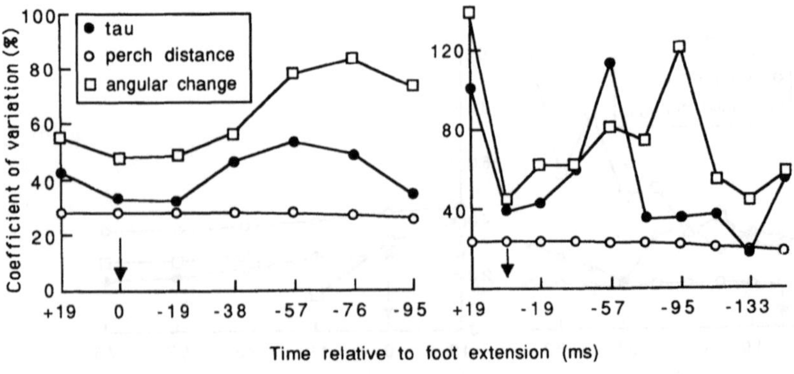

Fig. 16.2. Coefficients of variation for angular change, perch distance and tau at 19-ms intervals during landing flights of pigeons. Times are before (−) and after (+) the point of foot extension, marked by a *vertical arrow*. *Left* First landing flights made by newly fledged pigeons; *right* first flights made by adult pigeons to a novel perch (Davies 1989)

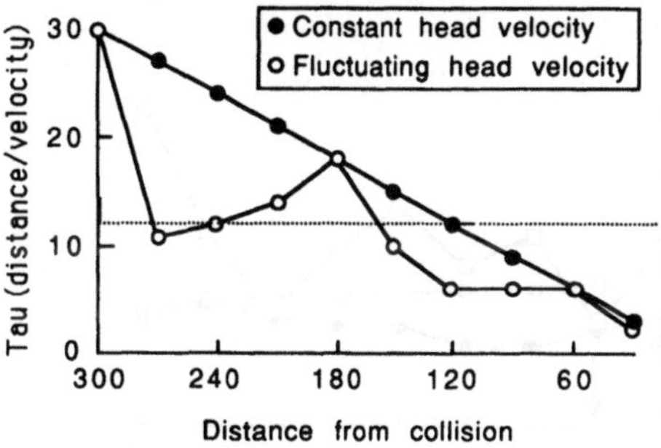

Fig. 16.3. Illustration of the relationship between tau and distance from a target for an observer travelling at constant body velocity, with head velocity either constant or fluctuating. Distance and tau are expressed in arbitrary units. The *horizontal line* shows how a particular critical value of tau (here, 12) provides ambiguous information about distance if head velocity fluctuates

indicates that when head velocity is smooth, the flow field variable tau can become effective in timing landing manoeuvres.

A later investigation attempted to dissect out the importance of binocular disparity in timing landing in the pigeon, by temporarily occluding one eye during landing (Davies and Green 1991a). The procedure was as described above, except that the birds flew either under monocular occlusion or sham occlusion. Temporary monocular occlusion was achieved by covering one eye with a protective layer of gelatin sponge, overlaid with black masking tape. Under the control occlusion condition, strips of masking tape were placed around one of the bird's eyes so as to form a circle. Care was taken to prevent occlusion of the visual field, particularly the frontal visual field.

Under the monocular occlusion condition, tau now showed *less* variation than distance, indicating that the visual parameter used to control landing had changed (see Fig. 16.4, left). The film data confirmed that the head-bobbing profiles were not distinguishable from those of birds flying unoccluded. These results seem to support the hierarchical model, suggesting that binocular disparity is the dominant cue and that tau is promoted from second place when disparity information is not available. However, the situation is not so straightforward when one considers the sham occlusion data. Sham occlusion was employed to control for any general motivational effects of placing masking tape close to the eye. Gross landing activity was unaffected, and the birds also head-bobbed normally. However, the visual cue that varied least 60 ms before foot extension was tau and not perch distance (see Fig. 16.4, right). The implication of this result is that even when disparity information is available, it is not necessarily used. Such a finding clearly cannot be handled by the hierarchical model.

The pigeon results so far described indicate that successful landing can be achieved using either a cue to distance such as binocular disparity, or tau, and that tau is favoured not only when a distance cue is unavailable, but also just under mildly stressful conditions. The fact that neither form of occlusion appeared to affect head-bobbing poses an anomaly; how can tau control the timing of landing flight when its value fluctuates rhythmically during landing flight (see Fig. 16.3)? One explanation is suggested by the finding that birds flying under either sham or monocular occlusion make slower approaches to the

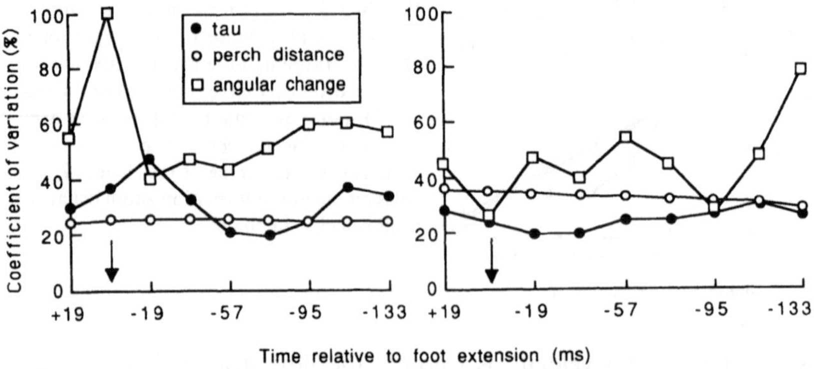

Fig. 16.4. Coefficients of variation for angular change, perch distance and tau at 19-ms intervals during landing flights of pigeons. Times are before (–) and after (+) the point of foot extension, marked by a *vertical arrow*. *Left* Monocular occlusion; *right* sham occlusion (After Davies and Green 1991a)

perch than birds with no tape attached to their heads. This suggests that birds landing under the occlusion conditions are able to make do with a relatively inaccurate depth estimate in order to time foot extension, as their slow approach gives them time to make adjustments to flight manoeuvres when they are near the perch.

These findings suggest that no single depth cue triggers foot extension in pigeons' landing flights in all circumstances. By apparently switching to tau when stressed by flying under either sham or monocular occlusion, birds may be using a visual parameter that requires less time and cognitive resources to process, so that more resources can be allocated to other tasks such as vigilance.

An alternative interpretation is possible; the tau function of the optical separation of the feet and the perch, unlike that of the optical size of the perch, does not fluctuate with the head-bobbing cycle, and could in principle be used to control foot extension in the same way as the tau function of the optical separation of a hand and a target can control reaching (see Sect. 13.2.4). Further analysis of pigeons' landings shows that, 60 ms before foot extension, the tau function of foot-perch distance has a lower coefficient of variation than eye-perch distance, which is consistent with the hypothesis that foot extension is controlled by this tau function (Lee et al. 1993). This pattern was found for birds flying normally and with one eye covered, but the opposite results were obtained for sham-occluded birds. The evidence therefore still suggests that a mildly stressful procedure, with no effect on the visual information available from the perch, causes a switch in the visual parameter controlling foot extension.

In conclusion, distance estimation by pigeons during landing flight has turned out to be far from a simple case of using a single depth cue to initiate a unitary landing manoeuvre. In addition, alighting from flight also involves problems of postural orientation (Green et al. 1992), control of braking (Lee et al. 1993; see Sect. 13.2.8) and foot placement (Davies and Green 1991b). In the case of foot extension, the effective cue can change with subtle changes in circumstances, demonstrating the inadequacy of the hierarchical model and providing evidence for the heterarchical model.

16.2.2.2 Depth Perception in the Visual Cliff

Experiments which we have recently carried out on depth perception in young chicks (Green et al. in press) also provide evidence that birds switch between different sources of depth information in different contexts. It is well known that chicks avoid stepping onto the deep side of a visual cliff, and prefer to step onto the shallow side when given a choice (Walk and Gibson 1961). In the second case, Walk et al. (1968) found that chicks showed a preference for the shallow side when the depth of the cliff exceeded a threshold of 4.5–5 cm. In a test where chicks had to step over the edge of a visual cliff in order to approach another chick, we found a similar threshold (between 4 and 8 cm), above which chicks were significantly slower to take a step. In contrast, if chicks were simply placed on the transparent floor over the deep side of the cliff, they were quicker to take a step at all depths, and the threshold above which stepping was significantly inhibited was higher, between 8 and 16 cm (see Fig. 16.5).

These results suggest that chicks do not use depth information in a single way. If there is a discontinuity in surface depth, such as a cliff edge, near their feet, stepping is strongly inhibited when the relative depth at the edge is greater than about 4 cm. If there is no such discontinuity near the chick, as when it is placed on the deep side of a visual cliff, then stepping is less strongly inhibited. There may be two different processes operating in the visual system of chicks, one to compute *differences* in depth at nearby surface discontinuities and one to compute the *absolute* depth of the visible surface immediately below the chick. The results of these two processes have different effects on behaviour, and cannot be treated as a unitary process of "depth perception".

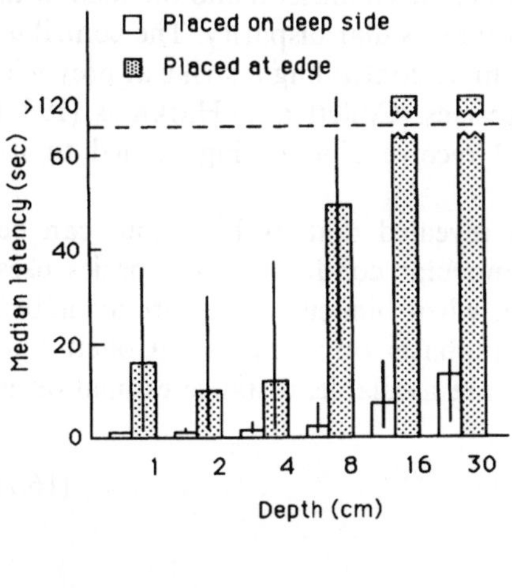

Fig. 16.5. Median latencies of chicks to take a step towards another chick in 2-min tests when placed on the deep side of a visual cliff of varying depths. *Vertical bars* show interquartile ranges. The differences between latencies in the two conditions are significant at all depths. Within conditions, the only significant differences at adjoining depths are between 8 and 16 cm (chicks placed on deep side) and between 4 and 8 cm (chicks placed at edge)

16.2.3 The Integration of Multiple Depth Cues

The heterarchical model, in common with the hierarchical model, assumes that at any one time only a single depth cue is effective in controlling behaviour. It would be surprising if this was generally true, since greater accuracy and precision could be achieved if some or all of the depth cues available in a particular context were combined. Rather than use a single sample (depth cue) to estimate the population (actual depth), more accurate and precise depth information could be obtained from several parallel sample estimates combined to describe the population (see Collett and Harkness 1982).

This problem has been considered previously (e.g. Freeman 1970) and has given rise to a small number of experimental investigations using psychometric techniques rather than behavioural performance to determine the relationships between different cues (Harker 1958; Jameson and Hurvich 1959; van der Meer 1979; Bruno and Cutting 1988; Landy et al. 1990). Only in recent years have there been attempts to model how multiple sources of depth information could be combined to improve the accuracy and precision of depth perception in the control of behaviour (e.g. Collett and Harkness 1982; Ellard et al. 1984, 1991; Cavallo and Laurent 1988; Coren and Ward 1989; Davies 1989; Maloney and Landy 1989; Curran and Johnston submitted).

Collett and Harkness' (1982) model of cue combination in the toad is one of the very few attempts to describe quantitatively the relationship between depth cues involved in controlling action, and the behavioural data gathered by Collett and his co-workers have also been used by Arbib and House (1987) to design their neural models of cue interaction and prey location. We will next look at these models in more detail.

16.2.3.1 The Toad – First Few Steps

The value of combining different sources of depth information is that each source is a separate estimate of the same real value. A critical factor in determining how cues are combined will be the degree of precision of depth estimation which each provides. For example, the chameleon and the toad both have the ability to use accommodation and binocular disparity. The contribution of each of these to the final depth estimate controlling a strike at prey will depend upon the weighting each cue receives. Collett and Harkness (1982) propose that each separate estimate will receive a weighting related to its precision.

Work on the toad *Bufo marinus* has revealed that its behaviour can be described by such a strategy. Under monocular conditions this species uses accommodation to estimate distance, but, when binocular cues are available, both are used together with a heavy emphasis on stereopsis (Collett and Harkness 1982). When both types of cue are available, distance estimation is described by the linear expression.

$$d_e = 0.06A + 0.94B, \quad (16.1)$$

where A and B are values of distance estimates from accommodation and binocular information respectively, and d_e is an overall estimate of distance. Equation (16.1) implies that the influence of binocular information on depth estimation is 16 times greater than that of accommodation in the experimental situation investigated.

In general, we would expect the weightings of two cues such as binocular disparity and accommodation to vary from one species to another, according to various structural and behavioural constraints. For example, an animal's size will affect its eye size and therefore its visual acuity (see Sect. 1.4.1), which in turn will affect the precision of depth estimation possible from accommodation and binocular disparity. Other constraints which are likely to be important are the shape of the binocular visual field and the costs of making an erroneous judgement. Within a species, weightings may also vary from one context to another. Factors likely to be important in determining weightings include the speed of movement of a target which must be captured and its location in the visual field.

Behavioural investigations into visuomotor control in anurans have led to recent attempts to model explicitly the decision making processes involved. Arbib and House's (1987) model is concerned specifically with the decisions made by anurans when faced with various forms of barrier between themselves and prey. Out of this work has developed the "cue interaction" model.

16.2.3.2 Cue Interaction Model

Anurans presented with prey which can be seen behind a barrier such as a fence are faced with the problem of detouring around the barrier. Behavioural data show that these animals do not have a single solution to the detour problem, but instead have a spread of preferred options. These usually take the form of preferred directions in which the animal moves to avoid the barrier and reach the prey.

In modelling the neural control of this behaviour, Arbib and House (1987) are concerned with the integration of depth information with motor control within the barrier environment. In the Cue Interaction model they propose that retinal information is processed into multiple cue-related depth maps, each covering the whole of the visual field. Accommodation and binocular disparity are explicitly considered, and the weightings identified by Collett and Harkness [see Eq. (16.1)] are incorporated into the interaction between the two maps.

As well as modelling Collett and Harkness' (1982) behavioural results, House (1984) developed a "prey localization" model, which also relies on multiple cues, but is concerned with the location of a single target rather than whole field depth map representation. The prey localization model incorporates further findings by Collett and Udin (1983), which show that if the neural pathway involved in processing disparity information is lesioned, the toad can use triangulation as a simpler binocular system. It is assumed that both eyes are focused to the same distance. The proximal image is then analyzed by the Imager for positional information, while the Pattern Analyzer processes the output and identifies the target. Once these two preliminary stages are complete, the Prey Selector selects from a number of potential targets the one which has the least blurred image; a decision strongly influenced by input from both eyes. The output feeds into the accommodation controller which converts weighted image coordinates into an estimate of depth through triangulation. The final step is for feedback to adjust the setting of the lens. It should be noted that this model cannot operate in the monocular mode.

Thus it is possible to model either the intact animal's behaviour with the Cue Interaction model, or the lesioned anuran with the Prey Localization model. Arbib and House (1987) postulate that instead of there being a single

"general depth-perception mechanism we have here a case of various neural strategies functioning either cooperatively or alternatively to cope with the vast array of visuo-motor tasks required of the freely functioning animal" (Arbib and House 1987, p. 144).

16.2.3.3 Cue Interaction Is Only One Side of the Coin

In Sect. 16.2.3.2 we reviewed what has been achieved so far in work on the concept of cue interaction. For the particular case of modelling the toad's visuomotor behaviour, such an approach is clearly more valuable than the heterarchical model, as it allows for the parallel computation of two or more depth cues. Even so, the cue interaction model has its own limitations. So far, only the physical environment and an animal's anatomy have been considered as sources of factors affecting the weighting of different cues. The cue integration approach has not considered the importance of internal organismic factors such as the allocation of cognitive resources, motivational factors and the effects of learning. Arbib and House (1987) have drawn attention to the probable role of motivation in their attempt to explain the spread of detour orientation produced by their computer model of the toad's actual behaviour. They argued that the variation in orientation obtained could be explained in terms of the modulation of tectal cell receptive field sizes by motivational state and experience.

However, before neurophysiological studies are undertaken to investigate the influence of motivation and other internal factors in the processing of sensory information, behavioural methods should be employed to determine how internal factors affect perceptuomotor performance. Behavioural studies into visuomotor control have made little effort to study the link between factors internal to the animal and depth estimation strategies. As a first step in exploring these links, a framework bringing together depth perception, organismic and external factors is presented below as the "multipotentialism" model of depth perception. A key assumption of this model is that cues are combined to produce an estimate of depth, each given a weighting which depends upon contextual factors at the time. Thus the animal should strive for an optimal solution to the problem of maximizing precision of visuomotor control, while minimizing the costs.

16.2.3.4 Multipotentialism Model of Depth Perception

The model presented here is not a neural model of the animal's decision making processes; instead it is an information processing model that attempts to place the cue interaction approach within a broader, ecological context and to guide experimental work. Cue interaction is represented by a series of analyzing components as seen in Fig. 16.6. To maintain continuity with Arbib and House's (1987) cue interaction model, the retinal image is processed in three stages. The directions of items in space are first computed, then items are categorized, and finally depth cues are combined to produce the most precise estimate of distance possible. The sequence of stages is not crucial for the model, which is concerned with this last stage.

Depth cues may be summed in a linear fashion, as suggested by Collett and Harkness (1982) and Bruno and Cutting (1988) or they may be integrated multiplicatively, or in some other way. Both the rule for cue integration, and the weightings apportioned to each cue estimate, will depend upon the species, environment and individual history of the animal. Nevertheless, the basic premise is that multiple cue estimates of depth are brought together to produce a more precise estimate of the actual distance of the target.

The full multipotentialism model is displayed in Fig. 16.7, which illustrates how various constraints influence the depth estimation process by affecting the weightings associated with each individual cue estimate. One constraint is learning; it is proposed that the relative reliability of cues over time can be used to alter weights. This could come about in two main ways. Feedback from the success or failure of actions could be used to measure the reliability of a cue, in a form of "supervised" learning. Alternatively, weights could be adjusted without the use of feedback from behaviour, for example by using the relative variances over a period of time in the depth estimates derived from different cues. In a particular context, the larger the variance in depth estimates associated with a

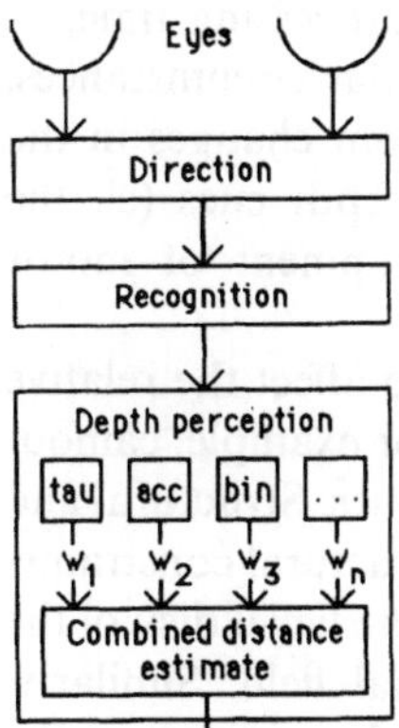

Fig. 16.6. Model of depth cue interaction. Visual input is processed to extract the direction, identity and distance of a target. In the last stage, the outputs of a number of processes computing different depth cues (e.g. tau, the state of accommodation of the eye, binocular disparity) are combined according to their weights $w_1 \ldots w_n$ to yield a distance estimate which controls motor output

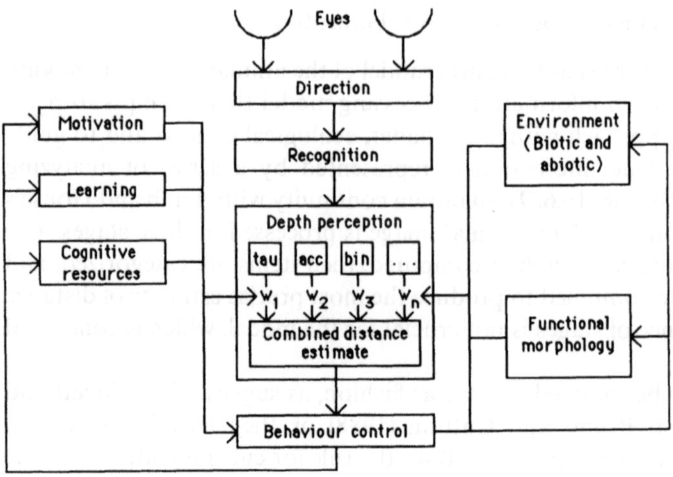

Fig. 16.7. A summary of the factors which influence the weightings of depth cues in the multipotentialism model of depth perception

particular cue, relative to those associated with other cues, the less reliable it is and the lower the weight assigned to it should be.

A second constraint is motivation, including current levels of fear, hunger, sexual arousal and so on. Let us consider hunger as an example of a motivational factor which may influence depth estimation in the visuomotor guidance of pecking. We would predict that the trade-off between accuracy and speed of pecking might vary with motivational state, with greater emphasis on accuracy in satiated than in hungry birds, where speed would be more important. This trade-off would also be influenced by social factors; the presence of other birds competing for scarce grain would be expected to cause greater emphasis on speed. If different cues make different demands on neural processing time, as well as achieving different degrees of precision under particular circumstances, then changes in the precision-speed trade-off could arise from changes in the weights attached to the estimates provided by different depth cues (cf. the discussion of speed-accuracy considerations in the development of motor control of pecking in Chap. 8).

In addition, non-motivational external factors could also affect the relative importance of individual cues, and thus their weightings. For example, camouflaged targets may enhance the efficiency of relative motion cues. Structural and behavioural constraints will also be relevant. Examples of structural constraints are eye geometry (see Chap. 1) and the beak and feathers of the facial disk of the tawny owl (Martin 1984) that affect the size of the retinal field. Similarly, variation in the form of the head between different breeds of pigeon alters the shape of the binocular visual field (Jahnke 1984). Behavioural constraints are well illustrated by head bobbing, which constrains the accuracies of sources of depth information during landing flight (see Sect. 16.2.2.1). Therefore, in the multipotentialism model, weightings of depth cues reflect not only their precision, as envisaged by Collett and Harkness (1982), but also non-perceptual parameters which have a bearing on the usefulness of a particular source of distance information in a specific context.

16.3 Conclusions

The question of how multiple sources of depth information interact to determine depth judgements, or to control visually guided actions, is an old one, although it has attracted relatively little experimental and theoretical attention. We have reviewed evidence to show that the problem arises in the context of visuomotor organization in birds, and we have argued that it should be considered in the wider context of optimality theory (McFarland and Houston 1981). Like any mechanisms controlling behaviour, those which combine sources of depth information to control visually guided actions can be considered as naturally selected means of achieving optimal solutions to a variety of environmental demands, which differ between species, individuals and contexts. The weightings assigned to different depth cues in the central process of integration are likely to depend not only on the precision associated with each cue, as determined by geometry and optics, but also on the relative importance at any one time of precision and speed of depth estimation. Our multipotentialism model provides a framework for incorporating these factors into explanations of depth perception, and an agenda for experimental work on this problem.

The visual control of actions in birds where distance or time to contact information is required, such as pecking (see Chaps. 2, 3, 8, 9 and 15) or perching (see Chap. 13 and Sect. 16.2.2.1), provides opportunities for further experiments. The first step should be to develop means of comparing the weightings of two or more depth cues, as Collett and Harkness (1982) were able to do in toads. The methods used by Martinoya et al. (1988) and McFadden (1990; see Sect. 3.3.2.3) to test the role of binocular information in pigeon depth perception are promising, as they provide a means of measuring the relative weights of binocular disparity and a set of other depth cues.

Given such methods, the next step should be to examine the effects on weightings of various manipulations of context and motivation. The effects of sham occlusion of the eyes on the control of landing manoeuvres (see Sect. 16.2.2.1) suggest that the effects of mild stress on weightings of depth cues are worth examining further. There is also evidence that stress caused by repeated landings affects another aspect of visuomotor control, the accuracy of foot placement (Davies and Green 1991b). The theoretical arguments developed earlier suggest further factors which are likely to influence the weighting of cues; these include hunger, the characteristics of food or other targets, and social competition. It should be possible to use knowledge of the speed and precision which can be achieved with different depth cues to make quantitative predictions about the effects of these variables.

In proposing that depth information is computed in different ways by multiple independent processes, the multipotentialism model resembles Bruno and Cutting's (1988) "minimodularity" theory of the combination of depth cues in determining psychophysical judgements. It can also be related to a broader range of current thinking about brain organization and artificial intelligence, which sees sensorimotor control as involving competitive interactions between many different subsystems, each extracting a particular limited kind of information from sensory receptors. Strong neurophysiological evidence for parallel processing of this kind in birds has been obtained from the pigeon (Frost et al. 1990; see Chap. 12), where local and whole-field image motion are processed separately in the tectofugal and accessory optic systems respectively. An exciting challenge for the future is the question of how these two systems interact in the control of locomotion.

In artificial intelligence, this trend is exemplified by the "situated action" approach in robotics (e.g. Brooks 1991), which builds robot control systems from independent activity-producing "layers". This approach has succeeded in building machines which can navigate and carry out simple tasks in biologically realistic time and in fairly unconstrained situations such as offices and corridors. The biological inspiration for these developments has so far come mainly from work on the visual control of insect behaviour. Behavioural and neurophysiological work on perception and motor control in birds will soon, we believe, reach a stage where it will contribute biological principles to guide the development of more complex robots built on these principles.

References

Arbib MA, House DH (1987) Depth and detours: an essay on visually guided behaviour. In: Arbib MA, Hanson AR (eds) Vision, brain and co-operative computation. MIT Press, Cambridge, pp 129–163

Brooks RA (1991) Intelligence without representation. Artif Intell 47:139–159

Bruce V, Green PR (1990) Visual perception: physiology, psychology and ecology, 2nd edn. Erlbaum, London

Bruno N, Cutting JE (1988) Minimodularity and the perception of layout. J Exp Psychol Gen 117:161–170

Cavallo V, Laurent M (1988) Visual information and skill level in time-to-collision estimation. Perception 17:623–632

Collett TS, Harkness LIK (1982) Depth vision in animals. In: Ingle DJ, Goodale MA, Mansfield RJW (eds) Analysis of visual behaviour. MIT Press, Cambridge, pp 111–176

Collett TS, Udin SB (1983) The role of the toad's nucleus isthmi in prey-catching behaviour. In: Proc 2nd worksh on visuomotor coordination in frog and toad: models and experiments. COINS Tech Rep 83–19, Computer and Information Science Dept, University of Massachusetts, Amherst

Coren S, Ward LM (1989) Sensation and perception, 3rd edn. Harcourt Brace Jovanovich, London

Curran W, Johnston A, Integration of shading and texture cues: testing the linear model. (submitted)

Davies MNO (1989) The perception of relative movement and the control of action. Thesis, University of Nottingham, UK

Davies MNO, Green PR (1988) Head-bobbing during walking, running and flying: relative motion perception in the pigeon. J Exp Biol 138:71–91

Davies MNO, Green PR (1990) Flow-field variables trigger landing in hawk but not in pigeons. Naturwissenschaften 77:142–144
Davies MNO, Green PR (1991a) The adaptability of visuomotor control in the pigeon during landing flight. Zool Jahrb Physiol 95:331–338
Davies MNO, Green PR (1991b) Footedness in pigeons, or simply sleight of foot? Anim Behav 42:311–312
Davies MNO, Green PR, Thorpe PH, Head-bobbing and head orientation during upwards landing flights of pigeons. (submitted)
Ellard CG, Goodale MA, Timney B (1984) Distance estimation in the Mongolian gerbil: the role of dynamic depth cues. Behav Brain Res 14: 29–39
Ellard CG, Chapman DG, Cameron KA (1991) Calibration of retinal image size with distance in the Mongolian gerbil: rapid adjustments of calibrations in different contexts. Percept Psychophys 49:38–42
Fitzke FW, Hayes BP, Hodos W, Holden AL (1985) Refractive sectors in the visual field of the pigeon eye. J Physiol 369:33–44
Freeman RB (1970) Theory of cues and the psychophysics of visual space perception. Psychon Monogr Suppl 3 (13, 45):171–181
Frost BJ (1978) The optokinetic basis of head-bobbing in the pigeon. J Exp Biol 74:187–195
Frost BJ, Wylie DR, Wang Y-C (1990) The processing of object and self-motion in the tectofugal and accessory optic pathways of birds. Vision Res 30:1677–1688
Goldstein EB (1989) Sensation and perception, 3rd edn. Wadsworth, Belmont, CA
Green PR, Davies MNO, Thorpe PH (1992) Head orientation in pigeons during landing flight. Vision Res 32:2229–2234
Green PR, Davies IB, Davies MNO, Interaction of visual and tactile information in the control of chicks' locomotion in the visual cliff. Perception (in press)
Harker GS (1958) Interaction of monocular and binocular acuities in the making of equidistance judgements. J Opt Soc Am 48:233–240
House DH (1984) Neural models of depth perception in frogs and toads. Thesis, University of Massachusetts, Amherst
Jahnke HK (1984) Binocular visual field differences among various breeds of pigeons. Bird Behav 5:96–102
Jameson D, Hurvich LM (1959) Note on factors influencing the relation between stereoscopic acuity and observation distance. J Opt Soc Am 49:639
Landy MS, Maloney LT, Young MJ (1990) Psychophysical estimation of the human depth combination rule. In: Schenker PS (ed) Sensor fusion III: 3-D perception and recognition. Proc SPIE – Int Soc Opt Eng 1383:247–254
Lee DN (1980) The optic flow field: the foundation of vision. Philos Trans R Soc Lond B 290: 169–179
Lee DN, Davies MNO, Green PR, van der Weel FR (1993) Visual control of velocity of approach by pigeons when landing. J Exp Biol 180:85–104
Maloney LT, Landy MS (1989) A statistical framework for robust fusion of depth information. Proc SPIE – Int Soc Opt Eng: visual communications and image processing, part 2, pp 1154–1163
Marr D, Poggio T (1979) A computational theory of human stereoscopic vision. Proc R Soc Lond B 204:301–328
Martin GR (1984) The visual fields of the tawny owl (*Strix aluco*). Vision Res 24:1739–1751
Matinoya C, Le Houezec J, Bloch S (1988) Depth resolution in the pigeon. J Comp Physiol 163:33–42
Mayhew JEW, Frisby JP (1980) The computation of binocular edges. Perception 9:69–86
McFadden SA (1990) Eye design for depth and distance perception in birds: an observer oriented perspective. J Comp Psychol: Comp Stud Percept Cognit 3:1–31
McFarland DJ, Houston A (1981) Quantitative ethology: the state space approach. Pitman, London
McKee SP, Levi DM, Bowne SF (1990) The imprecision of stereopsis. Vision Res 30:1763–1779
Schaeffel F, Howland HC (1987) Corneal accommodation in chick and pigeon. J Comp Physiol 160:375–384

van der Meer HC (1979) Interaction of the effects of binocular disparity and perspective cues on judgements of depth and height. Percept Psychophys 26:481–488
Wagner H (1982) Flow-field variables trigger landing in flies. Nature 297:147–148
Walk RD, Gibson EJ (1961) A comparative and analytical study of visual depth perception. Psychol Monogr 75 (15):44 pp
Walk RD, Falbo TL, Lebowitz C (1968) Differential visual depth threshold of the chick. Psychon Sci 12:197–198
White KD, Odom JV (1985) Temporal integration in global stereopsis. Percept Psychophys 37:139–144

Subject Index

Springer-Verlag and the Environment

We at Springer-Verlag firmly believe that an international science publisher has a special obligation to the environment, and our corporate policies consistently reflect this conviction.

We also expect our business partners – paper mills, printers, packaging manufacturers, etc. – to commit themselves to using environmentally friendly materials and production processes.

The paper in this book is made from low- or no-chlorine pulp and is acid free, in conformance with international standards for paper permanency.

MIX
Papier aus verantwortungsvollen Quellen
Paper from responsible sources
FSC® C105338

If you have any concerns about our products,
you can contact us on
ProductSafety@springernature.com

In case Publisher is established outside the EU,
the EU authorized representative is:
Springer Nature Customer Service Center GmbH
Europaplatz 3, 69115 Heidelberg, Germany

Printed by Libri Plureos GmbH
in Hamburg, Germany